Process Piping Design

Process

Volume 1
1. General Piping
2. Process Terms
3. Plant Arrangement and Storage Tanks
4. Process Unit Plot Plans
5. Piping Systems and Details
6. Pipe Fabrication
7. Vessels
8. Instrumentation

Volume 2
1. Pumps and Turbines
2. Compressors
3. Fired Heaters
4. Exchangers
5. Piping Flexibility

Piping Design
Volume 1

Rip Weaver

Gulf Publishing Company
Houston, Texas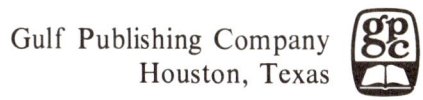

Books by the Author
Process Piping Drafting
Process Piping Drafting Workbook
Process Piping Design, 2 Volumes
Modern Basic Drafting
Modern Basic Drafting Workbook,
   Parts 1 and 2
Structural Drafting
Piper's Pocket Handbook

First Printing, March 1973
Second Printing, August 1974
Third Printing, December 1974
Fourth Printing, January 1977
Fifth Printing, February 1979
Sixth Printing, November 1981

Piping design is both demanding and strenuous. It makes relaxation imperative. This book is dedicated to the three men who have helped me relax the most: my brother, Joe Weaver, Joe Garrett and Kenneth Campbell.

Copyright © 1973 by Gulf Publishing Company, Houston, Texas. All rights reserved. Printed in the United States of America. This book, or parts thereof, may not be reproduced in any form without permission of the publisher.

Library of Congress Card Number: 72-84332

ISBN: 0-87201-759-1

Contents

Preface

Introduction

1 General Piping, 1
Piping Materials
Methods of Manufacturing Pipe
Pipe Diameters, Thicknesses and Schedules
Fittings and Flanges
Valves

2 Process Terms, 26
The Hydrocarbon
Hydrocarbon Structures
Fractionation
Piping Equilibrium Liquids
Two-Phase Flow
Hot Vapor By-Pass
Static Head
Steam Data

3 Plant Arrangement and Storage Tanks, 37
Site Data
Hilly Terrain
Block Plot Plan
The Process Block
Storage Tanks
Tank Dike Regulations
Storage Tank Design
A Refinery
Exercise
Dike Area Drainage
Storage Tank Piping
Foam Protection System

4 Process Unit Plot Plans, 57
Equipment Plot Plan
Preliminary Plot Plan Dimensions
Plot Plan Exercise
Foundation Location Plan
Excavation Plan
Flow Diagram Transportation
Piping Drawing Index
Equipment Setting

Contents, continued

5 Piping Systems and Details, 73
Underground Piping
Sewer System Terms
Sewer Flow Diagram
Cast Iron Soil Pipe and Fittings
Designing Systems
Student Exercise
Design Guidelines
Underground Pressurized Systems
Fire Water
Underground Cooling Water
Pump-Out system
Glycol Lines
Steam Tracing
Steam Tracing Design Practice
Tail Pipe Cuts

6 Pipe Fabrication, 92
Welding Shop Details
Pipe Bends
Miter Welds
Small Fittings
The Triangle
The Cutback

7 Vessels, 119
Definitions
Horizontal Vessels
Piping Arrangement for Elevated Vessels
Manhold Davits and Hinges
Vertical Vessels

8 Instrumentation, 149
Instrument Types
Instrument Functions
Dual Instruments
Transmitters
Thermowells
Other Temperature Instruments
Pressure Instruments
Flow Instruments
Orifice Flange Taps
Level Instruments
Level Gage
Control Valves

Appendix, 171
Conversions

Index, 209

Preface

This two-volume set on *Process Piping Design* has been written to supplement my first book, *Process Piping Drafting*. I have purposely used a very practical writing style for both of these efforts, applying my personal practical experience rather than trying to present the technical aspects of piping. In *Process Piping Design* I have elected to have each chapter self-sustaining. I have reproduced some charts and tables but only those required in learning the basics of piping design.

To become a competent piping designer requires many years of experience plus a talent for creative thinking. Piping designers must call on their knowledge for each design job but also must apply their own ingenuity daily. I like to call that *horse sense*.

Many piping designers have told me that their job is 25% knowledge, 25% experience and 50% horse sense. This book will try to deal with the first 25% and touch on the last 50%.

Fluor Corporation has been very farsighted by instigating piping drafting and piping design training classes and they have been very generous in allowing me to reproduce some of their instructional material. I wish to take this opportunity to thank Fluor for its assistance.

Introduction

In a refining or petrochemical complex, piping constitutes the major expenditure of all the design disciplines. Piping consumes about 50% of the design engineering manhours, 35% of the material cost of the plant and about 30% of the labor cost in the field. Inept piping design in the office can increase the cost of a plant.

The design engineering department of a contractor designing refinery or petrochemical complexes consists of four main functions: piping, structural, electrical and vessels. The piping section encompasses flow diagrams, model making, insulation and painting, piping material take-off, piping material control, instrument design and piping design itself.

Experienced piping designers know the functions of all of these groups. They also must have a broad knowledge of the structural, electrical and vessel sections. And they must know the many pieces of equipment that they must pipe up, the numerous details of piping, the materials necessary for various services, piping flexibility basics and field construction practices. Then comes a general knowledge of plant operation and maintenance. When all of this is considered, it is easy to see why there is always a shortage of experienced, competent piping designers.

Ironically, piping design is the one subject that has been neglected in the school systems. This is largely due to the fact that adequate textbooks were not available. No book can teach a person to become a piping designer but it can teach the fundamentals and how to apply them to become a designer.

Process Piping Design

1 General Piping

The dictionary describes pipe as a long tube of clay, concrete, metal, wood, etc., for conveying water, gas, oil or other fluids. A fluid is defined as any substance that can flow—liquid or gas.

Pipe has been with us for over 5000 years. Each year introduces new purposes for piping which create demands for new and improved material specifications and ingenuity on the part of the piping designer. Piping has advanced from its earlier function of transporting water through wooden conduits from the mountains' melting snow to the village below. Today piping can be found almost any place one looks, from the piping in an automobile to the complicated maze of piping in a process plant.

A piping designer is one who designs piping systems. Since piping is so widely used, the piping designers cannot learn the complete field of piping, and consequently they have become more specialized, going into the fields of process piping, pipe line piping, steam power plant piping or one of the many other categories.

Process piping design is, in the author's estimation, the most interesting specialization in the piping field. Each process piping unit presents a new challenge to the designer. Each plant is custom designed, and while there is some similarity between a crude unit of 50,000 barrels a day and one of 75,000 barrels a day, there is no duplication. A piping designer may work on two or three crude units in his entire career. There are literally hundreds of different types of units in a refinery or petrochemical complex. Each type of unit presents its own piping challenges.

How, then, can one learn process piping design from a book? It can't be done. A book can present the basic similarities that exist in many types of units and introduce guidelines for competent design. From there on, the process piping designer must utilize his personal judgment.

Piping Materials

The term *piping* means not only pipe but the fittings, flanges, valves and other items which form part of the overall piping system. Piping materials are divided into two basic classes, metallic and nonmetallic. Nonmetallic piping might be glass, ceramic, plastic, etc. The metallic piping is again divided into two classes, ferrous and nonferrous. Ferrous materials are those of, containing, or derived from iron and are most commonly used in process piping. Ferrous metals are carbon steel, stainless steel, chrome steel, cast iron, etc. Nonferrous metals include aluminum.

Table 1-1.
Some Steel Pipe Specifications

ASTM Number	Type	Material	Remarks
A-53	Gr. A, B	Carbon Steel	Manufactured in welded and seamless. Grade B is most commonly specified.
A-106	Gr. A, B	Carbon Steel	Seamless. Grade B is preferred and mostly used. In the modern plant, almost all carbon steel pipe is this specification.
A-333	Gr. 1	Carbon Steel	Used for sub-zero temperatures. Incorporates special testing. For use to -50°F.
A-335	P1	Carbon Moly	Basically a carbon steel with ½% molybdenum. Used in medium high temperature service.
A-335	P11	Chrome Moly	1¼% chrome, ½% molybdenum. Used in higher temperature, corrosive services.
A-335	P5	Chrome Moly	5% chrome, ½% molybdenum. Used in higher temperature, corrosive services.
A-335	P9	Chrome Moly	9% chrome, 1% molybdenum. Used in high temperature highly corrosive services.
A-312	304	Stainless	Used for temperatures below -50°F and for corrosive service at higher temperatures. Widely used for food product piping.
A-312	316	Stainless	Used for high temperature, highly corrosive service.
A-312	321	Stainless	Used for very high temperature, highly corrosive service.
A-312	347	Stainless	Used in harsher conditions than Type 321 stainless.
A-333	Gr. 3	Nickel	3½% nickel. Used for temperatures from -50°F to -150°F.

Table 1-1 lists some of the many hundreds of ferrous metals available for pipe. Those listed are the most commonly used in process units. For a complete listing see the ASTM book, *Ferrous Materials**.

Methods of Manufacturing Pipe

Pipe diameter, wall thickness, material specification and delivery requirements are determining factors in the selection of the manufacturing process. Steel piping is made by lap-welding, spiral welding, buttwelding and seamless methods. Welded pipe types are made from flat plates which are rolled to form round shapes; the edges are then welded together to form a longitudinal weld. The longitudinal weld reduces the pressure-containing characteristics of pipe and the ANSI (American National Standards Institute—formerly ASA) piping code reduces the allowable stress of this method of manufacture by imposing a "joint efficiency" of less than 100%. Seamless piping has a joint efficiency of 100% since there is no longitudinal joint. Welded pipe also can attain this 100% joint efficiency rating with special quality control procedures such as stress relieving and full x-ray examination. However, these add to the cost and may not be needed. In the smaller sizes, seamless piping is quite often as economical as welded if 100% joint efficiency is specified.

Whatever method of manufacture is specified, a "mill tolerance" must be added to the minimum calculated wall thickness. Plate is manufactured to

*Refer to ASTM Standard, Part 1: Steel Piping, Tubing and Fittings.

Table 1-2
Commercial Wrought Steel Pipe Data

Nominal Pipe Size Inches	Outside Diameter (D) Inches	Schedule No. See Note 1	Wall Thickness (t) Inches	Inside Diameter (d) Inches	Area of Metal (a) Square Inches	Transverse Internal Area		Moment of Inertia (I) Inches to 4th Power	Weight of Pipe Pounds per foot	Weight of Water Pounds per foot of pipe	External Surface Sq. Ft. per foot of pipe	Section Modulus $\left(2\dfrac{I}{D}\right)$
						Square Inches	See Note 2 Square Feet					
1/8	0.405	40s	.068	.269	.0720	.0568	.00040	.00106	.244	.025	.106	.00523
		80x	.095	.215	.0925	.0364	.00025	.00122	.314	.016	.106	.00602
1/4	0.540	40s	.088	.364	.1250	.1041	.00072	.00331	.424	.045	.141	.01227
		80x	.119	.302	.1574	.0716	.00050	.00377	.535	.031	.141	.01395
3/8	0.675	40s	.091	.493	.1670	.1910	.00133	.00729	.567	.083	.178	.02160
		80x	.126	.423	.2173	.1405	.00098	.00862	.738	.061	.178	.02554
1/2	0.840	40s	.109	.622	.2503	.3040	.00211	.01709	.850	.132	.220	.04069
		80x	.147	.546	.3200	.2340	.00163	.02008	1.087	.102	.220	.04780
		160	.187	.466	.3836	.1706	.00118	.02212	1.300	.074	.220	.05267
		...xx	.294	.252	.5043	.050	.00035	.02424	1.714	.022	.220	.05772
3/4	1.050	40s	.113	.824	.3326	.5330	.00371	.03704	1.130	.231	.275	.07055
		80x	.154	.742	.4335	.4330	.00300	.04479	1.473	.188	.275	.08531
		160	.218	.614	.5698	.2961	.00206	.05269	1.940	.128	.275	.10036
		...xx	.308	.434	.7180	.148	.00103	.05792	2.440	.064	.275	.11032
1	1.315	40s	.133	1.049	.4939	.8640	.00600	.08734	1.678	.375	.344	.1328
		80x	.179	.957	.6388	.7190	.00499	.1056	2.171	.312	.344	.1606
		160	.250	.815	.8365	.5217	.00362	.1251	2.840	.230	.344	.1903
		...xx	.358	.599	1.0760	.282	.00196	.1405	3.659	.122	.344	.2136
1 1/4	1.660	40s	.140	1.380	.6685	1.495	.01040	.1947	2.272	.649	.435	.2346
		80x	.191	1.278	.8815	1.283	.00891	.2418	2.996	.555	.435	.2913
		160	.250	1.160	1.1070	1.057	.00734	.2839	3.764	.458	.435	.3421
		...xx	.382	.896	1.534	.630	.00438	.3411	5.214	.273	.435	.4110
1 1/2	1.900	40s	.145	1.610	.7995	2.036	.01414	.3099	2.717	.882	.497	.3262
		80x	.200	1.500	1.068	1.767	.01225	.3912	3.631	.765	.497	.4118
		160	.281	1.338	1.429	1.406	.00976	.4824	4.862	.608	.497	.5078
		...xx	.400	1.100	1.885	.950	.00660	.5678	6.408	.42	.497	.5977
2	2.375	40s	.154	2.067	1.075	3.355	.02330	.6657	3.652	1.45	.622	.5606
		80x	.218	1.939	1.477	2.953	.02050	.8679	5.022	1.28	.622	.7309
		160	.343	1.689	2.190	2.241	.01556	1.162	7.440	.97	.622	.979
		...xx	.436	1.503	2.656	1.774	.01232	1.311	9.029	.77	.622	1.104
2 1/2	2.875	40s	.203	2.469	1.704	4.788	.03322	1.530	5.79	2.07	.753	1.064
		80x	.276	2.323	2.254	4.238	.02942	1.924	7.66	1.87	.753	1.339
		160	.375	2.125	2.945	3.546	.02463	2.353	10.01	1.54	.753	1.638
		...xx	.552	1.771	4.028	2.464	.01710	2.871	13.70	1.07	.753	1.997
3	3.500	40s	.216	3.068	2.228	7.393	.05130	3.017	7.58	3.20	.916	1.724
		80x	.300	2.900	3.016	6.605	.04587	3.894	10.25	2.86	.916	2.225
		160	.438	2.624	4.205	5.408	.03755	5.032	14.32	2.35	.916	2.876
		...xx	.600	2.300	5.466	4.155	.02885	5.993	18.58	1.80	.916	3.424
3 1/2	4.000	40s	.226	3.548	2.680	9.886	.06870	4.788	9.11	4.29	1.047	2.394
		80x	.318	3.364	3.678	8.888	.06170	6.280	12.51	3.84	1.047	3.140
4	4.500	40s	.237	4.026	3.174	12.73	.08840	7.233	10.79	5.50	1.178	3.214
		80x	.337	3.826	4.407	11.50	.07986	9.610	14.98	4.98	1.178	4.271
		120	.438	3.624	5.595	10.31	.0716	11.65	19.00	4.47	1.178	5.178
		160	.531	3.438	6.621	9.28	.0645	13.27	22.51	4.02	1.178	5.898
		...xx	.674	3.152	8.101	7.80	.0542	15.28	27.54	3.38	1.178	6.791
5	5.563	40s	.258	5.047	4.300	20.01	.1390	15.16	14.62	8.67	1.456	5.451
		80x	.375	4.813	6.112	18.19	.1263	20.67	20.78	7.88	1.456	7.431
		120	.500	4.563	7.953	16.35	.1136	25.73	27.10	7.09	1.456	9.250
		160	.625	4.313	9.696	14.61	.1015	30.03	32.96	6.33	1.456	10.796
		...xx	.750	4.063	11.340	12.97	.0901	33.63	38.55	5.61	1.456	12.090
6	6.625	40s	.280	6.065	5.581	28.89	.2006	28.14	18.97	12.51	1.734	8.50
		80x	.432	5.761	8.405	26.07	.1810	40.49	28.57	11.29	1.734	12.22
		120	.562	5.501	10.70	23.77	.1650	49.61	36.40	10.30	1.734	14.98
		160	.718	5.189	13.32	21.15	.1469	58.97	45.30	9.16	1.734	17.81
		...xx	.864	4.897	15.64	18.84	.1308	66.33	53.16	8.16	1.734	20.02
8	8.625	20	.250	8.125	6.57	51.85	.3601	57.72	22.36	22.47	2.258	13.39
		30	.277	8.071	7.26	51.16	.3553	63.35	24.70	22.17	2.258	14.69
		40s	.322	7.981	8.40	50.03	.3474	72.49	28.55	21.70	2.258	16.81
		60	.406	7.813	10.48	47.94	.3329	88.73	35.64	20.77	2.258	20.58
		80x	.500	7.625	12.76	45.66	.3171	105.7	43.39	19.78	2.258	24.51
		100	.593	7.439	14.96	43.46	.3018	121.3	50.87	18.83	2.258	28.14
		120	.718	7.189	17.84	40.59	.2819	140.5	60.63	17.59	2.258	32.58
		140	.812	7.001	19.93	38.50	.2673	153.7	67.76	16.68	2.258	35.65
		...xx	.875	6.875	21.30	37.12	.2578	162.0	72.42	16.10	2.258	37.56
		160	.906	6.813	21.97	36.46	.2532	165.9	74.69	15.80	2.258	38.48

Source: Crane Co.

(Table 1-2 continued on following page)

Note 1: The letters s, x, and xx in the column of Schedule Numbers indicate Standard, Extra Strong, and Double Extra Strong Pipe, respectively.

Note 2: The values shown in square feet for the Transverse Internal Area also represent the volume in cubic feet per foot of pipe length.

(Table 1-2 continued)

Nominal Pipe Size Inches	Outside Diameter (D) Inches	Schedule No. See Note 1	Wall Thickness (t) Inches	Inside Diameter (d) Inches	Area of Metal (a) Square Inches	Transverse Internal Area Square Inches	Transverse Internal Area See Note 2 Square Feet	Moment of Inertia (I) Inches to 4th Power	Weight of Pipe Pounds per foot	Weight of Water Pounds per foot of pipe	External Surface Sq. Ft. per foot of pipe	Section Modulus $\left(2\frac{I}{D}\right)$
10	10.750	20	.250	10.250	8.24	82.52	.5731	113.7	28.04	35.76	2.814	21.15
		30	.307	10.136	10.07	80.69	.5603	137.4	34.24	34.96	2.814	25.57
		40s	.365	10.020	11.90	78.86	.5475	160.7	40.48	34.20	2.814	29.90
		60x	.500	9.750	16.10	74.66	.5185	212.0	54.74	32.35	2.814	39.43
		80	.593	9.564	18.92	71.84	.4989	244.8	64.33	31.13	2.814	45.54
		100	.718	9.314	22.63	68.13	.4732	286.1	76.93	29.53	2.814	53.22
		120	.843	9.064	26.24	64.53	.4481	324.2	89.20	27.96	2.814	60.32
		140	1.000	8.750	30.63	60.13	.4176	367.8	104.13	26.06	2.814	68.43
		160	1.125	8.500	34.02	56.75	.3941	399.3	115.65	24.59	2.814	74.29
12	12.75	20	.250	12.250	9.82	117.86	.8185	191.8	33.38	51.07	3.338	30.2
		30	.330	12.090	12.87	114.80	.7972	248.4	43.77	49.74	3.338	39.0
		..s	.375	12.000	14.58	113.10	.7854	279.3	49.56	49.00	3.338	43.8
		40	.406	11.938	15.77	111.93	.7773	300.3	53.53	48.50	3.338	47.1
		..x	.500	11.750	19.24	108.43	.7528	361.5	65.42	46.92	3.338	56.7
		60	.562	11.626	21.52	106.16	.7372	400.4	73.16	46.00	3.338	62.8
		80	.687	11.376	26.03	101.64	.7058	475.1	88.51	44.04	3.338	74.6
		100	.843	11.064	31.53	96.14	.6677	561.6	107.20	41.66	3.338	88.1
		120	1.000	10.750	36.91	90.76	.6303	641.6	125.49	39.33	3.338	100.7
		140	1.125	10.500	41.08	86.59	.6013	700.5	133.68	37.52	3.338	109.9
		160	1.312	10.126	47.14	80.53	.5592	781.1	160.27	34.89	3.338	122.6
14	14.00	10	.250	13.500	10.80	143.14	.9940	255.3	36.71	62.03	3.665	36.6
		20	.312	13.376	13.42	140.52	.9758	314.4	45.68	60.89	3.665	45.0
		30s	.375	13.250	16.05	137.88	.9575	372.8	54.57	59.75	3.665	53.2
		40	.438	13.124	18.66	135.28	.9394	429.1	63.37	58.64	3.665	61.3
		..x	.500	13.000	21.21	132.73	.9217	483.8	72.09	57.46	3.665	69.1
		60	.593	12.814	24.98	128.96	.8956	562.3	84.91	55.86	3.665	80.3
		80	.750	12.500	31.22	122.72	.8522	687.3	106.13	53.18	3.665	98.2
		100	.937	12.126	38.45	115.49	.8020	824.4	130.73	50.04	3.665	117.8
		120	1.093	11.814	44.32	109.62	.7612	929.6	150.67	47.45	3.665	132.8
		140	1.250	11.500	50.07	103.87	.7213	1027.0	170.22	45.01	3.665	146.8
		160	1.406	11.188	55.63	98.31	.6827	1117.0	189.12	42.60	3.665	159.6
16	16.00	10	.250	15.500	12.37	188.69	1.3103	383.7	42.05	81.74	4.189	48.0
		20	.312	15.376	15.38	185.69	1.2895	473.2	52.36	80.50	4.189	59.2
		30s	.375	15.250	18.41	182.65	1.2684	562.1	62.58	79.12	4.189	70.3
		40x	.500	15.000	24.35	176.72	1.2272	731.9	82.77	76.58	4.189	91.5
		60	.656	14.688	31.62	169.44	1.1766	932.4	107.50	73.42	4.189	116.6
		80	.843	14.314	40.14	160.92	1.1175	1155.8	136.46	69.73	4.189	144.5
		100	1.031	13.938	48.48	152.58	1.0596	1364.5	164.83	66.12	4.189	170.5
		120	1.218	13.564	56.56	144.50	1.0035	1555.8	192.29	62.62	4.189	194.5
		140	1.438	13.124	65.78	135.28	.9394	1760.3	223.64	58.64	4.189	220.0
		160	1.593	12.814	72.10	128.96	.8956	1893.5	245.11	55.83	4.189	236.7
18	18.00	10	.250	17.500	13.94	240.53	1.6703	549.1	47.39	104.21	4.712	61.1
		20	.312	17.376	17.34	237.13	1.6467	678.2	59.03	102.77	4.712	75.5
		..s	.375	17.250	20.76	233.71	1.6230	806.7	70.59	101.18	4.712	89.6
		30	.438	17.124	24.17	230.30	1.5990	930.3	82.06	99.84	4.712	103.4
		..x	.500	17.000	27.49	226.98	1.5763	1053.2	92.45	98.27	4.712	117.0
		40	.562	16.876	30.79	223.68	1.5533	1171.5	104.75	96.93	4.712	130.1
		60	.750	16.500	40.64	213.83	1.4849	1514.7	138.17	92.57	4.712	168.3
		80	.937	16.126	50.23	204.24	1.4183	1833.0	170.75	88.50	4.712	203.8
		100	1.156	15.688	61.17	193.30	1.3423	2180.0	207.96	83.76	4.712	242.3
		120	1.375	15.250	71.81	182.66	1.2684	2498.1	244.14	79.07	4.712	277.6
		140	1.562	14.876	80.66	173.80	1.2070	2749.0	274.23	75.32	4.712	305.5
		160	1.781	14.438	90.75	163.72	1.1369	3020.0	308.51	70.88	4.712	335.6
20	20.00	10	.250	19.500	15.51	298.65	2.0740	756.4	52.73	129.42	5.236	75.6
		20s	.375	19.250	23.12	290.04	2.0142	1113.0	78.60	125.67	5.236	111.3
		30x	.500	19.000	30.63	283.53	1.9690	1457.0	104.13	122.87	5.236	145.7
		40	.593	18.814	36.15	278.00	1.9305	1703.0	122.91	120.46	5.236	170.4
		60	.812	18.376	48.95	265.21	1.8417	2257.0	166.40	114.92	5.236	225.7
		80	1.031	17.938	61.44	252.72	1.7550	2772.0	208.87	109.51	5.236	277.1
		100	1.281	17.438	75.33	238.83	1.6585	3315.2	256.10	103.39	5.236	331.5
		120	1.500	17.000	87.18	226.98	1.5762	3754.0	296.37	98.35	5.236	375.5
		140	1.750	16.500	100.33	213.82	1.4849	4216.0	341.10	92.66	5.236	421.7
		160	1.968	16.064	111.49	202.67	1.4074	4585.5	379.01	87.74	5.236	458.5
24	24.00	10	.250	23.500	18.65	433.74	3.0121	1315.4	63.41	187.95	6.283	109.6
		20s	.375	23.250	27.83	424.56	2.9483	1942.0	94.62	183.95	6.283	161.9
		..x	.500	23.000	36.91	415.48	2.8853	2549.5	125.49	179.87	6.283	212.5
		30	.562	22.876	41.39	411.00	2.8542	2843.0	140.80	178.09	6.283	237.0
		40	.687	22.626	50.31	402.07	2.7921	3421.3	171.17	174.23	6.283	285.1
		60	.968	22.064	70.04	382.35	2.6552	4652.8	238.11	165.52	6.283	387.7
		80	1.218	21.564	87.17	365.22	2.5362	5672.0	296.36	158.26	6.283	472.8
		100	1.531	20.938	108.07	344.32	2.3911	6849.9	367.40	149.06	6.283	570.8
		120	1.812	20.376	126.31	326.08	2.2645	7825.0	429.39	141.17	6.283	652.1
		140	2.062	19.876	142.11	310.28	2.1547	8625.0	483.13	134.45	6.283	718.9
		160	2.343	19.314	159.41	292.98	2.0346	9455.9	541.94	126.84	6.283	787.9

a tolerance of 0.01". Pipe made from plate (all pipe with a longitudinal seam) will have 0.01" added to its calculated minimum thickness for this mill tolerance. Seamless pipe is made by a process that requires a tolerance of 12½%.

Seamless pipe is made from hot, round solid billets of steel. A mandrel is centered and penetrates the hot billet, expanding the solid piece to a hollow pipe. This method of manufacturing can cause some possible thin spots in the pipe wall; consequently, the 12½% tolerance is imposed.

Special manufacturing of seamless pipe, such as centrifugally cast or special forging, is specified for very thick requirements. This type is cast or forged to a thicker wall and is precision machined, inside and out, and the tolerance is usually nil.

The piping engineer or designer must recognize the method of manufacture and its related mill tolerance before calculating the minimum wall thickness required for his piping.

The various methods of manufacture also determine the length of the delivered pipe. Commonly, pipe is made in "random length" which is ±20'-0", and in "double random length" which is ±40'-0". Unless double random length is specified, the manufacturer will ship single random. For long, straight runs of piping, considerable savings can be made by utilizing the longer pieces, saving buttwelding.

Centrifugally cast and special forged pipe will be shipped in 6-12' lengths. Since these are machined inside and out, the lengths are very short. This adds numerous buttwelds to long runs of pipe, increasing the system cost. Consequently, this method of manufacture is utilized only where the other methods cannot produce the desired pipe more economically. And any economic evaluation must consider the number of buttwelds.

Pipe Diameters, Thicknesses and Schedules

Table 1-2 lists pipe data for most commercially available sizes. Normally, sizes 1¼", 2½", 3½" and 5" are considered as noncommercially manufactured and are not specified by a piping designer. Equipment manufacturers will employ these sizes and the piping designer will have to attach a flange or reducer to this connection but should immediately increase to the next larger size for his piping.

Pipe and tubing are not the same. Tubing is specified by its outside diameter; 4" tubing is 4" OD, 4" pipe is 4.5"OD. This is usually specified as 4" IPS (Iron Pipe Size) for pipe and can also be defined by specifying 4" schedule 40. The schedule number defines the OD and the "Nominal" wall thickness for IPS piping.

Nominal wall thickness is the average wall of the pipe—not the minimum wall. To ascertain the minimum wall, the mill tolerance must be subtracted.

Fittings and Flanges

Welding fittings are manufactured to match the companion pipe. However, it is not mandatory that the fitting and the pipe have the same thickness. While pipe of several schedules is available, fittings are not stocked for all schedules. Fittings are usually specified as standard weight, extra strong, schedule 160 and double extra strong. It is usually advantageous to specify the fitting thickness of the next higher available weight if the pipe wall thickness is not standard, extra strong, etc. As an example, with 14" schedule 10 (0.250" wall) the standard weight fitting would be specified, which is 0.375" wall. For 14" schedule 40 (0.438" wall) the extra strong fitting would be specified, which is 0.500" wall.

For pipe sizes 2" and below, welding fittings are usually not used. For low pressure, noncritical service, the screwed fitting is specified, while for higher pressures and most process systems, the socketwelding fittings are employed.

Figure 1-1 describes forged steel screwed fittings and Figure 1-2 gives dimensional data for them.

Figure 1-3 describes forged steel socketwelding fittings and Figure 1-4 gives their dimensions. Starting in 1970, the 2000-pound series was discontinued by most manufacturers. The 3000-pound series would be used with schedule 40 and schedule 80 pipe.

Socketwelding costs a small amount more than screwed fittings. This cost reverses into an overall installed savings as socketwelded systems withstand the hydrostatic testing and remain a leak-free system for years.

Figure 1-5 describes steel buttwelding fittings and Figure 1-6 gives their dimensions.

The fittings depicted in Figure 1-5 are the ones that may be specified by a piping designer. However, due to delivery and cost, additional practical data is:

90° Elbow
No. 240, 2000-Pound
No. 380, 3000-Pound
No. 660, 6000-Pound

Tee
No. 241, 2000-Pound
No. 381, 3000-Pound
No. 661, 6000-Pound

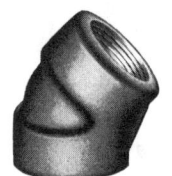
45° Elbow
No. 242, 2000-Pound
No. 382, 3000-Pound
No. 662, 6000-Pound

Cross
No. 243, 2000-Pound
No. 383, 3000-Pound
No. 663, 6000-Pound

90° Street Elbow
No. 384, 3000-Pound
No. 664, 6000-Pound

Coupling
No. 386, 3000-Pound
No. 666, 6000-Pound

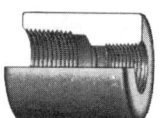
Reducer
No. 387, 3000-Pound
No. 667, 6000-Pound

Half Coupling
No. 388, 3000-Pound
No. 668, 6000-Pound

Cap
No. 389, 3000-Pound

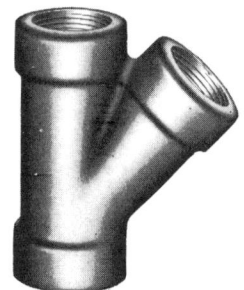
45° Y-Bend
No. 245, 2000-Pound
No. 665, 6000-Pound

Working Pressures
Steam, Water, Oil, Oil Vapor, Gas, or Air

Temp. Deg. Fahr.	Psi, Non-Shock Carbon Steel ASTM A105, Grade II		
	2000 Pound W.O.G.	3000 Pound W.O.G.	6000 Pound W.O.G.
100°	2000	3000	6000
150	1970	2955	5915
200	1940	2915	5830
250	1915	2875	5750
300	1895	2845	5690
350	1875	2810	5625
400	1850	2775	5550
450	1810	2715	5430
500	1735	2605	5210
550	1640	2460	4925
600	1540	2310	4620
650	1430	2150	4300
700	1305	1960	3920
750	1180	1775	3550
800	1015	1525	3050
850[1]	830	1250	2500
875[1]	725	1090	2180
900[1]	615	925	1855
925[1,2]	520	785	1570
950[1,2]	425	640	1285
975[1,2]	330	500	1000
1000[1,2]	235	355	715

[1] Product used within the jurisdiction of Section 1, Power Boilers, of the ASME Boiler and Pressure Vessel Code is subject to the same maximum temperature limitations placed upon the material in Table P7, 1959 edition thereof.

[2] Product used within the jurisdiction of Section 1, Power Piping, of the ASA Code for Pressure Piping B31.1 is subject to the same maximum temperature limitations placed upon piping of the same general composition in Table 2a, 1955 edition thereof.

Recommendations: These are unusually strong, rugged fittings. They are ideally suited for high pressure hydraulic lines and for high pressure-temperature service in oil refineries, oil and gas fields, central power stations, and industrial and chemical plants.

The 2000-Pound W.O.G. Fittings, exceptionally compact and light in weight, are intended for services beyond the temperature range of malleable iron fittings and for many relatively low pressure installations where the extra strength and safety afforded by steel fittings are desired.

Materials and design: Elbows, tees, crosses, and Y-bends are forged solid; the caps, couplings, reducers, plugs, and bushings are machined from solid steel. Carbon steel billets or bar stock used in the manufacturing process are subject to rigid specifications for strength, toughness, and resistance to temperature and shock.

The fittings feature liberal metal sections throughout and have an ample factor of safety over the recommended working pressures. All openings are drilled; on forged fittings, each opening is reinforced with a wide band which completely surrounds the thread chamber, extending beyond the last thread. The design provides the requisite strength, adds to the compact, neat appearance, and permits a sure wrench grip.

Threads: Threads are long and are accurately cut to gauge. All openings are in true alignment and chamfered to permit easy entrance of pipe.

MSS ratings: Working pressures agree with those in the MSS Standard for Forged Steel Screwed Fittings, No. SP-49-1956.

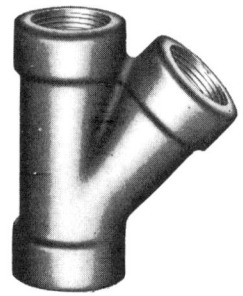

Round Head Plug
No. 308, 3000-Pound

Square Head Plug
No. 309, 3000-Pound

Hexagon Head Plug
No. 602, 6000-Pound

Face Bushing
No. 601, 6000-Pound

Hexagon Bushing
No. 600, 6000-Pound

Figure 1-1. Forged steel socketwelding fittings (2000, 3000, 4000 and 6000 pound W.O.G. Reprinted courtesy of Crane Co.

General Piping

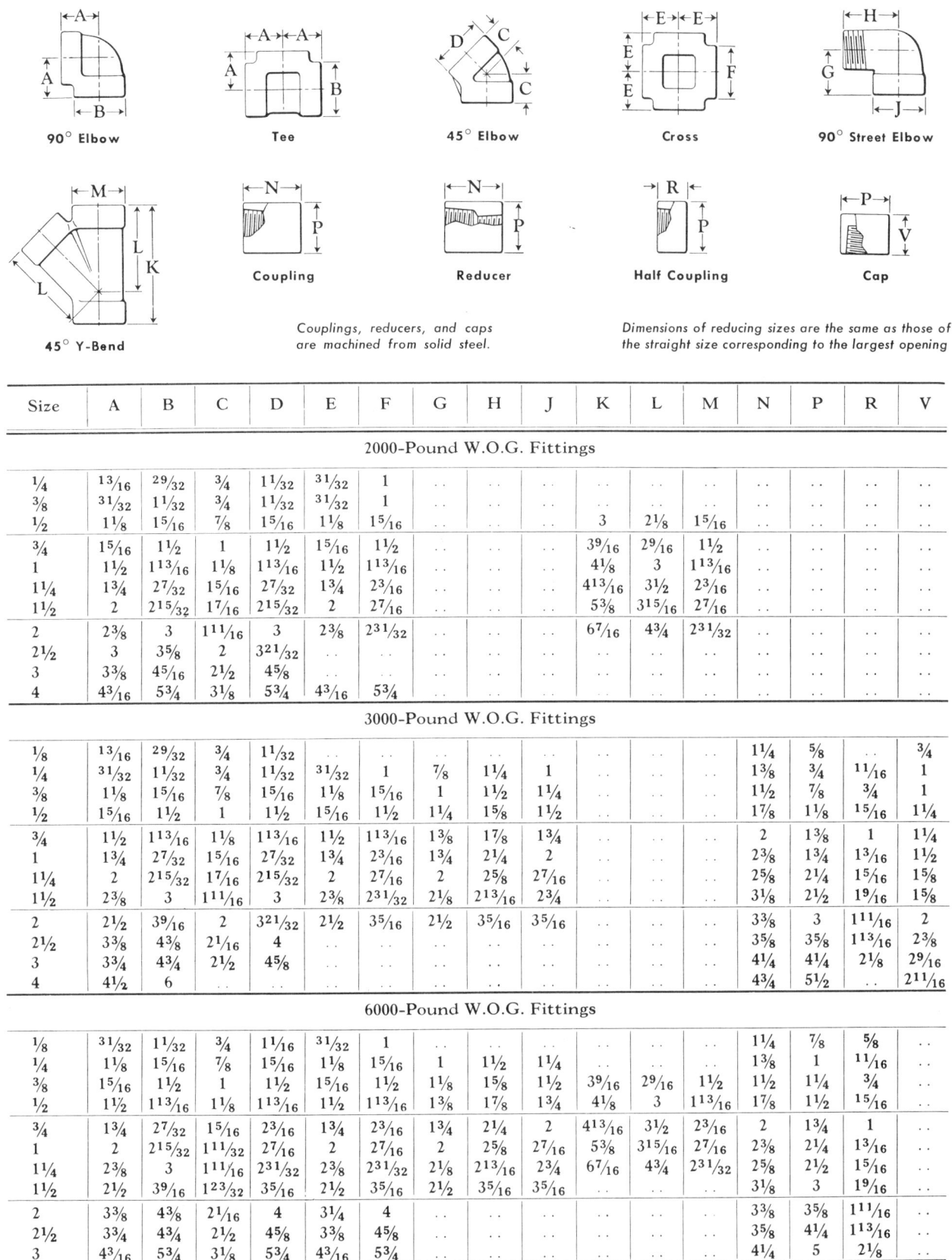

Figure 1-2. Forged steel screwed fittings (dimensions in inches).

Process Piping Design

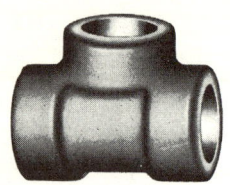

Tee
No. 1241, 2000-Pound WOG
No. 1381, 3000-Pound WOG
No. 1461, 4000-Pound WOG
No. 1661, 6000-Pound WOG

45° Elbow
No. 1242, 2000-Pound WOG
No. 1382, 3000-Pound WOG
No. 1462, 4000-Pound WOG
No. 1662, 6000-Pound WOG

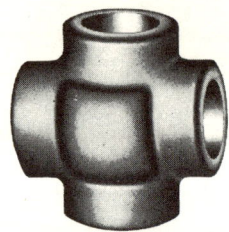

Cross
No. 1243, 2000-Pound WOG
No. 1383, 3000-Pound WOG
No. 1463, 4000-Pound WOG
No. 1663, 6000-Pound WOG

90° Elbow
No. 1240, 2000-Pound WOG
No. 1380, 3000-Pound WOG
No. 1460, 4000-Pound WOG
No. 1660, 6000-Pound WOG

2000-Pound WOG Fittings are for use with Schedule 40 or Standard pipe
3000-Pound WOG Fittings are for use with Schedule 80 or Extra Strong pipe
4000-Pound WOG Fittings are for use with Schedule 160 pipe
6000-Pound WOG Fittings are for use with Double Extra Strong pipe

45° Y-Bend
No. 1245, 2000-Pound WOG
No. 1385, 3000-Pound WOG
No. 1465, 4000-Pound WOG
No. 1665, 6000-Pound WOG

Coupling
No. 1246, 2000-Pound WOG
No. 1386, 3000-Pound WOG
No. 1466, 4000-Pound WOG
No. 1666, 6000-Pound WOG

Reducer
No. 1247, 2000-Pound WOG
No. 1387, 3000-Pound WOG
No. 1467, 4000-Pound WOG
No. 1667, 6000-Pound WOG

Cap
No. 1249, 2000-Pound WOG
No. 1389, 3000-Pound WOG
No. 1469, 4000-Pound WOG
No. 1669, 6000-Pound WOG

Recommendations: These unusually rugged, durable fittings are ideal for small (up to and including 4") welded lines on relatively low pressure service, for high pressure hydraulic lines, or for high pressure-temperature service.

The 2000-Pound WOG Fittings are for use with Schedule 40 or Standard pipe... the 3000-Pound, with Schedule 80 or Extra Strong pipe... the 4000-Pound, with Schedule 160 pipe... and the 6000-Pound, with Double Extra Strong pipe.

Design: Elbows, tees, crosses, and Y-bends are forged solid; their openings are reinforced with a wide band which completely surrounds the socket chamber, extends well beyond the back of the socket, and meets recognized requirements for socket-weld dimensions. Reducer inserts, couplings, reducers, and caps are machined from solid steel. Openings of all fittings are drilled and the ends are bored to slip over pipe.

Materials: The fittings are made from high grade carbon steel (ASTM A 105, Grade II) of unusual strength and toughness. It is particularly suitable for fusion welding.

American Standard: These fittings conform to the American Standard for Steel Socket-Welding Fittings (B16.11-1946). This Standard includes elbows, tees, crosses, and couplings in sizes 3-inch and smaller for use with Schedule 40, Schedule 80, and Schedule 160 pipe.

*Note: When pipe is rated in accordance with the Code for Pressure Piping or any other Code, these fittings may be used for the same pressures and temperatures as the pipe even though such ratings exceed those in the table above.

The fittings, of course, must be made of a material having chemical and physical properties comparable to the pipe, and must be of suitable weight, as indicated by the schedule numbers.

Working Pressures*
Steam, Water, Oil, Oil Vapor, Gas, or Air

Material	Temp. Deg. Fahr.	2000 Pound WOG	3000 Pound WOG	4000 Pound WOG	6000 Pound WOG
		Pounds, Non-Shock			
Carbon Steel ASTM A 105 Grade II	100°	2000	3000	4000	6000
	150	1970	2955	3940	5915
	200	1940	2915	3885	5830
	250	1915	2875	3830	5750
	300	1895	2845	3790	5690
	350	1875	2810	3750	5625
	400	1850	2775	3700	5550
	450	1810	2715	3620	5430
	500	1735	2605	3470	5210
	550	1640	2460	3280	4925
	600	1540	2310	3080	4620
	650	1430	2150	2865	4300
	700	1305	1960	2610	3920
	750	1180	1775	2365	3550
	800	1015	1525	2030	3050
	850	830	1250	1665	2500
	875	725	1090	1450	2180
	900	615	925	1235	1855
	925	520	785	1045	1570
	950	425	640	855	1285
	975	330	500	665	1000
	1000	235	355	475	715

Figure 1-3. Forged steel socketwelding fittings (2000, 3000, 4000 and 6000 pound W.O.G.).

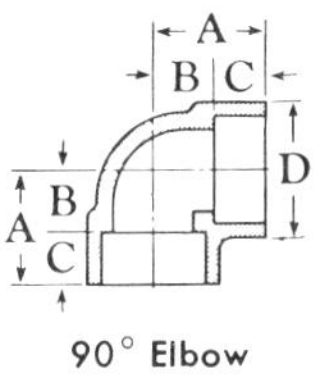

Dimensions of reducing sizes are the same as those of the straight size corresponding to the largest opening.

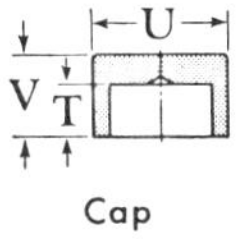

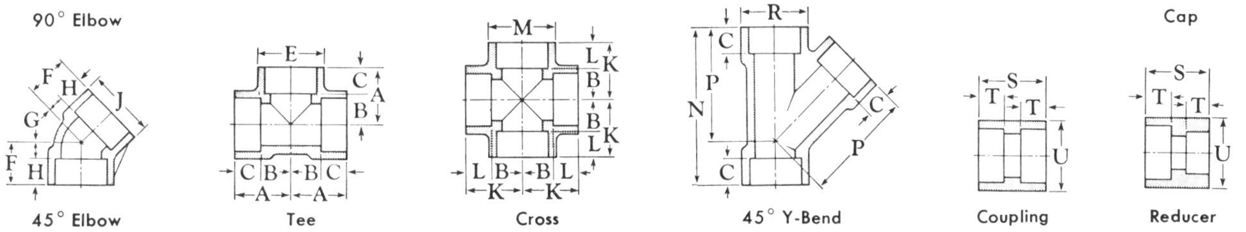

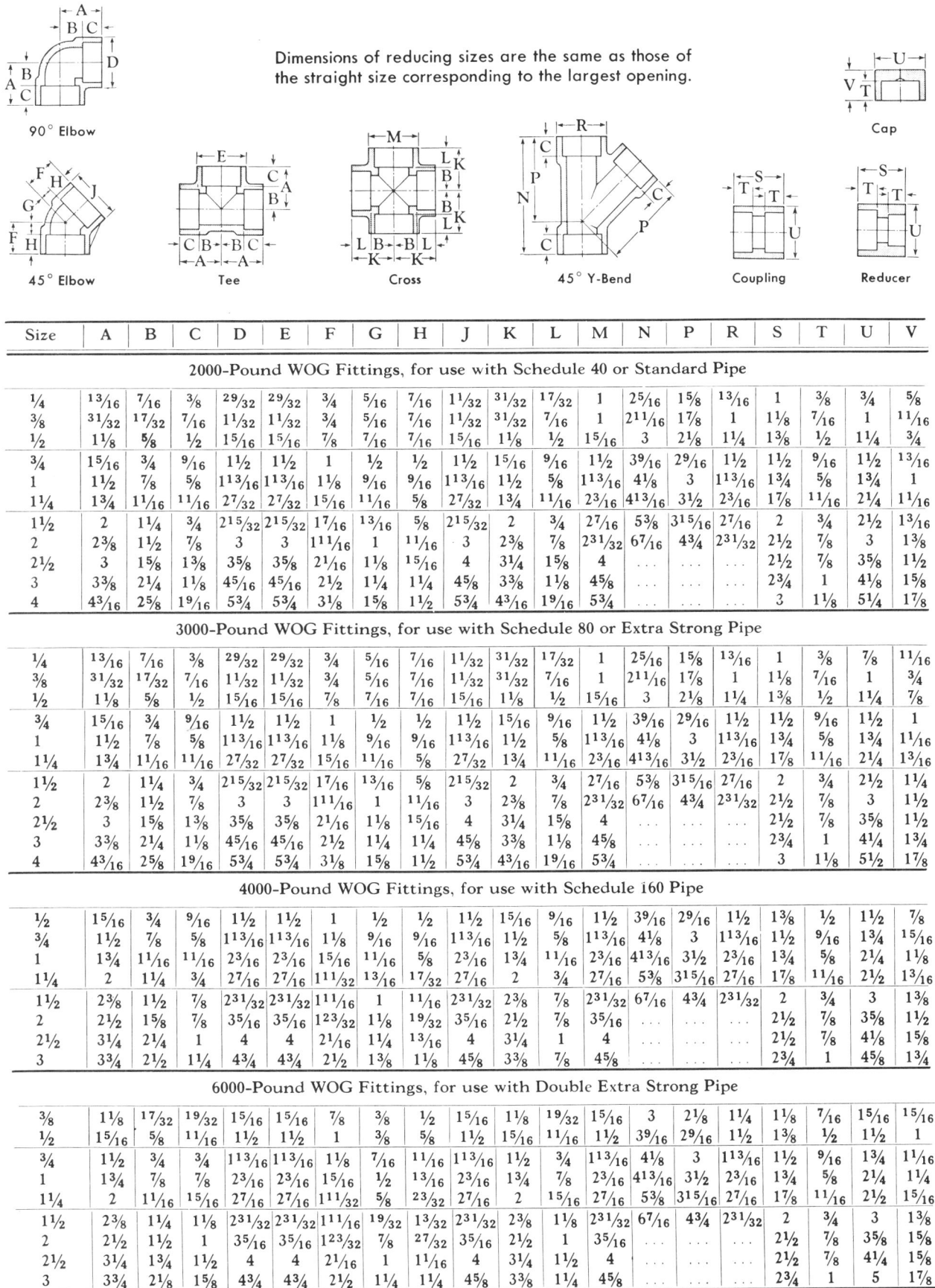

Size	A	B	C	D	E	F	G	H	J	K	L	M	N	P	R	S	T	U	V
2000-Pound WOG Fittings, for use with Schedule 40 or Standard Pipe																			
1/4	13/16	7/16	3/8	29/32	29/32	3/4	5/16	7/16	1 1/32	31/32	17/32	1	2 5/16	1 5/8	13/16	1	3/8	3/4	5/8
3/8	31/32	17/32	7/16	1 1/32	1 1/32	3/4	5/16	7/16	1 1/32	31/32	7/16	1	2 11/16	1 7/8	1	1 1/8	7/16	1	11/16
1/2	1 1/8	5/8	1/2	1 5/16	1 5/16	7/8	7/16	7/16	1 5/16	1 1/8	1/2	1 5/16	3	2 1/8	1 1/4	1 3/8	1/2	1 1/4	3/4
3/4	1 5/16	3/4	9/16	1 1/2	1 1/2	1	1/2	1/2	1 1/2	1 5/16	9/16	1 1/2	3 9/16	2 9/16	1 1/2	1 1/2	9/16	1 1/2	13/16
1	1 1/2	7/8	5/8	1 13/16	1 13/16	1 1/8	9/16	9/16	1 13/16	1 1/2	5/8	1 13/16	4 1/8	3	1 13/16	1 3/4	5/8	1 3/4	1
1 1/4	1 3/4	1 1/16	11/16	2 7/32	2 7/32	1 5/16	11/16	5/8	2 7/32	1 3/4	11/16	2 3/16	4 13/16	3 1/2	2 3/16	1 7/8	11/16	2 1/4	1 1/16
1 1/2	2	1 1/4	3/4	2 15/32	2 15/32	1 7/16	13/16	5/8	2 15/32	2	3/4	2 7/16	5 3/8	3 15/16	2 7/16	2	3/4	2 1/2	1 3/16
2	2 3/8	1 1/2	7/8	3	3	1 11/16	1	11/16	3	2 3/8	7/8	2 31/32	6 7/16	4 3/4	2 31/32	2 1/2	7/8	3	1 3/8
2 1/2	3	1 5/8	1 3/8	3 5/8	3 5/8	2 1/16	1 1/8	15/16	4	3 1/4	1 5/8	4	...	...	...	2 1/2	7/8	3 5/8	1 1/2
3	3 3/8	2 1/4	1 1/8	4 5/16	4 5/16	2 1/2	1 1/4	1 1/4	4 5/8	3 3/8	1 1/8	4 5/8	...	...	...	2 3/4	1	4 1/8	1 5/8
4	4 3/16	2 5/8	1 9/16	5 3/4	5 3/4	3 1/8	1 5/8	1 1/2	5 3/4	4 3/16	1 9/16	5 3/4	...	...	...	3	1 1/8	5 1/4	1 7/8
3000-Pound WOG Fittings, for use with Schedule 80 or Extra Strong Pipe																			
1/4	13/16	7/16	3/8	29/32	29/32	3/4	5/16	7/16	1 1/32	31/32	17/32	1	2 5/16	1 5/8	13/16	1	3/8	7/8	11/16
3/8	31/32	17/32	7/16	1 1/32	1 1/32	3/4	5/16	7/16	1 1/32	31/32	7/16	1	2 11/16	1 7/8	1	1 1/8	7/16	1	3/4
1/2	1 1/8	5/8	1/2	1 5/16	1 5/16	7/8	7/16	7/16	1 5/16	1 1/8	1/2	1 5/16	3	2 1/8	1 1/4	1 3/8	1/2	1 1/4	7/8
3/4	1 5/16	3/4	9/16	1 1/2	1 1/2	1	1/2	1/2	1 1/2	1 5/16	9/16	1 1/2	3 9/16	2 9/16	1 1/2	1 1/2	9/16	1 1/2	1
1	1 1/2	7/8	5/8	1 13/16	1 13/16	1 1/8	9/16	9/16	1 13/16	1 1/2	5/8	1 13/16	4 1/8	3	1 13/16	1 3/4	5/8	1 3/4	1 1/16
1 1/4	1 3/4	1 1/16	11/16	2 7/32	2 7/32	1 5/16	11/16	5/8	2 7/32	1 3/4	11/16	2 3/16	4 13/16	3 1/2	2 3/16	1 7/8	11/16	2 1/4	1 3/16
1 1/2	2	1 1/4	3/4	2 15/32	2 15/32	1 7/16	13/16	5/8	2 15/32	2	3/4	2 7/16	5 3/8	3 15/16	2 7/16	2	3/4	2 1/2	1 1/4
2	2 3/8	1 1/2	7/8	3	3	1 11/16	1	11/16	3	2 3/8	7/8	2 31/32	6 7/16	4 3/4	2 31/32	2 1/2	7/8	3	1 1/2
2 1/2	3	1 5/8	1 3/8	3 5/8	3 5/8	2 1/16	1 1/8	15/16	4	3 1/4	1 5/8	4	...	...	...	2 1/2	7/8	3 5/8	1 1/2
3	3 3/8	2 1/4	1 1/8	4 5/16	4 5/16	2 1/2	1 1/4	1 1/4	4 5/8	3 3/8	1 1/8	4 5/8	...	...	...	2 3/4	1	4 1/4	1 3/4
4	4 3/16	2 5/8	1 9/16	5 3/4	5 3/4	3 1/8	1 5/8	1 1/2	5 3/4	4 3/16	1 9/16	5 3/4	...	...	...	3	1 1/8	5 1/2	1 7/8
4000-Pound WOG Fittings, for use with Schedule 160 Pipe																			
1/2	1 5/16	3/4	9/16	1 1/2	1 1/2	1	1/2	1/2	1 1/2	1 5/16	9/16	1 1/2	3 9/16	2 9/16	1 1/2	1 3/8	1/2	1 1/2	7/8
3/4	1 1/2	7/8	5/8	1 13/16	1 13/16	1 1/8	9/16	9/16	1 13/16	1 1/2	5/8	1 13/16	4 1/8	3	1 13/16	1 1/2	9/16	1 3/4	15/16
1	1 3/4	1 1/16	11/16	2 3/16	2 3/16	1 5/16	11/16	5/8	2 3/16	1 3/4	11/16	2 3/16	4 13/16	3 1/2	2 3/16	1 3/4	5/8	2 1/4	1 1/8
1 1/4	2	1 1/4	3/4	2 7/16	2 7/16	1 11/32	13/16	17/32	2 7/16	2	3/4	2 7/16	5 3/8	3 15/16	2 7/16	1 7/8	11/16	2 1/2	1 3/16
1 1/2	2 3/8	1 1/2	7/8	2 31/32	2 31/32	1 11/16	1	11/16	2 31/32	2 3/8	7/8	2 31/32	6 7/16	4 3/4	2 31/32	2	3/4	3	1 3/8
2	2 1/2	1 5/8	7/8	3 5/16	3 5/16	1 23/32	1 1/8	19/32	3 5/16	2 1/2	7/8	3 5/16	...	...	...	2 1/2	7/8	3 5/8	1 1/2
2 1/2	3 1/4	2 1/4	1	4	4	2 1/16	1 1/4	13/16	4	3 1/4	1	4	...	...	...	2 1/2	7/8	4 1/8	1 5/8
3	3 3/4	2 1/2	1 1/4	4 3/4	4 3/4	2 1/2	1 3/8	1 1/8	4 5/8	3 3/8	7/8	4 5/8	...	...	...	2 3/4	1	4 5/8	1 3/4
6000-Pound WOG Fittings, for use with Double Extra Strong Pipe																			
3/8	1 1/8	17/32	19/32	1 5/16	1 5/16	7/8	3/8	1/2	1 5/16	1 1/8	19/32	1 5/16	3	2 1/8	1 1/4	1 1/8	7/16	1 5/16	15/16
1/2	1 5/16	5/8	11/16	1 1/2	1 1/2	1	3/8	5/8	1 1/2	1 5/16	11/16	1 1/2	3 9/16	2 9/16	1 1/2	1 3/8	1/2	1 1/2	1
3/4	1 1/2	3/4	3/4	1 13/16	1 13/16	1 1/8	7/16	11/16	1 13/16	1 1/2	3/4	1 13/16	4 1/8	3	1 13/16	1 1/2	9/16	1 3/4	1 1/16
1	1 3/4	7/8	7/8	2 3/16	2 3/16	1 5/16	1/2	13/16	2 3/16	1 3/4	7/8	2 3/16	4 13/16	3 1/2	2 3/16	1 3/4	5/8	2 1/4	1 1/4
1 1/4	2	1 1/16	15/16	2 7/16	2 7/16	1 11/32	5/8	23/32	2 7/16	2	15/16	2 7/16	5 3/8	3 15/16	2 7/16	1 7/8	11/16	2 1/2	1 5/16
1 1/2	2 3/8	1 1/4	1 1/8	2 31/32	2 31/32	1 11/16	19/32	1 3/32	2 31/32	2 3/8	1 1/8	2 31/32	6 7/16	4 3/4	2 31/32	2	3/4	3	1 3/8
2	2 1/2	1 1/2	1	3 5/16	3 5/16	1 23/32	7/8	27/32	3 5/16	2 1/2	1	3 5/16	...	...	...	2 1/2	7/8	3 5/8	1 5/8
2 1/2	3 1/4	1 3/4	1 1/2	4	4	2 1/16	1	1 1/16	4	3 1/4	1 1/2	4	...	...	...	2 1/2	7/8	4 1/4	1 5/8
3	3 3/4	2 1/8	1 5/8	4 3/4	4 3/4	2 1/2	1 1/4	1 1/4	4 5/8	3 3/8	1 1/4	4 5/8	...	...	...	2 3/4	1	5	1 7/8

Figure 1-4. Forged steel socketwelding fittings (dimensions in inches).

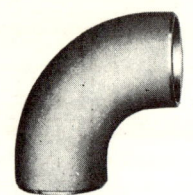

No. 352 E
90° Long Radius Elbow
Straight and Reducing

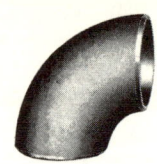

No. 331 E
90° Short Radius Elbow

No. 354 E
45° Long Radius Elbow

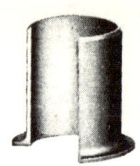

No. 574
Cranelap Stub End

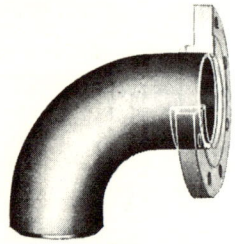

No. 335 E
90° Long Radius Elbow
Long Tangent on One End
(flange is not included)

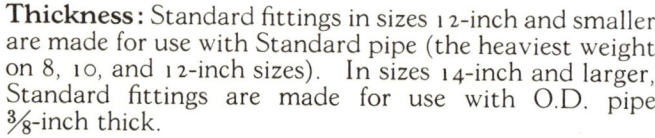

No. 353 E
Tee
Straight and Reducing

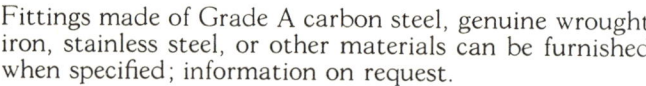
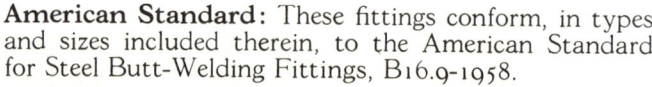

No. 336 E
Cross
Straight and Reducing

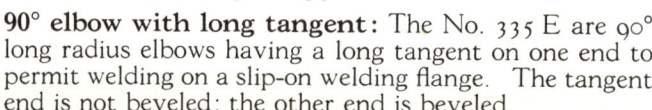

Return Bend
No. 372 E, Short Radius
No. 373 E, Long Radius

No. 350 E
90° Shaped Nipple

No. 351 E
45° Shaped Nipple

No. 357 E
Cap

Concentric Reducer

Eccentric Reducer

Thickness: Standard fittings in sizes 12-inch and smaller are made for use with Standard pipe (the heaviest weight on 8, 10, and 12-inch sizes). In sizes 14-inch and larger, Standard fittings are made for use with O.D. pipe ⅜-inch thick.

Materials: Unless otherwise specified, the fittings are made of carbon steel conforming to requirements of ASTM Specification A 234, Grade B.

Fittings made of Grade A carbon steel, genuine wrought iron, stainless steel, or other materials can be furnished when specified; information on request.

American Standard: These fittings conform, in types and sizes included therein, to the American Standard for Steel Butt-Welding Fittings, B16.9-1958.

The Standard does not include sizes smaller than 1-inch, nor does it include 90° elbows with a long tangent on one end, short radius 90° elbows, crosses, short radius return bends, or shaped nipples.

90° elbow with long tangent: The No. 335 E are 90° long radius elbows having a long tangent on one end to permit welding on a slip-on welding flange. The tangent end is not beveled; the other end is beveled.

Cranelap stub ends: Cranelap stub ends, made of Grade B seamless steel pipe lapped to the full thickness of the pipe wall, and Cranelap flanges afford an ideal method of installing flanged equipment in a welded line. The swivel flange eliminates the difficulty of aligning bolt holes and permits installing the equipment at any angle.

Shaped nipples: Shaped nipples eliminate the use of templates when saddling one pipe upon another; they save erection time and assure an accurate fit. Both ends are beveled for welding. When ordering, be sure to specify both the pipe size and the nominal size of the header on which the nipple will be used; header sizes which the nipples are shaped to fit are included in the upper table on the facing page.

Prices: Prices are furnished on request.

Ordering reducing tees and crosses: When ordering reducing tees and crosses, specify the size of openings in the sequence of the lower case letters (a and b) shown on their illustrations at the left.

Figure 1-5. Steel buttwelding fittings for use with standard pipe. Reprinted courtesy of Crane Co.

General Piping

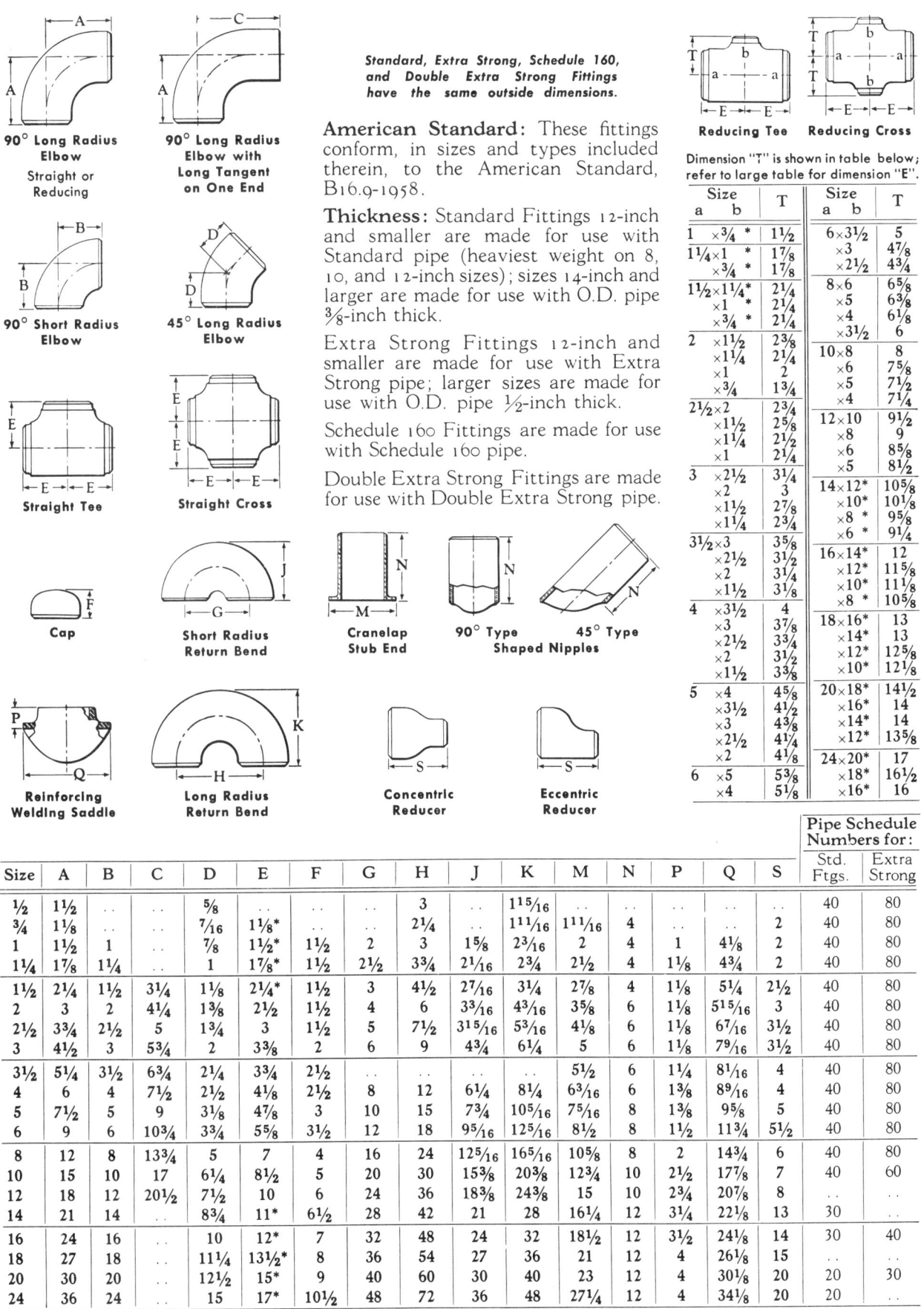

Figure 1-6. Steel buttwelding fittings (dimensions in inches). Courtesy of Crane Co.

Welded flanged joints can be furnished in the types illustrated here. The Cranelap stub ends with Cranelap flange, also illustrated, afford an auxiliary flanged connection for welding.

Application: Any of the welded flanged joints shown at the right can be applied to straight pipe, pipe bends, the ends and nozzles of welded headers, and the flanged ends of welded assemblies. Special shop equipment assures the perfect alignment of flange faces on all Crane Welded Flanged Joints.

Welding: The shop welding of these flanged joints is performed by Crane welders working under approved procedure control.

Special piping materials: These types of welded flanged joints can be furnished on many special piping materials, including numerous alloy steels, with facilities for heat-treating after fabrication.

Complete information and prices will be furnished on application.

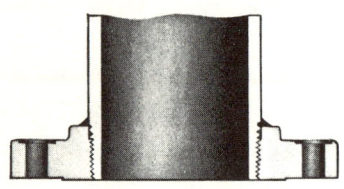

Screwed Flange
Seal-Welded and Refaced

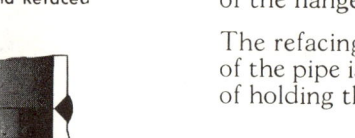

Welding Neck Flange
Butt-Welded to Pipe

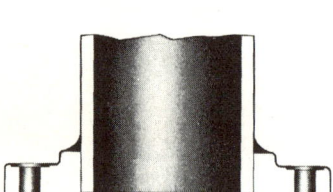

Type No. 1
Slip-On Welding Flange
Welded Front and Back

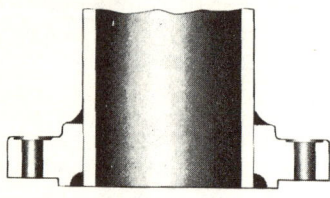

Type No. 2
Slip-On Welding Flange
Welded Front and Back and Refaced

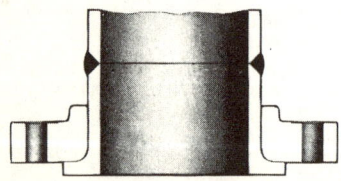

Cranelap Flange with
Cranelap Stub End
Butt-Welded to Pipe

Forged Steel Screwed Flange, Seal-Welded

A Crane Forged Steel Screwed Flange is used in this joint. The pipe and the flange are accurately threaded; the flange is made up tight on the pipe, seal-welded, and then refaced. The joint is sealed by fillet-welding the back of the flange to the pipe, thus assuring no leakage through the threads.

The refacing assures perfect alignment of the flange faces, and that the end of the pipe is flush with the face of the flange. The threads retain the function of holding the flange securely on the pipe, hence there is no shearing action.

Forged Steel Welding Neck Flange

Crane Welding Neck Flanges are of forged steel. They are machined with a beveled end and bored to match the inside diameter of the pipe to which they are applied. A butt-weld is used to attach the welding neck flange to the pipe, which is also machine beveled.

Forged Steel Slip-On Welding Flange

Crane Forged Steel Slip-On Welding Flanges are bored for a snug fit on the pipe and, when applied to fabricated piping, are welded at the front and back through the two methods defined below and illustrated at the right.

Type No. 1: Type No. 1 is Crane standard for welded flanged joints using Forged Steel Slip-On Welding Flanges. Regular flanges are utilized with the end of the pipe set back from the face of the flange and the flange welded to the pipe both in front and back.

Type No. 2: Type No. 2 is furnished on special order only; slip-on flanges with a special front groove for welding are used. The pipe is flush with the flange face; this is accomplished by refacing, after both the front and back of the flange are welded to the pipe.

Code limitation: When piping must comply with the American Standard Code for Pressure Piping or the ASME Boiler and Pressure Vessel Code, the use of the slip-on flanged joint is permissible on all sizes of flanges listed under primary service pressure ratings up to and including the 900-pound class, and in sizes 2½-inch and smaller of the 1500-pound class, of the American Steel Flange Standard (ASA B16.5-1957).

Cranelap Stub Ends and Cranelap Flange

The Cranelap stub end with Cranelap flange can be applied to fabricated piping. Both the stub end and the pipe are machine beveled. A butt-weld is used to complete the joint.

This type of joint has all of the advantages of the regular Cranelap joint. In most cases, piping can be fabricated with Cranelap joints applied directly, which eliminates the weld necessary for the application of the Cranelap stub end with Cranelap flange.

Figure 1-7. Flanging processes—welded flanged joints. Reprinted courtesy of Crane Co.

General Piping

The Crane line of Forged Steel Flanges comprises the complete assortment of straight and reducing types illustrated on this page. Made in seven different pressure classes.... 150, 300, 400, 600, 900, 1500, and 2500-Pound.... they are available in a variety of materials and with various flange facings, providing a correct type for any service requirement.

Reducing Screwed Flange

No. 558½,	150-Pound
No. 292 E,	300-Pound
No. 658 E,	400-Pound
No. 857 E,	600-Pound
No. 1263 E,	900-Pound
No. 1558 E,	1500-Pound

Screwed Flange

No. 556,	150-Pound
No. 291 E,	300-Pound
No. 651 E,	400-Pound
No. 856 E,	600-Pound
No. 1266 E,	900-Pound
No. 1556 E,	1500-Pound

Materials: Crane flanges are made of carbon steel forgings having a highly refined grain structure and generally excellent physical properties well in excess of recognized minimum requirements.

In the 150 and 300-pound pressure classes, the flanges are regularly made of carbon steel conforming to ASTM Specification A 181, Grade II; on special order, they can be furnished heat-treated (normalized or annealed) to conform to ASTM Specification A 105, Grade II.

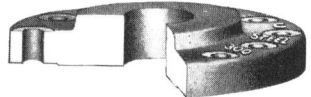

Reducing Slip-On Welding Flange

No. 554½,	150-Pound
No. 290 E,	300-Pound
No. 693 E,	400-Pound
No. 853 E,	600-Pound
No. 1295 E,	900-Pound
No. 1595 E,	1500-Pound

Slip-On Welding Flange

No. 554,	150-Pound
No. 294 E,	300-Pound
No. 694 E,	400-Pound
No. 854 E,	600-Pound
No. 1294 E,	900-Pound
No. 1594 E,	1500-Pound

In the 400-pound and higher pressure classes, the flanges are regularly made of carbon steel conforming to ASTM Specification A 105, Grade II.

In addition, flanges in 300-pound and higher pressure classes can be made to order of Crane No. 5 Chrome-Molybdenum Forged Steel (ASTM A 182, Grade F5a).

American Standard: The dimensions and drilling of all flanges conform to the American Steel Flange Standard B16.5-1957, for their respective pressure class.

This Standard does not include slip-on welding flanges of the 2500-pound class nor sizes 3-inch and larger of the 1500-pound class; in such classes and sizes, Crane slip-on welding flanges have the same dimensions as American Standard Steel Screwed Flanges, being bored instead of threaded.

Flange facings: The 150 and 300-Pound Screwed, Slip-On Welding, Welding Neck, and Blind Flanges are regularly furnished with an American Standard 1/16-inch raised face.

The aforementioned flanges, in 400-pound and higher pressure classes, are regularly furnished with an American Standard 1/4-inch male face (large male).

Other types of facings such as ring joint, female, tongue, groove, etc., can be furnished; see pages 332 to 335 for complete information.

In addition, flanges of any pressure class are available with a flat face (raised or male face removed); the flat face will have a spiral serrated finish.

Finish of flange faces: The 1/16-inch raised faces and the 1/4-inch large male faces are regularly furnished with a serrated finish. A smooth finish can be furnished when specified.

Drilling: The flanges are regularly furnished faced, drilled, and spot faced to the corresponding pressure class of the American Standard. They can be furnished faced only, when specified.

Reducing flanges: The Reducing Screwed and Reducing Slip-On Welding Flanges, illustrated above, are available in any size reduction; prices are based on the outside diameter of the flange. For ordering information, see page 311.

Reducing Welding Neck Flanges and Eccentric Reducing Screwed or Slip-On Welding Flanges can be made to order; information on request.

Reducing Cranelap Flanges are not recommended and, consequently, are not manufactured. Another type of flanged joint or connection should be used.

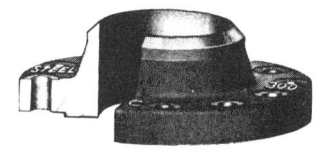

Welding Neck Flange
(For boring, see page 314.)

No. 568,	150-Pound
No. 296 E,	300-Pound
No. 656 E,	400-Pound
No. 855 E,	600-Pound
No. 1265 E,	900-Pound
No. 1565 E,	1500-Pound

Cranelap Flange

No. 572,	150-Pound
No. 496 E,	300-Pound
No. 664 E,	400-Pound
No. 862 E,	600-Pound
No. 1262 E,	900-Pound
No. 1562 E,	1500-Pound

Blind Flange

No. 556½,	150-Pound
No. 297 E,	300-Pound
No. 657 E,	400-Pound
No. 858 E,	600-Pound
No. 1267 E,	900-Pound
No. 1557 E,	1500-Pound

Figure 1-8. Forged steel flanges. Reprinted courtesy of Crane Co.

Steam, Water, Oil, Oil Vapor, Gas, or Air

Metal	Temp. Deg. Fahr.	Pounds per Square Inch, Non-Shock						
		150 Lb.	300 Lb.	400 Lb.	600 Lb.	900 Lb.	1500 Lb.	2500 Lb.
[1]Carbon Steel Flanges	100°	275	720	960	1440	2160	3600	6000
	150	255	710	945	1420	2130	3550	5915
	200	240	700	930	1400	2100	3500	5830
	250	225	690	920	1380	2070	3450	5750
	300	210	680	910	1365	2050	3415	5690
	350	195	675	900	1350	2025	3375	5625
	400	180	665	890	1330	2000	3330	5550
	450	165	650	870	1305	1955	3255	5430
	500	150	625	835	1250	1875	3125	5210
	550	140	590	790	1180	1775	2955	4925
	600	130	555	740	1110	1660	2770	4620
	650	120	515	690	1030	1550	2580	4300
	700	110	470	635	940	1410	2350	3920
	750	100	425	575	850	1275	2125	3550
	800	92	365	490	730	1100	1830	3050
	850	[2]82	[2]300	[2]400	[2]600	[2]900	[2]1500	[2]2500
	875	[2]75	[2]260	[2]350	[2]525	[2]785	[2]1305	[2]2180
	900	[2]70	[2]225	[2]295	[2]445	[2]670	[2]1115	[2]1855
	925	[2]60[3]	[2]190[3]	[2]250[3]	[2]375[3]	[2]565[3]	[2]945[3]	[2]1570[3]
	950	[2]55[3]	[2]155[3]	[2]205[3]	[2]310[3]	[2]465[3]	[2]770[3]	[2]1285[3]
	975	[2]50[3]	[2]120[3]	[2]160[3]	[2]240[3]	[2]360[3]	[2]600[3]	[2]1000[3]
	1000	[2]40[3]	[2]85[3]	[2]115[3]	[2]170[3]	[2]255[3]	[2]430[3]	[2]715[3]

At temperatures lower than 700 F, ratings are the same as those for Carbon Steel Flanges.

Metal	Temp.	150 Lb.	300 Lb.	400 Lb.	600 Lb.	900 Lb.	1500 Lb.	2500 Lb.
No. 5 Chromium- Molybdenum Alloy Steel Flanges (made to order)	700°	...	485	645	965	1450	2415	4025
	750	...	450	600	900	1350	2250	3745
	800	...	415	555	835	1250	2080	3470
	850	...	385	510	765	1150	1915	3190
	875	...	365	490	735	1100	1830	3055
	900	...	350	465	700	1050	1750	2915
	925	...	335	445	665	1000	1665	2775
	950	...	315	420	635	950	1585	2640
	975	...	300	400	600	900	1500	2500
	1000	...	250	335	500	750	1250	2085
	1025	...	215	285	430	645	1070	1785
	1050	...	180	240	355	535	890	1485
	1075	...	145	195	290	435	730	1215
	1100	...	115	150	225	340	565	945
	1125[4]	...	95	125	190	285	470	785
	1150[4]	...	75	100	150	225	375	630
	1175[4]	...	65	85	125	190	315	530
	1200[4]	...	50	70	105	155	255	430

ASA and API Standards: Crane pressure-temperature ratings conform to those listed in the American Steel Flange Standard, ASA B16.5-1957, and in the American Petroleum Institute (API) Standard No. 600, Fourth Edition, 1958.

Cold service: For temperatures between minus 20 F and plus 100 F, the ratings shown in the table for 100 F will apply.

For temperatures below minus 20 F, steels with suitable impact strength must be used; pressure ratings for such steels will be the same as shown in the table for 100 F.

Gaskets: The use of these ratings requires gaskets conforming to requirements set forth in American Standard B16.5-1957.

The user is responsible for selecting gaskets of dimensions and materials capable of withstanding the required bolt loading without injurious crushing, as well as being suitable for the service conditions in all other respects.

Flange facings: Unless otherwise ordered, Crane screwed, slip-on welding, welding neck, and blind flanges of the 150 and 300-pound classes are furnished with a 1/16-inch raised face. In the 400-pound and higher pressure classes, these flanges are furnished with a 1/4-inch large male face.

Cranelap flanges and joints: These ratings also apply to Cranelap flanges, the rating being dependent upon the type of facing applied to the lapped pipe end. Ratings for Cranelap joints are contingent upon the use of pipe of proper material having an equal or higher rating.

---

[1]Where welded construction is used, consideration should be given to the possibility of graphite formation on carbon steel at temperatures above 775 F.

[2]Product used within the jurisdiction of Section 1, Power Boilers, of the ASME Boiler and Pressure Vessel Code, is subject to the same maximum temperature limitations placed upon the material in Table P7, 1959 edition thereof.

[3]Product used within the jurisdiction of Section 1, Power Piping, of the ASA Code for Pressure Piping, B31.1, is subject to the same maximum temperature limitations placed upon piping of the same general composition in Table 2a, 1955 edition thereof.

[4]Consideration should be given to the possibility of excessive oxidation (scaling) when No. 5 Chromium-Molybdenum Steel is used at temperatures above 1100 F.

Figure 1-9. Forged steel flanges—working pressures. Reprinted courtesy of Crane Co.

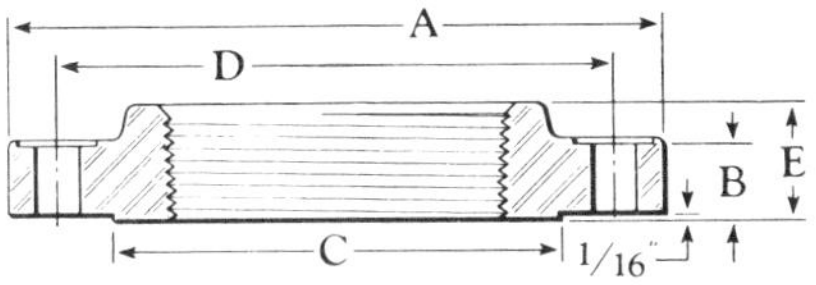

Screwed Flange
150 and 300-Pound

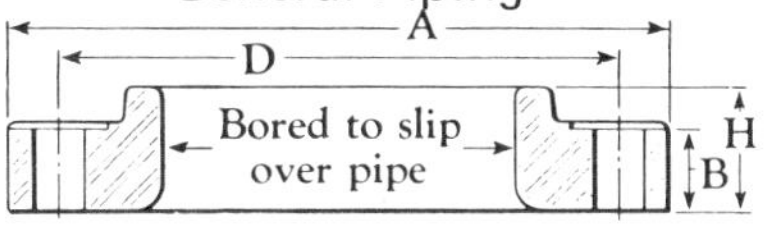

Cranelap Flange
400, 600, 900, 1500, and 2500-Pound

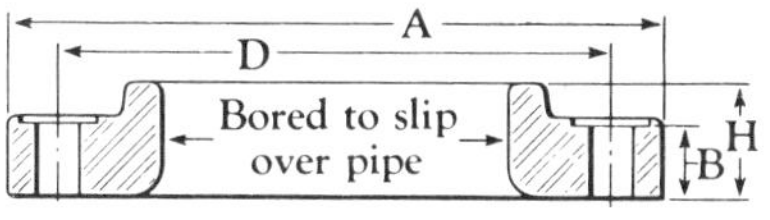

Cranelap Flange
150 and 300-Pound

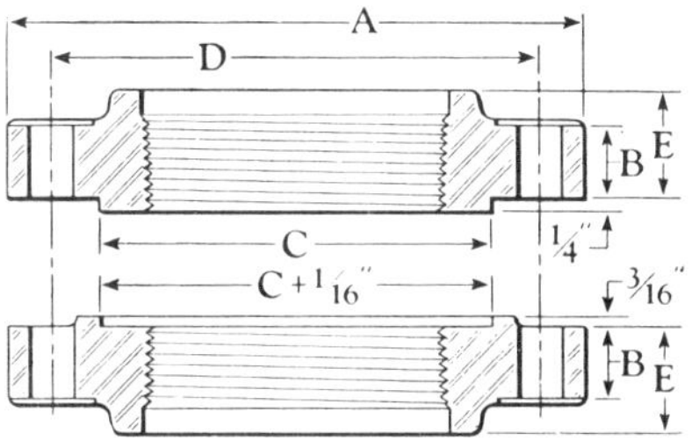

Screwed Flange
400, 600, 900, 1500, and 2500-Pound

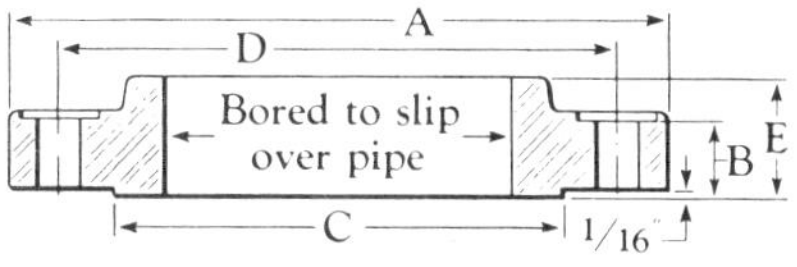

Slip-On Welding Flange
150 and 300-Pound

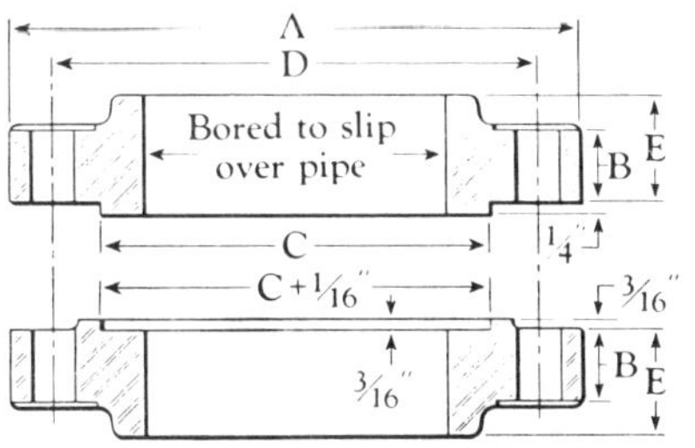

Slip-On Welding Flange
400, 600, 900, and 1500-Pound

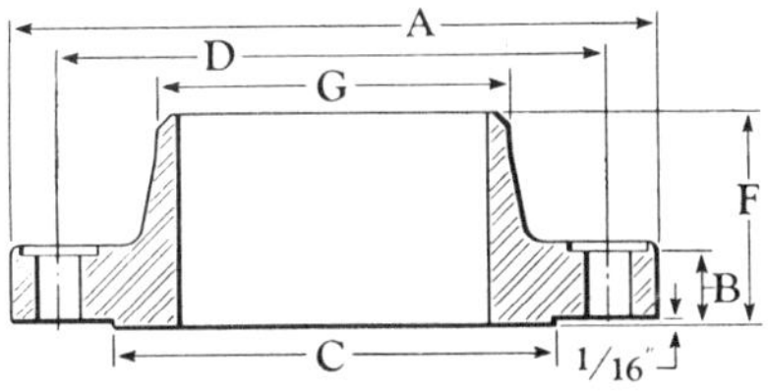

Welding Neck Flange
150 and 300-Pound

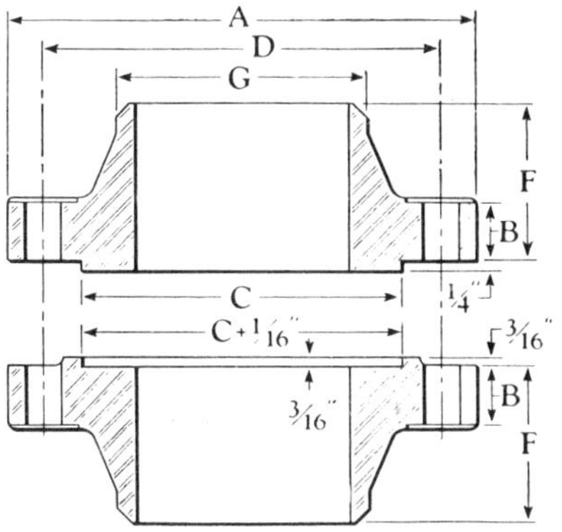

Welding Neck Flange
400, 600, 900, 1500, and 2500-Pound

Class	Pipe Size	A	B	C	D	Bolts No.	Bolts Dia.	E	F	G	H
150 Pound	1/2	3 1/2	7/16	1 3/8	2 3/8	4	1/2	5/8	1 7/8	0.84	5/8
	3/4	3 7/8	1/2	1 11/16	2 3/4	4	1/2	5/8	2 1/16	1.05	5/8
	1	4 1/4	9/16	2	3 1/8	4	1/2	11/16	2 3/16	1.32	11/16
	1 1/4	4 5/8	5/8	2 1/2	3 1/2	4	1/2	13/16	2 1/4	1.66	13/16
	1 1/2	5	11/16	2 7/8	3 7/8	4	1/2	7/8	2 7/16	1.90	7/8
	2	6	3/4	3 5/8	4 3/4	4	5/8	1	2 1/2	2.38	1
	2 1/2	7	7/8	4 1/8	5 1/2	4	5/8	1 1/8	2 3/4	2.88	1 1/8
	3	7 1/2	15/16	5	6	4	5/8	1 3/16	2 3/4	3.50	1 3/16
	3 1/2	8 1/2	15/16	5 1/2	7	8	5/8	1 1/4	2 13/16	4.00	1 1/4
	4	9	15/16	6 3/16	7 1/2	8	5/8	1 5/16	3	4.50	1 5/16
	5	10	15/16	7 5/16	8 1/2	8	3/4	1 7/16	3 1/2	5.56	1 7/16
	6	11	1	8 1/2	9 1/2	8	3/4	1 9/16	3 1/2	6.63	1 9/16
	8	13 1/2	1 1/8	10 5/8	11 3/4	8	3/4	1 3/4	4	8.63	1 3/4
	10	16	1 3/16	12 3/4	14 1/4	12	7/8	1 15/16	4	10.75	1 15/16
	12	19	1 1/4	15	17	12	7/8	2 3/16	4 1/2	12.75	2 3/16
	14	21	1 3/8	16 1/4	18 3/4	12	1	2 1/4	5	14.00	3 1/8
	16	23 1/2	1 7/16	18 1/2	21 1/4	16	1	2 1/2	5	16.00	3 7/16
	18	25	1 9/16	21	22 3/4	16	1 1/8	2 11/16	5 1/2	18.00	3 13/16
	20	27 1/2	1 11/16	23	25	20	1 1/8	2 7/8	5 11/16	20.00	4 1/16
	24	32	1 7/8	27 1/4	29 1/2	20	1 1/4	3 1/4	6	24.00	4 3/8
300 Pound	1/2	3 3/4	9/16	1 3/8	2 5/8	4	1/2	7/8	2 1/16	0.84	...
	3/4	4 5/8	5/8	1 11/16	3 1/4	4	5/8	1	2 1/4	1.05	1
	1	4 7/8	11/16	2	3 1/2	4	5/8	1 1/16	2 7/16	1.32	1 1/16
	1 1/4	5 1/4	3/4	2 1/2	3 7/8	4	5/8	1 1/16	2 9/16	1.66	1 1/16
	1 1/2	6 1/8	13/16	2 7/8	4 1/2	4	3/4	1 3/16	2 11/16	1.90	1 3/16
	2	6 1/2	7/8	3 5/8	5	8	5/8	1 5/16	2 3/4	2.38	1 5/16
	2 1/2	7 1/2	1	4 1/8	5 7/8	8	3/4	1 1/2	3	2.88	1 1/2
	3	8 1/4	1 1/8	5	6 5/8	8	3/4	1 11/16	3 1/8	3.50	1 11/16
	3 1/2	9	1 3/16	5 1/2	7 1/4	8	3/4	1 3/4	3 3/16	4.00	...
	4	10	1 1/4	6 3/16	7 7/8	8	3/4	1 7/8	3 3/8	4.50	1 7/8
	5	11	1 3/8	7 5/16	9 1/4	8	3/4	2	3 7/8	5.56	2
	6	12 1/2	1 7/16	8 1/2	10 5/8	12	3/4	2 1/16	3 7/8	6.63	2 1/16
	8	15	1 5/8	10 5/8	13	12	7/8	2 7/16	4 3/8	8.63	2 7/16
	10	17 1/2	1 7/8	12 3/4	15 1/4	16	1	2 5/8	4 5/8	10.75	3 3/4
	12	20 1/2	2	15	17 3/4	16	1 1/8	2 7/8	5 1/8	12.75	4
	14	23	2 1/8	16 1/4	20 1/4	20	1 1/8	3	5 5/8	14.00	4 3/8
	16	25 1/2	2 1/4	18 1/2	22 1/2	20	1 1/4	3 1/4	5 3/4	16.00	4 3/4
	18	28	2 3/8	21	24 3/4	24	1 1/4	3 1/2	6 1/4	18.00	5 1/8
	20	30 1/2	2 1/2	23	27	24	1 1/4	3 3/4	6 3/8	20.00	5 1/2
	24	36	2 3/4	27 1/4	32	24	1 1/2	4 3/16	6 5/8	24.00	6
400 Pound (For smaller sizes, use 600-Pound)	4	10	1 3/8	6 3/16	7 7/8	8	7/8	2	3 1/2	4.50	2
	5	11	1 1/2	7 5/16	9 1/4	8	7/8	2 1/8	4	5.56	2 1/8
	6	12 1/2	1 5/8	8 1/2	10 5/8	12	7/8	2 1/4	4 1/16	6.63	2 1/4
	8	15	1 7/8	10 5/8	13	12	1	2 11/16	4 5/8	8.63	2 11/16
	10	17 1/2	2 1/8	12 3/4	15 1/4	16	1 1/8	2 7/8	4 7/8	10.75	4
	12	20 1/2	2 1/4	15	17 3/4	16	1 1/4	3 1/8	5 3/8	12.75	4 1/4
	14	23	2 3/8	16 1/4	20 1/4	20	1 1/4	3 5/16	5 7/8	14.00	4 5/8
	16	25 1/2	2 1/2	18 1/2	22 1/2	20	1 3/8	3 11/16	6	16.00	5
	18	28	2 5/8	21	24 3/4	24	1 3/8	3 7/8	6 1/2	18.00	5 3/8
	20	30 1/2	2 3/4	23	27	24	1 1/2	4	6 5/8	20.00	5 3/4
	24	36	3	27 1/4	32	24	1 3/4	4 1/2	6 7/8	24.00	6 1/4

Figure 1-10. Forged steel flanges (dimensions in inches). Courtesy of Crane Co.

(Figure 1-10 continued on following page)

Regular Facings

In 150 and 300-pound pressure classes, the screwed, slip-on welding, welding neck, and blind flanges are furnished with a 1/16-inch raised face.

In 400-pound and higher pressure classes, the aforementioned flanges have a ¼-inch male face (large male).

American Standard

The dimensions and drilling of flanges conform to the American Steel Flange Standard, B16.5-1957, for their respective pressure class. This Standard does not include slip-on welding flanges in the 2500-pound class nor sizes 3-inch and larger of the 1500-pound class. Crane flanges of this type have the same dimensions as American Standard Steel Screwed Flanges, being bored instead of threaded.

Cranelap Flanges

Cranelap flanges also are recommended for use in combination with Cranelap stub ends.

3-inch Cranelap Joints
(300 and 600-Pound)

When 3-inch 300 or 600-pound flanges with ring joint facing are to be bolted to Cranelap joints, orders must so specify; they require a groove of special pitch diameter.

Galvanizing

Galvanized flanges can be furnished to order.

Dimensions, in Inches — continued

Class	Pipe Size	A	B	C	D	Bolts No.	Bolts Dia.	E	F	G	H
600 Pound	½	3¾	9/16	1⅜	2⅝	4	½	⅞	2 1/16	0.84	⅞
	¾	4⅝	⅝	1 11/16	3¼	4	⅝	1	2¼	1.05	1
	1	4⅞	11/16	2	3½	4	⅝	1 1/16	2 7/16	1.32	1 1/16
	1¼	5¼	13/16	2½	3⅞	4	⅝	1⅛	2⅝	1.66	1⅛
	1½	6⅛	⅞	2⅞	4½	4	¾	1¼	2¾	1.90	1¼
	2	6½	1	3⅝	5	8	⅝	1 7/16	2⅞	2.38	1 7/16
	2½	7½	1⅛	4⅛	5⅞	8	¾	1⅝	3⅛	2.88	1⅝
	3	8¼	1¼	5	6⅝	8	¾	1 13/16	3¼	3.50	1 13/16
	4	10¾	1½	6 3/16	8½	8	⅞	2⅛	4	4.50	2⅛
	5	13	1¾	7 5/16	10½	8	1	2⅜	4½	5.56	2⅜
	6	14	1⅞	8½	11½	12	1	2⅝	4⅝	6.63	2⅝
	8	16½	2 3/16	10⅝	13¾	12	1⅛	3	5¼	8.63	3
	10	20	2½	12¾	17	16	1¼	3⅜	6	10.75	4⅜
	12	22	2⅝	15	19¼	20	1¼	3⅝	6⅛	12.75	4⅝
	14	23¾	2¾	16¼	20¾	20	1⅜	3 11/16	6½	14.00	5
	16	27	3	18½	23¾	20	1½	4 3/16	7	16.00	5½
	18	29¼	3¼	21	25¾	20	1⅝	4⅝	7¼	18.00	6
	20	32	3½	23	28½	24	1⅝	5	7½	20.00	6½
	24	37	4	27¼	33	24	1⅞	5½	8	24.00	7¼
900 Pound (For smaller sizes, use 1500 Pound)	3	9½	1½	5	7½	8	⅞	2⅛	4	3.50	2⅛
	4	11½	1¾	6 3/16	9¼	8	1⅛	2¾	4½	4.50	2¾
	5	13¾	2	7 5/16	11	8	1¼	3⅛	5	5.56	3⅛
	6	15	2 3/16	8½	12½	12	1⅛	3⅜	5½	6.63	3⅜
	8	18½	2½	10⅝	15½	12	1⅜	4	6⅜	8.63	4½
	10	21½	2¾	12¾	18½	16	1⅜	4¼	7¼	10.75	5
	12	24	3⅛	15	21	20	1⅜	4⅝	7⅞	12.75	5⅝
	14	25¼	3⅜	16¼	22	20	1½	5⅛	8⅜	14.00	6⅛
	16	27¾	3½	18½	24¼	20	1⅝	5¼	8½	16.00	6½
	18	31	4	21	27	20	1⅞	6	9	18.00	7½
	20	33¾	4¼	23	29½	20	2	6¼	9¾	20.00	8¼
	24	41	5½	27¼	35½	20	2½	8	11½	24.00	10½
1500 Pound	½	4¾	⅞	1⅜	3¼	4	¾	1¼	2⅜	0.84	1¼
	¾	5⅛	1	1 11/16	3½	4	¾	1⅜	2¾	1.05	1⅜
	1	5⅞	1⅛	2	4	4	⅞	1⅝	2⅞	1.32	1⅝
	1¼	6¼	1⅛	2½	4⅜	4	⅞	1⅝	2⅞	1.66	1⅝
	1½	7	1¼	2⅞	4⅞	4	1	1¾	3¼	1.90	1¾
	2	8½	1½	3⅝	6½	8	⅞	2¼	4	2.38	2¼
	2½	9⅝	1⅝	4⅛	7½	8	1	2½	4⅛	2.88	2½
	3	10½	1⅞	5	8	8	1⅛	2⅞	4⅝	3.50	2⅞
	4	12¼	2⅛	6 3/16	9½	8	1¼	3 9/16	4⅞	4.50	3 9/16
	5	14¾	2⅞	7 5/16	11½	8	1½	4⅛	6⅛	5.56	4⅛
	6	15½	3¼	8½	12½	12	1⅜	4 11/16	6¾	6.63	4 11/16
	8	19	3⅝	10⅝	15½	12	1⅝	5⅝	8⅜	8.63	5⅝
	10	23	4¼	12¾	19	12	1⅞	6¼	10	10.75	7
	12	26½	4⅞	15	22½	16	2	7⅛	11⅛	12.75	8⅝
	14	29½	5¼	16¼	25	16	2¼	. . .	11¾	14.00	9½
2500 Pound	½	5¼	1 3/16	1⅜	3½	4	¾	1 9/16	2⅞	0.84	1 9/16
	¾	5½	1¼	1 11/16	3¾	4	¾	1 11/16	3⅛	1.05	1 11/16
	1	6¼	1⅜	2	4¼	4	⅞	1⅞	3½	1.32	1⅞
	1¼	7¼	1½	2½	5⅛	4	1	2 1/16	3¾	1.66	2 1/16
	1½	8	1¾	2⅞	5¾	4	1⅛	2⅜	4⅜	1.90	2⅜
	2	9¼	2	3⅝	6¾	8	1	2¾	5	2.38	2¾
	2½	10½	2¼	4⅛	7¾	8	1⅛	3⅛	5⅝	2.88	3⅛
	3	12	2⅝	5	9	8	1¼	3⅝	6⅝	3.50	3⅝
	4	14	3	6 3/16	10¾	8	1½	4¼	7½	4.50	4¼
	5	16½	3⅝	7 5/16	12¾	8	1¾	5⅛	9	5.56	5⅛
	6	19	4¼	8½	14½	8	2	6	10¾	6.63	6
	8	21¾	5	10⅝	17¼	12	2	7	12½	8.63	7
	10	26½	6½	12¾	21¼	12	2½	9	16½	10.75	9
	12	30	7¼	15	24⅜	12	2¾	10	18¼	12.75	10

Figure 1-10 continued)

General Piping

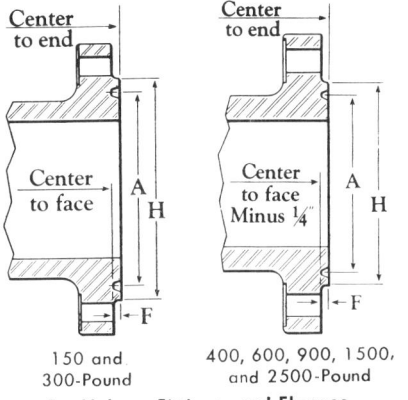

150 and 300-Pound

400, 600, 900, 1500, and 2500-Pound

For Valves, Fittings, and Flanges

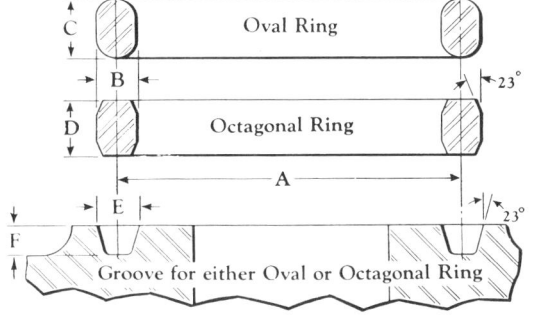

Oval rings fit grooves having either a flat or round bottom; octagonal rings only fit grooves having a flat bottom.

("Z" represents pipe thickness.)

Cranelap Joints

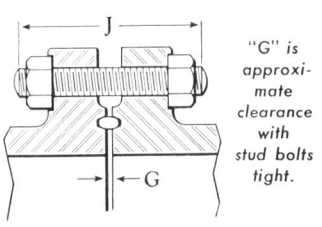

"G" is approximate clearance with stud bolts tight.

Assembled Ring Joint

*Dimension "J" does not apply to Cranelap Joints; see

†**Caution:** 3-inch 300 and 600-Pound Cranelap Ring Joints use Ring No. R 30, having a pitch diameter of 4 5/8 inches. When 3-inch 300 or 600-Pound ring joint valves, fittings, or flanges are to be bolted to Cranelap joints, orders must specify; they will be machined special.

Class	Size	Ring No.	A	B	C	D	E	F	G	H	*J	Stud Bolts No.	Dia.
150 Pound	1	R 15	1 7/8	5/16	9/16	1/2	11/32	1/4	5/32	2 1/2	3	4	1/2
	1 1/4	R 17	2 1/4	5/16	9/16	1/2	11/32	1/4	5/32	2 7/8	3	4	1/2
	1 1/2	R 19	2 9/16	5/16	9/16	1/2	11/32	1/4	5/32	3 1/4	3 1/4	4	1/2
	2	R 22	3 1/4	5/16	9/16	1/2	11/32	1/4	5/32	4	3 1/2	4	5/8
	2 1/2	R 25	4	5/16	9/16	1/2	11/32	1/4	5/32	4 3/4	3 3/4	4	5/8
	3	R 29	4 1/2	5/16	9/16	1/2	11/32	1/4	5/32	5 1/4	4	4	5/8
	3 1/2	R 33	5 3/16	5/16	9/16	1/2	11/32	1/4	5/32	6 1/16	4	8	5/8
	4	R 36	5 7/8	5/16	9/16	1/2	11/32	1/4	5/32	6 3/4	4	8	5/8
	5	R 40	6 3/4	5/16	9/16	1/2	11/32	1/4	5/32	7 5/8	4 1/4	8	3/4
	6	R 43	7 5/8	5/16	9/16	1/2	11/32	1/4	5/32	8 5/8	4 1/4	8	3/4
	8	R 48	9 3/4	5/16	9/16	1/2	11/32	1/4	5/32	10 3/4	4 1/2	8	3/4
	10	R 52	12	5/16	9/16	1/2	11/32	1/4	5/32	13	5	12	7/8
	12	R 56	15	5/16	9/16	1/2	11/32	1/4	5/32	16	5	12	7/8
	14	R 59	15 5/8	5/16	9/16	1/2	11/32	1/4	1/8	16 3/4	5 1/2	12	1
	16	R 64	17 7/8	5/16	9/16	1/2	11/32	1/4	1/8	19	5 3/4	16	1
	18	R 68	20 3/8	5/16	9/16	1/2	11/32	1/4	1/8	21 1/2	6 1/4	16	1 1/8
	20	R 72	22	5/16	9/16	1/2	11/32	1/4	1/8	23 1/2	6 1/2	20	1 1/8
	24	R 76	26 1/2	5/16	9/16	1/2	11/32	1/4	1/8	28	7 1/4	20	1 1/4

Class	Size	Ring No.	A	B	C	D	E	F	G 300 Lb.	G 400 Lb.	G 600 Lb.	H	*J 300 Lb.	*J 400 Lb.	*J 600 Lb.	No. Stud Bolts 300 Lb.	No. Stud Bolts 400 Lb.	No. Stud Bolts 600 Lb.	Dia. Stud Bolts 300 Lb.	Dia. Stud Bolts 400 Lb.	Dia. Stud Bolts 600 Lb.
†300, 400, and †600 Pound	1/2	R 11	1 11/32	1/4	7/16	3/8	9/32	7/32	1/8	...	1/8	2	3	...	3	4	...	4	1/2	...	1/2
	3/4	R 13	1 11/16	5/16	9/16	1/2	11/32	1/4	5/32	...	5/32	2 1/2	3 1/4	...	3 1/4	4	...	4	5/8	...	5/8
	1	R 16	2	5/16	9/16	1/2	11/32	1/4	5/32	...	5/32	2 3/4	3 1/2	...	3 1/2	4	...	4	5/8	...	5/8
	1 1/4	R 18	2 3/8	5/16	9/16	1/2	11/32	1/4	5/32	...	5/32	3 1/8	3 1/2	...	3 3/4	4	...	4	5/8	...	5/8
	1 1/2	R 20	2 11/16	5/16	9/16	1/2	11/32	1/4	5/32	...	5/32	3 9/16	4	...	4	4	...	4	3/4	...	3/4
	2	R 23	3 1/4	7/16	11/16	5/8	15/32	5/16	7/32	...	3/16	4 1/4	4	...	4 1/4	8	...	8	5/8	...	5/8
	2 1/2	R 26	4	7/16	11/16	5/8	15/32	5/16	7/32	...	3/16	5	4 1/2	...	4 3/4	8	...	8	3/4	...	3/4
	†3	†R 31	†4 7/8	7/16	11/16	5/8	15/32	5/16	7/32	...	3/16	5 3/4	4 3/4	...	5	8	...	8	3/4	...	3/4
	3 1/2	R 34	5 3/16	7/16	11/16	5/8	15/32	5/16	7/32	...	3/16	6 1/4	5	...	5 1/2	8	...	8	3/4	...	7/8
	4	R 37	5 7/8	7/16	11/16	5/8	15/32	5/16	7/32	7/32	3/16	6 7/8	5	5 1/2	5 3/4	8	8	8	3/4	7/8	7/8
	5	R 41	7 1/8	7/16	11/16	5/8	15/32	5/16	7/32	7/32	3/16	8 1/4	5 1/4	5 3/4	6 1/2	8	8	8	3/4	7/8	1
	6	R 45	8 5/16	7/16	11/16	5/8	15/32	5/16	7/32	7/32	3/16	9 1/2	5 1/2	6	6 3/4	12	12	12	3/4	7/8	1
	8	R 49	10 5/8	7/16	11/16	5/8	15/32	5/16	7/32	7/32	3/16	11 7/8	6	6 3/4	7 3/4	12	12	12	7/8	1	1 1/8
	10	R 53	12 3/4	7/16	11/16	5/8	15/32	5/16	7/32	7/32	3/16	14	6 3/4	7 1/2	8 1/2	16	16	16	1	1 1/8	1 1/4
	12	R 57	15	7/16	11/16	5/8	15/32	5/16	7/32	7/32	3/16	16 1/4	7 1/4	8	8 3/4	16	16	20	1 1/8	1 1/4	1 1/4
	14	R 61	16 1/2	7/16	11/16	5/8	15/32	5/16	7/32	7/32	3/16	18	7 1/2	8 1/4	9 1/4	20	20	20	1 1/8	1 1/4	1 3/8
	16	R 65	18 1/2	7/16	11/16	5/8	15/32	5/16	7/32	7/32	3/16	20	8	8 3/4	10	20	20	20	1 1/4	1 3/8	1 1/2
	18	R 69	21	7/16	11/16	5/8	15/32	5/16	7/32	7/32	3/16	22 5/8	8 1/4	9	10 3/4	24	24	20	1 1/4	1 3/8	1 5/8
	20	R 73	23	1/2	3/4	11/16	17/32	3/8	7/32	7/32	3/16	25	8 3/4	9 3/4	11 1/2	24	24	24	1 1/4	1 1/2	1 5/8
	24	R 77	27 1/4	5/8	7/8	13/16	21/32	7/16	1/4	1/4	7/32	29 1/2	10	11	13 1/4	24	24	24	1 1/2	1 3/4	1 7/8

Figure 1-11. Ring joint facing and rings, American standard (dimensions in inches). Reprinted courtesy of Crane Co.

"Flex Gate" Valves
(Sizes 2 to 12-inch regular; sizes 14 to 36-inch on order)

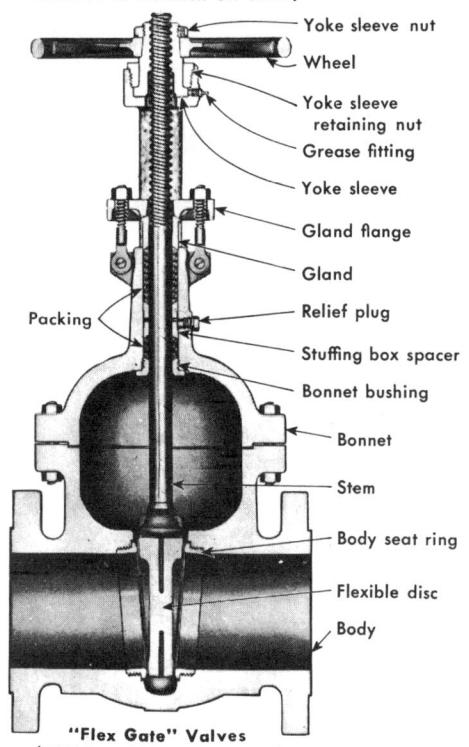

"Flex Gate" Valves
(300-Pound Valve illustrated)

Regularly furnished in sizes 2 to 12-inch; available on order in sizes 14-inch to 36-inch.

Crane's patented one-piece flexible disc . . . solid through the center only . . . permits each disc face to move independently of the other.

Crane "Flex Gate" Valves offer a host of benefits. The design effects easy operation; less torque is required to seat and unseat . . . . the disc will not stick in the closed position, even if closed while hot and allowed to cool . . . . the resiliency of the construction compensates for minor misalignment of seats due to pipeline deflection . . . . and the valves are tight over a wide range of pressures on both the inlet seat and the outlet seat.

A complete line . . . featuring Crane's patented "Flex Gate" design in sizes 2 to 12-inch

The term "Flex Gate" is a Crane Co. trademark; registration pending in the U.S. Patent Office.

Crane 150 and 300-Pound Cast Steel Wedge Gate Valves offer dependable service in steam, water, oil, and oil vapor lines. Quality materials and fine workmanship combined with tested designs assure high utility in severe service. A variety of trim materials are furnished.

The line, in the popular 2 to 12-inch size range, introduces Crane's "Flex Gate" Valves . . . . with patented one-piece flexible wedge disc . . . . a major step forward in fine valve construction.

"Flex Gate" Valves: Crane "Flex Gate" Valves feature a new concept in valve design . . . . a flexible wedge disc. Instead of being made solid with both seating faces maintained in the same rigid position, flexibility . . . . or resiliency . . . is attained by having the two faces separated from each other except fo a small section at the center. See the two illustrations at the left.

The shape of the flexible disc can be likened to two wheels on a very short axle. The "axle" or spud at the center of the disc is amply strong to carry the two halves of the disc together at all times . . . . and yet, it permits a degree of action between them. It is this "flexibility" that makes the disc tight on both faces over a wide range of pressures . . . . prevents sticking during temperature changes, and assures minimum operating torque.

Although each disc face can move independently of the other . . . . up to two full degrees . . . . the construction is one-piece. There are no loose parts to cause harmful vibration.

Solid Disc Valves: Crane Solid Wedge Disc Valves, illustrated at the upper right, are regularly furnished in the 1½-inch size and in sizes 14 to 24-inch; they are optional in sizes 2 to 12-inch. As in the "Flex Gate" design, careful engineering and workmanship are combined to produce a quality product highly dependable in severe service.

The disc is the solid web type. The facings are smoothly and accurately machined, and are then ground to a mirror-like finish. The disc is carefully fitted into the valve so that an even, wide, and true contact is made with the corresponding faces of the body seat rings.

Disc guides; stem connection: Both the flexible disc and the solid disc have long, machined guide slots which engage the guide ribs in the body to maintain true alignment of the disc throughout its travel. The seating faces do not contact each other until the valve is virtually closed. A tee-head disc-stem connection prevents lateral strains on the stem.

Solid Disc Valves
(Sizes 1½ to 24-inch regular; sizes 30 and 36-inch on order)

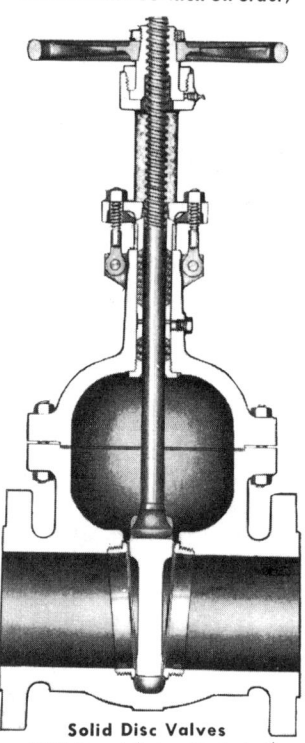

Solid Disc Valves
(300-Pound Valve illustrated)

Solid wedge disc valves, except for the disc, are the same design as the "Flex Gate" Valves.

Figure 1-12. Cast steel wedge gate valves—150 to 300 pound. Reprinted courtesy of Crane Co.

General Piping

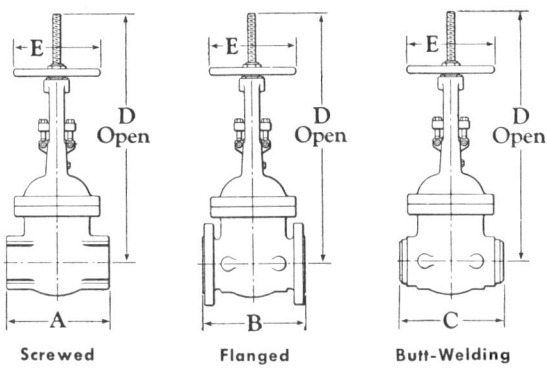

Face to face: Flanged valves of the 150 and 300-pound pressure classes are regularly furnished with a 1/16-inch raised face; those of the 400, 600, 900, 1500, and 2500-pound classes are regularly furnished with 1/4-inch high large male facing; face to face dimensions include these facings.

All flanged and butt-welding valves conform to the American Standard for Face-to-Face Dimensions of Ferrous Flanged and Welding End Valves, B16.10-1957, for their respective pressure class. This Standard does not include 3½-inch steel valves.

Dimensions, in Inches

Class	Size of Valve	A	B	C	D	E	Class	Size of Valve	B	C	D	E
150-Pound	2	6¼	7	8½	15⅝	8	600-Pound	2	11½	11½	18¼	8
	2½	7	7½	9½	16½	8		2½	13	13	22¼	9
	3	7⅜	8	11⅛	20¾	9		3	14	14	25¾	10
	3½	...	8½	...	23	9		3½	15	...	32	14
	4	8	9	12	25¾	10		4	17	17	31½	14
	5	...	10	15	30½	12		5	20	20	36¾	16
	6	...	10½	15⅞	35¼	14		6	22	22	42¾	20
	8	...	11½	16½	44	16		8	26	26	52¼	24
	10	...	13	18	52½	18		10	31	31	62¼	27
	12	...	14	19¾	60½	18		12	33	33	70	27
	14	...	15	22½	70¼	22		14	35	35	77¼	30
	16	...	16	24	79¾	24		16	39	39	83¾	30
	18	...	17	26	89	27		18	43	43	93¾	36
	20	...	18	28	97¼	30		20	47	47	104½	36
	24	...	20	32	112¾	30		24	55	55	126	42
300-Pound	1½	...	7½	...	16¾	8	900-Pound	3	15	15	27¼	12
	2	7	8½	8½	18	8		4	18	18	31½	14
	2½	8	9½	9½	19	8		5	22	22	36¾	16
	3	9	11⅛	11⅛	23¼	9		6	24	24	42¾	20
	4	11	12	12	28¼	10		8	29	29	52½	24
	5	...	15	15	33½	12		10	33	33	62¼	27
	6	...	15⅞	15⅞	38½	14		12	38	38	73½	30
	8	...	16½	16½	47	16		14	40½	40½	77¼	30
	10	...	18	18	56½	20		16	44½	44½	85¾	36
	12	...	19¾	19¾	64¼	20	1500-Pound	1	10	10	16	8
	14	...	30	30	75¼	27		1¼	11	11	16½	8
	16	...	33	33	81	27		1½	12	12	20	9
	18	...	36	36	91½	30		2	14½	14½	22⅛	10
	20	...	39	39	99¾	36		2½	16½	16½	26⅜	12
	24	...	45	45	120½	36		3	18½	18½	28	14
400-Pound	4	...	16	16	30¾	12		4	21½	21½	33	16
	5	...	18	18	35	14		5	26½	26½	38¾	20
	6	...	19½	19½	40¼	16		6	27¾	27¾	47	24
	8	...	23½	23½	50½	20		8	32¾	32¾	55	27
	10	...	26½	26½	59¾	24	2500-Pound	\multicolumn{4}{c}{Dimensions of 2500-Pound Valves are furnished on application.}			
	12	...	30	30	67¾	24						
	14	...	32½	32½	74¾	27						
	16	...	35½	35½	80¾	27						

Figure 1-13. Cast steel wedge gate valves—150 to 1500 pound dimensions. Reprinted courtesy of Crane Co.

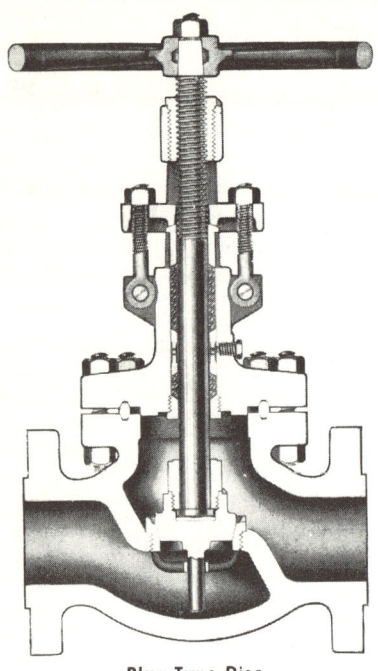

Plug Type Disc

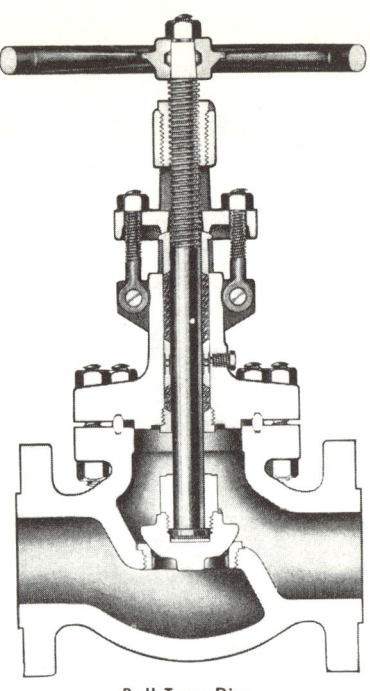

Ball Type Disc

Crane Steel Globe and Angle Valves embody many refinements in design and materials.

Disc and seat: The "XR" trimmed valves (for steam, water, or general service) and "U" trimmed valves (for steam, water, oil, or oil vapor service) in sizes 6-inch and smaller have a plug type disc and seat (illustrated at left); the 8-inch size has a flat disc and seat (not illustrated). The 2-inch valves do not have a disc stem guide.

All sizes of "X" trimmed valves (for oil or oil vapor service) are furnished with a 35° taper seat and a ball shaped seating face on the disc (illustrated at right).

Body seat ring: All valves have the shoulder-type screwed-in body seat ring for utmost tightness and security; in "U" trimmed valves, the rings are also seal brazed or seal welded.

Body and bonnet: The body and bonnet have heavy metal sections with liberal reinforcement at points subjected to greatest stress. The bonnet is fitted with a stem hole bushing.

Bonnet joint: A ring-type bonnet joint holds pressure easily on the 400, 600, 900, 1500, and 2500-pound valves, assuring tightness and maximum strength. On 150 and 300-pound valves, a close-fitting male and female bonnet joint retains the gasket and accurately centers the working parts.

The 300-pound and higher pressure valves have through stud bolts in the bonnet joint. The 150-pound valves employ studs, threaded into the bonnet flange on the body.

Stuffing box: The stuffing box on all valves is deep, assuring tightness and long packing life. The stuffing box is the lantern-type on all except the 150-pound valves. When wide open, the valves can be repacked while under pressure.

Gland: A two-piece ball-type gland and gland flange assure even pressure on the packing without binding on the stem. The gland flange is held in place by swinging eye bolts; the bolts will not loosen in service.

Stem: The stem is of liberal diameter and has unusual strength. Threads are clean and accurately cut and have long engagement with the yoke bushing. The stem and disc are held together by a disc stem ring, which permits the disc to swivel.

Drilling: Flanged valves of each pressure class are furnished with the end flanges faced, drilled, and spot faced (FD & SF) unless otherwise ordered. When orders so specify, flanged valves can be furnished faced only.

Flange facings: The 150 and 300-pound flanged valves are regularly furnished with an American Standard $\frac{1}{16}$-inch raised face on the end flanges; the 400, 600, 900, 1500, and 2500-pound flanged valves regularly have a $\frac{1}{4}$-inch male face (large male).

When so ordered, valves can be furnished with other types of facings, such as ring joint, female, tongue, groove, etc.

Finish of flange faces: The $\frac{1}{16}$-inch raised faces of the 150 and 300-pound valves and the $\frac{1}{4}$-inch male faces of the 400-pound and higher pressure class valves are regularly furnished with a serrated finish.

A smooth finish can be furnished on the raised or male faces, when specified.

American Standard: In design and materials, Crane Cast Steel Globe and Angle Valves exceed the requirements of Standards issued by the American Standards Association.

The butt-welding valve ends and the dimensions and drilling of end flanges on flanged valves conform to the American Steel Flange Standard, B16.5-1957, for their respective pressure class.

Flanged and butt-welding valves conform to the American Standard for Face-to-Face and End-to-End Dimensions of Ferrous Flanged and Welding End Valves, B16.10-1957, for their respective pressure class. This Standard does not include 3½-inch steel valves.

Figure 1-14. Cast steel globe and angle valves. Reprinted courtesy of Crane Co.

General Piping

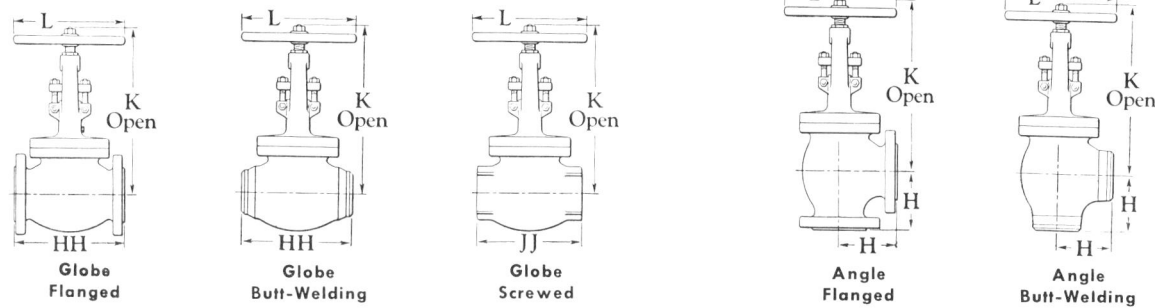

Dimensions, in Inches

All dimensions shown below apply to valves without gears; dimensions "HH" and "H" apply also to valves with gears. For sizes regularly furnished with gears, see asterisked (*) note at right.

Class	Size	Globe Valves						Angle Valves		All Valves
		Flanged		Butt-Welding		Screwed		Flanged or Butt-Welding†		
		HH	K	HH	K	JJ	K	H	K	L
150 Pound	2	8	13¾	8	13¾	8	13¾	4	12½	8
	2½	8½	14½	8½	14½	...	...	4¼	13	8
	3	9½	16½	9½	16½	...	...	4¾	15	9
	3½	10½	17¼	...	...	...	...	...	...	9
	4	11½	19¾	11½	19¾	...	...	5¾	17¾	10
	5	14	23	14	23	...	...	7	20¾	10
	6	16	24½	16	24½	...	...	8	21¾	12
	8	19½	26	19½	26	...	...	9¾	23½	16
300 Pound	2	10½	17¾	10½	17¾	...	...	5¼	17¾	9
	2½	11½	19	11½	19	...	...	5¾	19	10
	‡3	12½	20½	12½	20½	...	...	6¼	20½	10
	3½	13¼	22½	...	...	...	...	6⅝	22½	12
	4	14	24¾	14	24¾	...	...	7	24¾	14
	5	15¾	26½	...	...	...	...	7⅞	26½	16
	6	17½	29¾	17½	29¾	...	...	8¾	29¾	18
	8	22	36½	22	36½	...	...	11	36½	24
400 Pound	4	16	25¼	16	25¼	...	...	8	25¼	14
	5	18	28½	18	28½	...	...	9	28½	18
	6	19½	31¼	19½	31¼	...	...	9¾	31¼	20
	8	23½	38¼	23½	38¼	...	...	11¾	38¼	27
600 Pound	2	11½	19	11½	19	...	...	...	...	10
	2½	13	21¼	13	21¼	...	...	6½	21¼	12
	‡3	14	23½	14	23½	...	...	7	23½	12
	3½	15	25	...	...	...	...	7½	25	14
	4	17	27½	17	27½	...	...	8½	27½	18
	5	20	30¾	20	30¾	...	...	10	30¾	20
	6	22	35	22	35	...	...	11	35	24
900 Pound	3	15	24	15	24	...	...	7½	24	12
	4	18	29½	18	29½	...	...	9	29½	20
	6	24	37¾	24	37¾	...	...	12	37¾	27
1500 Pound	2	14½	25⅛	14½	25⅛	...	...	...	...	14
	2½	16½	28⅛	16½	28⅛	...	...	...	...	18
	3	18½	33½	18½	33½	...	...	...	...	24

†Angle butt-welding valves are made only in the 600-pound class in sizes 2½, 3, 4, 5, and 6-inch.

‡When 3-inch 300 and 600-pound flanged valves with ring joint facing are to be bolted to Cranelap Joints, orders must so specify; a groove of special pitch diameter is required.

*Ball-bearing yoke; gearing: Crane Cast Steel Globe and Angle Valves, in the larger sizes of the 300-pound and higher pressure classes, are regularly furnished with a ball-bearing yoke and spur or bevel gears, as follows:

    300-Pound............8-inch
    400-Pound......6 and 8-inch
    600-Pound......5 and 6-inch
    900-Pound......4 and 6-inch
    1500-Pound...........3-inch

Orders must state whether spur or bevel gears are wanted; see page 149 for description.

When specified, the above valves can be furnished without gears (plain bearing yoke).

Note: All dimensions apply to valves without gears. Face to face (HH) and center to face (H) dimensions also apply to valves with gears; for additional dimensions of geared valves; see page 149.

Face to face: The 150 and 300-pound flanged valves are regularly furnished with a 1/16-inch raised face; valves of the 400-pound and higher pressure classes have a ¼-inch high large male face. The face to face (HH) and center to face (H) dimensions include this facing.

Flanged and butt-welding valves conform to the American Standard for Face-to-Face Dimensions of Ferrous Flanged and Welding End Valves, B16.10-1957. This Standard does not include steel valves in the 3½-inch size.

Figure 1-15. Cast steel globe and angle valves. Reprinted courtesy of Crane Co.

22 Process Piping Design

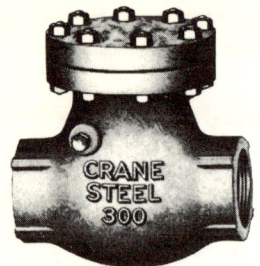

Screwed
For Oil, Oil Vapor,
Steam, or Water
No. 148 X, 150-Pound
No. 158 X, 300-Pound

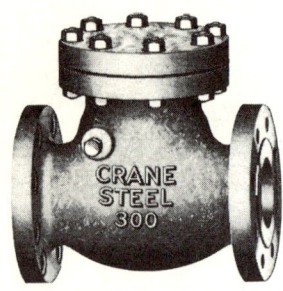

Flanged
For Oil, Oil Vapor,
Steam, or Water
No. 147 X,   150-Pound
No. 159 X,   300-Pound
No. 169 X,   400-Pound
No. 175 X,   600-Pound
No. 187 X,   900-Pound
No. 199 X,   1500-Pound

Butt-Welding
For Oil, Oil Vapor,
Steam, or Water
No. 147½ X,   150-Pound
No. 159½ X,   300-Pound
No. 169½ X,   400-Pound
No. 175½ X,   600-Pound
No. 187½ X,   900-Pound
No. 199½ X,   1500-Pound

A rugged line . . . designed for severe service on oil, oil vapor, steam, and water lines.

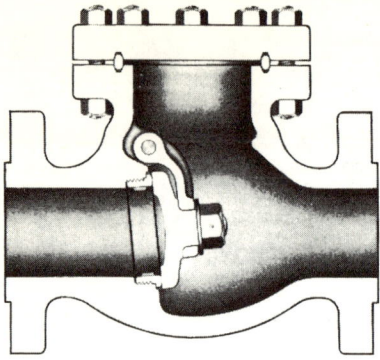

**Cross Section of
Cast Steel Swing Check Valve**

Note: The 150-pound valves in sizes 14 and 16-inch (not illustrated) have a bottom seated body seat ring . . . . and the complete disc, hinge, and hinge pin assembly is suspended from a hinge bracket; the bracket is securely fastened to a pad which is cast integral with the body.

Crane Cast Steel Swing Check Valves, described on these facing pages, embody the many refinements in design and materials necessary to withstand severe service.

Materials: These valves, in all pressure classes, are regularly furnished with a body and cap made of Crane Carbon Steel conforming to requirements of ASTM A 216, Grade WCB.

Seating materials are Exelloy to Exelloy (Class "X" trim), suitable for steam, water, oil, oil vapor, air, or gas.

Design: On flanged and butt-welding valves the full port area is maintained without pockets, from the inlet port to the valve seat, to avoid turbulence. On the outlet side of the valve seat, the body is of generous proportions, allowing full swing of the disc and minimizing erosion and flow resistance.

Body seat ring: A shoulder-type screwed-in body seat ring provides maximum tightness and security.

Cap joint: Valves of the 150 and 300-pound pressure classes have a male and female type cap joint.

Valves of the 400, 600, 900, 1500, and 2500-pound pressure classes have a ring type cap joint.

Crane Triplex Steel studs and stud bolts assure an unusually strong and tight joint. The 150-pound valves are equipped with studs; all other valves have through stud bolts.

Flange facings: The 150 and 300-pound flanged valves are regularly furnished with an American Standard ¹⁄₁₆-inch raised face on the end flanges.

The 400, 600, 900, 1500, and 2500-pound flanged valves are regularly furnished with a ¼-inch male face (large male).

When so ordered, flanged valves can be furnished with other types of facings, such as ring joint, female, tongue, groove, etc.

Finish of flange faces: The ¹⁄₁₆-inch raised faces and the ¼-inch male faces are regularly furnished with a serrated finish.

A smooth finish can be furnished on raised or male faces, when specified.

Standards: In design and materials, Crane Cast Steel Swing Check Valves exceed the requirements of Standards issued by the American Standards Association and the American Petroleum Institute.

The end flanges on flanged valves as well as the dimensions of butt-welding valve ends conform to the American Steel Standard, B16.5-1957, for their respective pressure class.

Flanged and butt-welding valves of all classes, in sizes 12-inch and smaller, conform to the American Standard for Face-to-Face and End-to-End Dimensions of Ferrous Flanged and Welding End Valves, B16.10-1957, for their respective pressure class. This Standard does not include 3½-inch steel valves.

Flanged and butt-welding valves of all classes conform also to the API Standard for Pipe Line Valves, No. 6-D, Ninth Edition, April, 1960. This Standard does not include a 3½ or 5-inch size.

Figure 1-16. Cast steel swing check valves. Reprinted courtesy of Crane Co.

General Piping

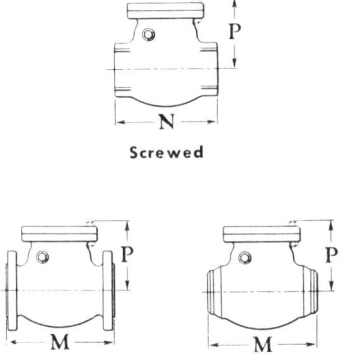

Screwed

Flanged   Butt-Welding

When ordering, specify catalog number and suffix; see the preceding page.

Drilling: Flanged valves are regularly furnished with end flanges faced, drilled and spot faced (FD & SF); they are drilled to the corresponding pressure class of the American Standard; they can be furnished faced only, when specified.

Face to face: Face to face dimensions (M) of flanged valves include the $\frac{1}{16}$-inch raised face on the 150 and 300-pound pressure classes and the $\frac{1}{4}$-inch high large male face on the 400-pound and higher pressure classes.

Butt-welding valves: Unless otherwise ordered, 150 and 300-pound butt-welding valves are bored to match the inside diameter of standard pipe (heaviest weight on the 8, 10, and 12-inch sizes). For all other pressure classes, orders must specify the diameter of the bore (I.D. of pipe).

Smaller size 400 and 900-pound valves: For smaller size 400-pound valves, use the 600-pound valves. For smaller size 900-pound valves, use the 1500-pound valves.

2500-pound valves: Prices, weights, and dimensions of 2500-pound valves are furnished on request. For sizes and general description.

Weights and Dimensions — Prices on Request

Pressure Class	Size Inches	Pounds, Each			Dimensions, in Inches			
		Screwed Valves	Flanged Valves FD & SF	Butt-Welding Valves	Screwed		Flanged or Butt-Welding	
					N	P	M	P
150 Pound	2	27	34	25	8	5	8	5
	2½	40	50	30	8½	5½	8½	5½
	3	50	65	50	9½	6	9½	6
	3½	...	94	...	...	...	10½	6½
	4	96	100	100	11½	7	11½	7
	5	...	140	120	...	...	13	8
	6	...	200	160	...	...	14	9
	8	...	390	360	...	...	19½	10¼
	10	...	510	...	...	...	24½	12⅛
	12	...	775	...	...	...	27½	13¾
	14	...	1200	...	...	...	35	on
	16	...	1450	...	...	...	39	request
300 Pound	2	40	62	47	9½	6¾	10½	6¾
	2½	70	80	60	10¾	8	11½	8
	*3	100	120	80	11¾	8½	12½	8½
	4	...	180	130	...	...	14	9¾
	5	...	250	240	...	...	15¾	10¾
	6	...	330	260	...	...	17½	11¾
	8	...	620	510	...	...	21	14
	10	...	920	760	...	...	24½	15
	12	...	1290	1015	...	...	28	16¾
400 Pound	4	...	200	190	...	...	16	10
	5	...	270	265	...	...	18	12
	6	...	395	310	...	...	19½	12½
	8	...	680	580	...	...	23½	14½
	10	...	900	820	...	...	26½	15¼
	12	...	1250	1150	...	...	30	16⅞
600 Pound	1¼	...	38	32	...	...	9	6¼
	1½	...	58	40	...	...	9½	6¾
	2	...	70	55	...	...	11½	7
	2½	...	105	70	...	...	13	8¼
	*3	...	140	100	...	...	14	9
	4	...	260	170	...	...	17	10¼
	5	...	400	300	...	...	20	12¾
	6	...	530	420	...	...	22	13½
	8	...	900	740	...	...	26	15¼
	10	...	1440	880	...	...	31	18¾
	12	...	1970	1200	...	...	33	21½
900 Pound	3	...	180	155	...	...	15	9½
	4	...	340	240	...	...	18	11
	6	...	640	500	...	...	24	13¾
	8	...	1180	890	...	...	29	16½
1500 Pound	1½	...	110	80	...	...	12	8¼
	2	...	160	130	...	...	14½	9¾
	2½	...	245	170	...	...	16½	10½
	3	...	280	210	...	...	18½	11¼
	4	...	630	390	...	...	21½	13¼
	5	...	950	480	...	...	26½	15¼
	6	...	1360	780	...	...	27¾	15¾
	8	...	2100	1320	...	...	32¾	18¼

*3-inch Cranelap Joints: When 3-inch 300 and 600-pound flanged valves with ring joint facing are to be bolted to Cranelap Joints, orders must so specify. A groove of special pitch diameter is required.

Figure 1-17. Cast steel swing check valves. Reprinted courtesy of Crane Co.

1. Avoid the use of the cross. Cost and delivery are undesirable.
2. Use eccentric reducers only where absolutely necessary. They cost almost twice as much as concentric reducers.
3. Reducing elbows are a cost saving (and are generally available) for sizes 8" and below for the large end in carbon steel materials. Avoid their use in alloy materials.
4. Avoid the use of the 90° elbow with long tangent on one end.
5. The short-radius elbow causes additional pressure drop in a piping system. Use it only where close connections are needed.
6. Avoid the shaped nipples. Stub the pipe directly into the header.

Figure 1-7 shows how flanges are joined with pipe. Figure 1-8 describes the seven basic pressure classes. Figure 1-9 gives their allowable working pressures and Figure 1-10 supplies flange dimensional data. Figure 1-11 gives dimensions for ring joint facing and ring numbers.

Valves

The three basic valve types are gate, globe and check. Figure 1-12 describes gate valves. Note that the stem rises up, raising the disc into the bonnet and offering relatively smooth fluid flow through the valve body. Flow can enter either end of the gate body.

Figure 1-13 gives dimensions for gate valves. When drawing valves always show the stem in the open position, dimension D. The handwheel is located about one pipe size below the top of stem. Always show the handwheel diameter, dimension E, to scale. Gate valve handwheels should not be oriented below the horizontal. It is preferable that they be vertical, however horizontal installation will keep the bonnet cavity relatively free of loose debris. It can easily be seen that if the bonnet cavity is below the flowing fluid, a trap is created for collection of debris, condensate, etc.

Figure 1-14 describes the globe valve. The globe body is designed for throttling. Flow must enter the valve and flow up, against the seat, and change direction again to the outlet end. This turbulent flow causes pressure drop and pressure drop is money in a piping system. The globe body pattern is also more expensive to make than the gate pattern. Consequently, globe body valves should be specified only where throttling is required and must be used *often* for throttling.

It is the author's opinion that specifying globe valves for control valve by-passes is too costly for nine out of ten installations. Some people will say that the valve is for throttling so a globe body is needed. However, this valve is closed almost all the time and a gate valve would be able to handle the necessary throttling, would be cheaper, and, because of the body design, a gate valve could be used which is *one size smaller* than the required globe valve.

Figure 1-15 gives dimensions for globe and angle valves. On a globe valve, the handwheel is fixed to the stem and as the valve is opened the handwheel rises with the stem. When drawing the globe valve, always show the handwheel in the open position, dimension K, and show the handwheel diameter, dimension L, to scale. The angle valve is quite expensive and should be avoided. Globe valves larger than 6" are seldom justified.

Figure 1-16 depicts cast steel swing check valves. The check valve is designed to permit flow in only one direction. In Figure 1-16 the flow enters the valve on the left side and the pressure raises the disc, allowing flow through the valve. Velocity of the fluid keeps the disc raised up in the cavity. As flow subsides or stops, the disc spring forces the disc down on the seat, preventing flow from reversing.

The swing check valve is the most commonly specified type of check valve. It is very efficient for constant flowing fluids. It should *never* be used for pulsating flow such as reciprocating pump or compressor installations. For pulsating flow, the piston or ball type check valve, spring loaded, should be specified.

The swing check valve should not be installed with flow in the down position. The weight of the fluid upstream of the disc would tend to open the disc when the line pressure was less upstream, thereby permitting the flow to reverse, destroying the "check" feature of the valve. Check valves should be installed horizontally or with the flow going up. Piston checks are generally for horizontal installation only.

Figure 1-17 gives dimensional data for cast steel swing check valves.

Chapter 1
Review Test

1. Define "ferrous" metal. _____

2. Name one "non-ferrous" metal. _____

3. Name four pipe manufacturing methods. _____

4. Define ANSI. _____

5. What is meant by "joint efficiency?" _____

6. What is meant by "mill tolerance?" _____

7. The mill tolerance for plate is _____

8. The mill tolerance for seamless pipe is _____

9. Random length pipe is ± _____ feet long.

10. Supply the following outsdie diameters for IPS pipe:

Pipe Size	OD
3"	_____
6"	_____
10"	_____
12"	_____
16"	_____

11. The OD of a 6" tube is _____.

12. Define the difference between "minimum" and "nominal" wall. _____
_____

13. Why is socketwelding preferred over screwed systems? _____

14. An eccentric reducer costs almost _____ the cost of a concentric reducer.

15. What type of valve shall be specified for control valve by-passes? _____
_____

2 Process Terms

To understand the complete piping problem, the piping designer must realize certain process basics. When handling utilities, the designer knows that steam cools and forms condensate which must be drawn off. This is accomplished with a "steam trap" installed at selected low points in the steam system. But his education must also include how to handle "two-phase flow," "equilibrium liquids," "hot vapor by-passes" and other special conditions.

The Hydrocarbon

Hydrocarbon is the term applied to most process fluid designations. A hydrocarbon is a mixture of the atoms hydrogen and carbon. How they are mixed and the quantities mixed are very important. Whether the hydrocarbon is in a liquid or vapor state must be known by the designer.

Crude oil is a hydrocarbon which may be "sweet" or "sour," depending on its percent of "mercaptans," a sulfur-bearing compound. A sweet stream is one with very low or no sulfur content while the sour stream ranks high in mercaptans.

Sour streams must be sweetened by removing the mercaptans or the piping and equipment will be subject to corrosion by contact with the sulfur. These mercaptans are separated out and sent to a unit for disposal. One type of disposal unit is the Merox unit, shortened from mercaptan oxidation, licensed by UOP, Universal Oil Products.

Crude oil also contains salt, which must be removed. This is accomplished by running the crude through a desalter, usually located in the crude unit of a refinery.

Hydrocarbon Structures

Methane is the lightest hydrocarbon, having one atom of carbon and four of hydrogen. This is stated CH_4. The next lightest hydrocarbon is ethane which contains two atoms of carbon. Table 2-1 lists the lightest 20 hydrocarbons by weight.

Hydrocarbons consist of only two elements, but the number of the hydrogen and carbon atoms and their arrangement in the molecule cause the wide variety of products we know. The arrangement generally is divided into two classes, the straight chain (aliphatic) and the ring (napthenic and aromatic).

Atoms combine to form a chemical molecule and each carbon atom holds or can contain four hydrogen atoms or their equivalent. When this happens the carbon atom is said to be saturated. The force holding the atoms together is called the *bond* and is indicated by a small straight line.

Table 2-1
Nomenclature of Hydrocarbons

No. of C Atoms	Paraffin Name	No. of C Atoms	Paraffin Name
1	Methane	11	Undecane
2	Ethane	12	Duodecane
3	Propane	13	Tridecane
4	Butane	14	Tetradecane
5	Pentane	15	Pentradecane
6	Hexane	16	Hexadecane
7	Heptane	17	Heptadecane
8	Octane	18	Octadecane
9	Nonane	19	Nonadecane
10	Decane	20	Eicosane

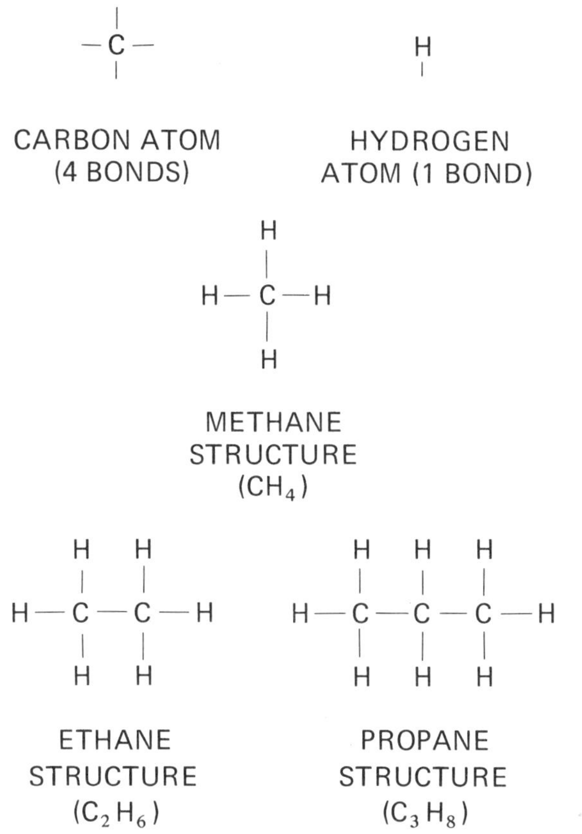

Figure 2-1 shows the chemical composition of some hydrocarbons and how these chains are built. These are the straight chain series of paraffins.

Of the 20 hydrocarbons listed in Table 2-1, the first four are normally gases and have a very low boiling point. The next 13 are liquids and the last three are solids. As a CH_2 is added, the hydrocarbon is heavier and the boiling point and melting point rise.

Fractionation

Fractionation is the most widely used operation in process plants today and has been for hundreds of years. Crude oil is sent to the crude tower for fractionation as its first processing step. Any process unit will have several fractionation (sometimes referred to as distillation) towers.

Fractionation is the separation of lighter from heavier fluid components, which occurs while these fluids are in contact with each other and are in equilibrium.

Equilibrium liquid is liquid that is at it's boiling point due to pressure and temperature.

Assume that two pure separate liquids are placed in a container. One liquid, say pentane, is lighter than the other, say heptane (refer to Table 2-1). By application of an exact amount of heat, the lighter fluid, pentane, can be vaporized while the heavier fluid remains a liquid. Then the two fluids are separated into two containers; one fluid is a liquid and one a vapor. As the vapor cools it returns to its liquid state and is pure pentane. This is simple fractionation.

This structure can continue for the complete hydrocarbon series. Note that adding one atom of carbon adds two hydrogen.

Figure 2-1. Composition of the hydrocarbon molecules.

Multiple fractionation occurs in a fractionating tower. Each tray has liquid of different weight and temperature. As the hotter vapors rise up the tower they make physical contact with the cooler liquid on each tray. As the vapor cools, the heavier hydrocarbons drop out in liquid form and become part of the tray's liquid. Conversely, as the hotter vapors heat the tray liquid, some of the lighter hydrocarbons vaporize and join the vapor on its upward journey. As a result, fractionation is occurring on every tray in a tower. Also, each tray contains liquid that is in equilibrium, liquid that is at its boiling point.

As liquid floods the uppermost tray, meeting the warmer vapors, fractionation occurs. As the liquid runs down the tower, the lighter hydrocarbons

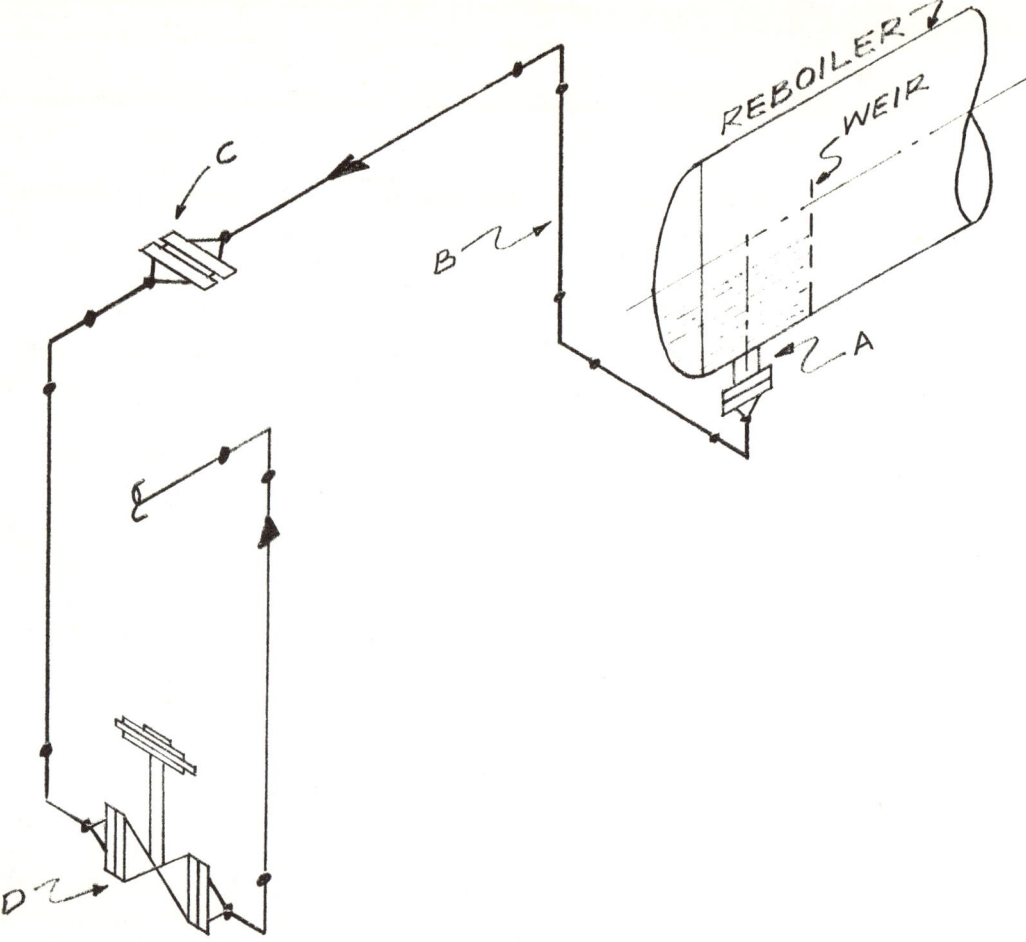

Figure 2-2. Incorrect equilibrium piping.

are fractionated out, leaving a heavier liquid. Each tray down, then, contains a heavier liquid. The heavier the liquid, the hotter the temperature must be to vaporize hydrocarbons.

Hot vapors from a reboiler enter the tower's vapor space below the bottom tray, making contact with the heavy hydrocarbons on the bottom tray. As the vapors rise, a calculated temperature drop occurs at each tray, maintaining just enough temperature to vaporize lighter hydrocarbons and dropping the vapor temperature enough so that the heavier hydrocarbons will condense out.

Piping Equilibrium Liquids

Equilibrium liquids require the piping designer's special attention. By definition, the liquid, at its pressure and temperature, is at its boiling point. Should the piper induce a very small amount of pressure drop in the piping system, the equilibrium liquid will start flashing, resulting in two-phase flow, increased line velocity and a fluid that is difficult to control and impossible to measure.

In some instances flashing does no harm, but the piping designer must first recognize what liquids are in equilibrium, and second, when flashing can be tolerated.

The main streams that contain equilibrium liquids are any tray draw-off, tower bottoms, two-phase flow and reboiler liquid draw-offs. The biggest piping problem occurs at reboiler liquid draw-offs.

Figure 2-2 shows incorrect piping of reboiler liquid draw-off. The piping designer has the prob-

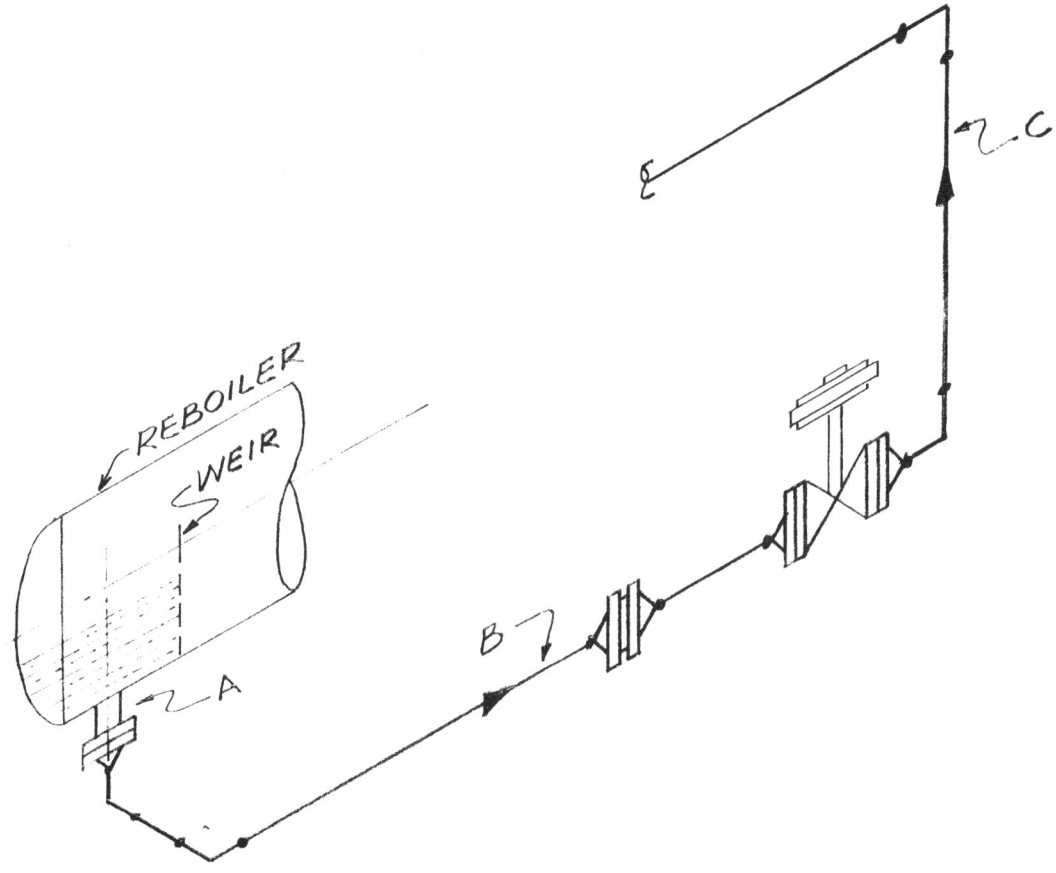

Figure 2-3. Correct equilibrium piping.

lem of routing the reboiler liquid from the area behind the weir, through a pair of orifice flanges (meter run), through a level control valve and into the main pipeway or rack. Liquid is drawn off at nozzle A, runs horizontally, and rises at B. Because the liquid must push against the head of liquid in riser B, pressure drop is induced in the system and flashing will start. The meter run, C, cannot properly measure two-phase flow. The control valve, D, cannot control properly. If this installation were fabricated and installed, it would have to be dismantled and rebuilt in the field, a costly mistake.

Figure 2-3 is one correct method of piping equilibrium liquid. Liquid is drawn off through nozzle A, stays horizontal through meter run B and the control valve, then rises vertically at C.

Now, flashing occurs after the measuring and control functions and if excessive flashing occurs, the line size can be increased to keep the velocity low. By keeping both the meter run and the control valve *below* the liquid level in the reboiler, flashing is prevented in this run.

This reboiler liquid is often a product going to storage. It usually must be cooled before going to a product tank. Cooling will drop this liquid below its equilibrium point and condense any flashed vapors. Then it is preferable to locate the meter run and the control valve downstream of the cooler.

The basic rule for piping equilibrium or any liquid subject to flashing is to keep pressure drop to a minimum and have no vertical risers before measuring or control devices.

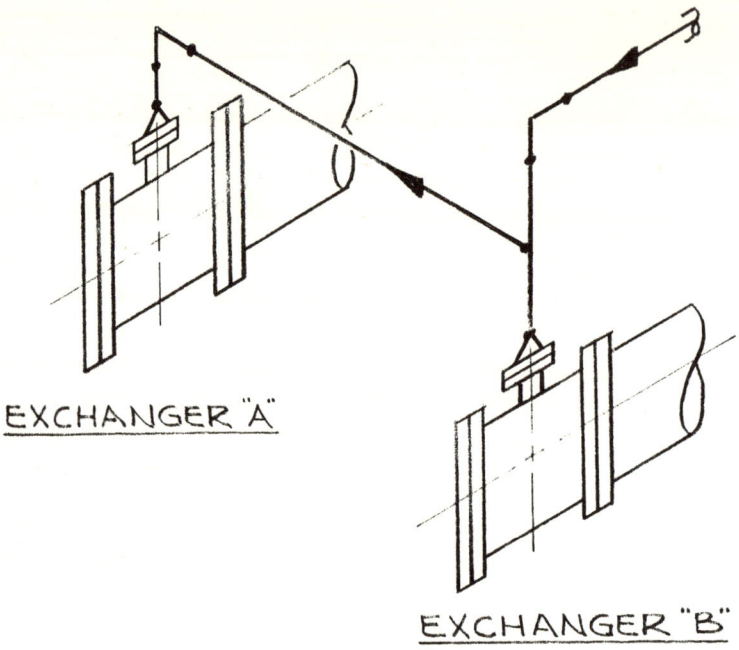

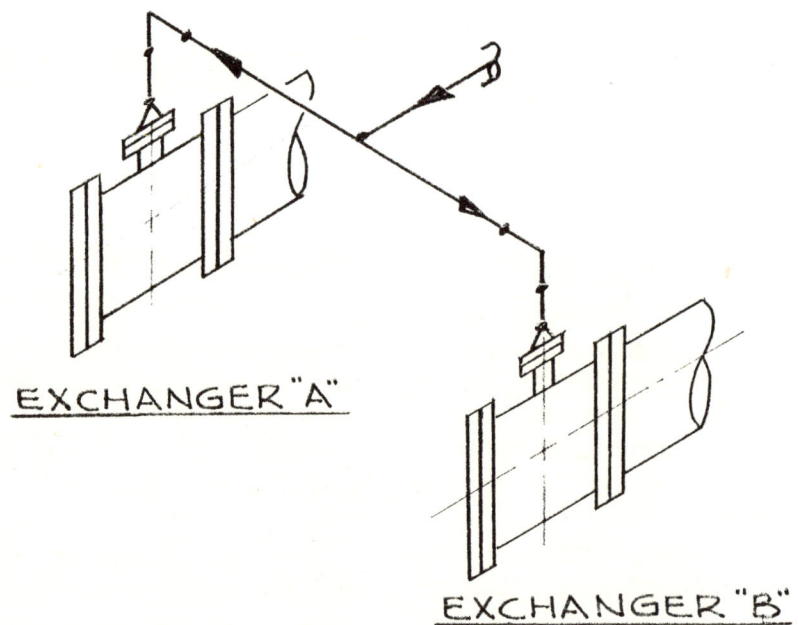

Figure 2-4. Proper and improper two-phase piping.

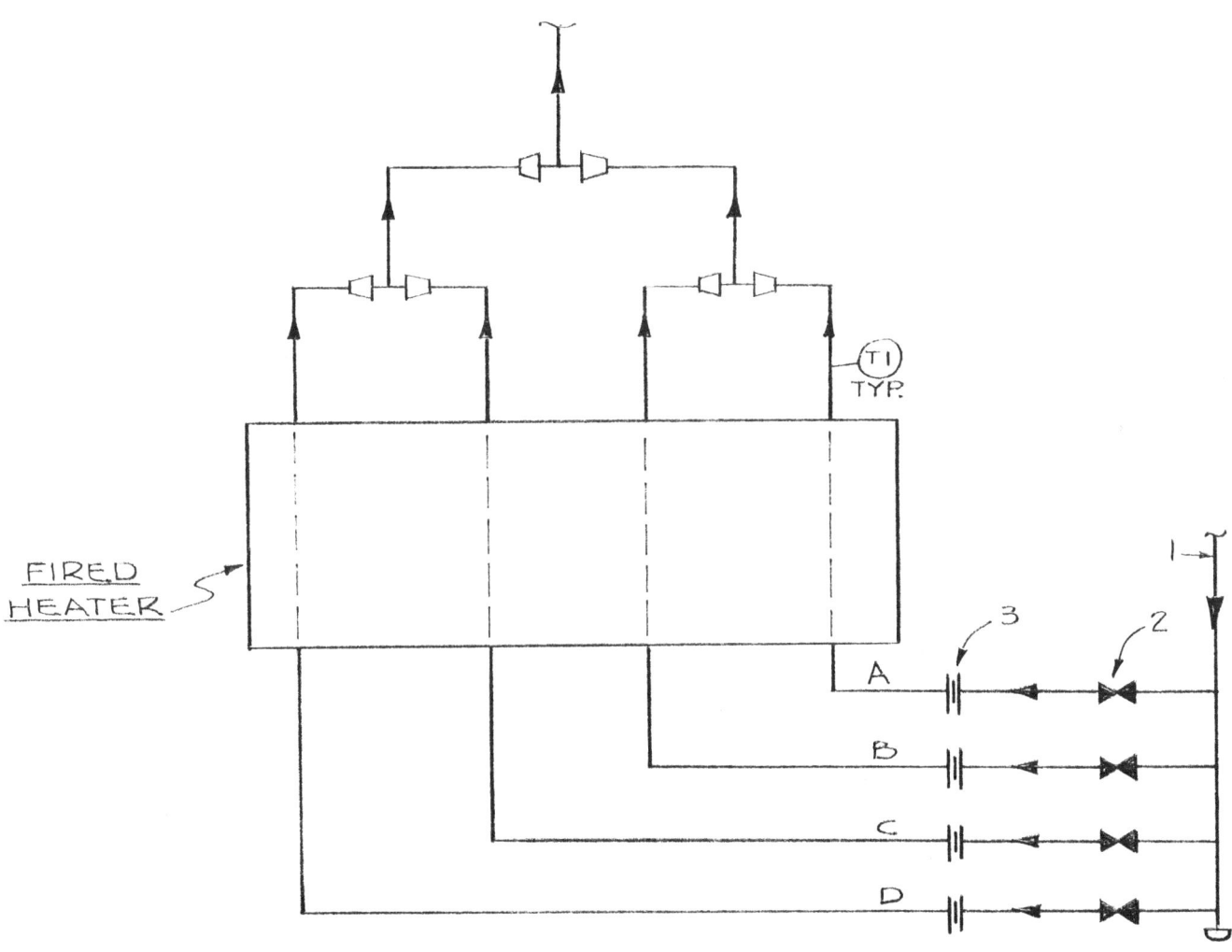

Figure 2-5. Piping at fired heaters.

Two-Phase Flow

Next to adequate flexibility, two-phase flow causes piping designers the most problems. Two-phase flow occurs when both liquid and vapor are in the same pipe, flowing together. Since the pipe's friction reacts more on the liquid portion, the vapors tend to flow at a greater velocity. The real problem occurs when two-phase flow must be divided equally into separate piping systems.

Figure 2-4 shows improper and proper two-phase piping at exchanger inlets. In the improper illustration, the major portion of the flow would be routed to exchanger B. This would cause exchanger A to have less pressure drop and exchanger B to have more when both exchangers were designed for the same pressure drop, as well as for the same heat transfer duty. Since A has less pressure drop and the liquid's velocity will direct it to B, another problem is evident. The vapor of the two-phase flow will take the path of least resistance and most of it will go to exchanger A. Now exchanger A is getting the vapor and B is getting the liquid. But the heat transfer rate of both exchangers is designed for equal flow of liquid and vapor. So they will not transfer heat as designed. And because these exchangers will not perform as designed, the process unit will not perform and must be shut down for correction. And this is because a piping designer did not recognize he was piping two-phase flow, or if he knew he was not knowledgeable enough to design the piping.

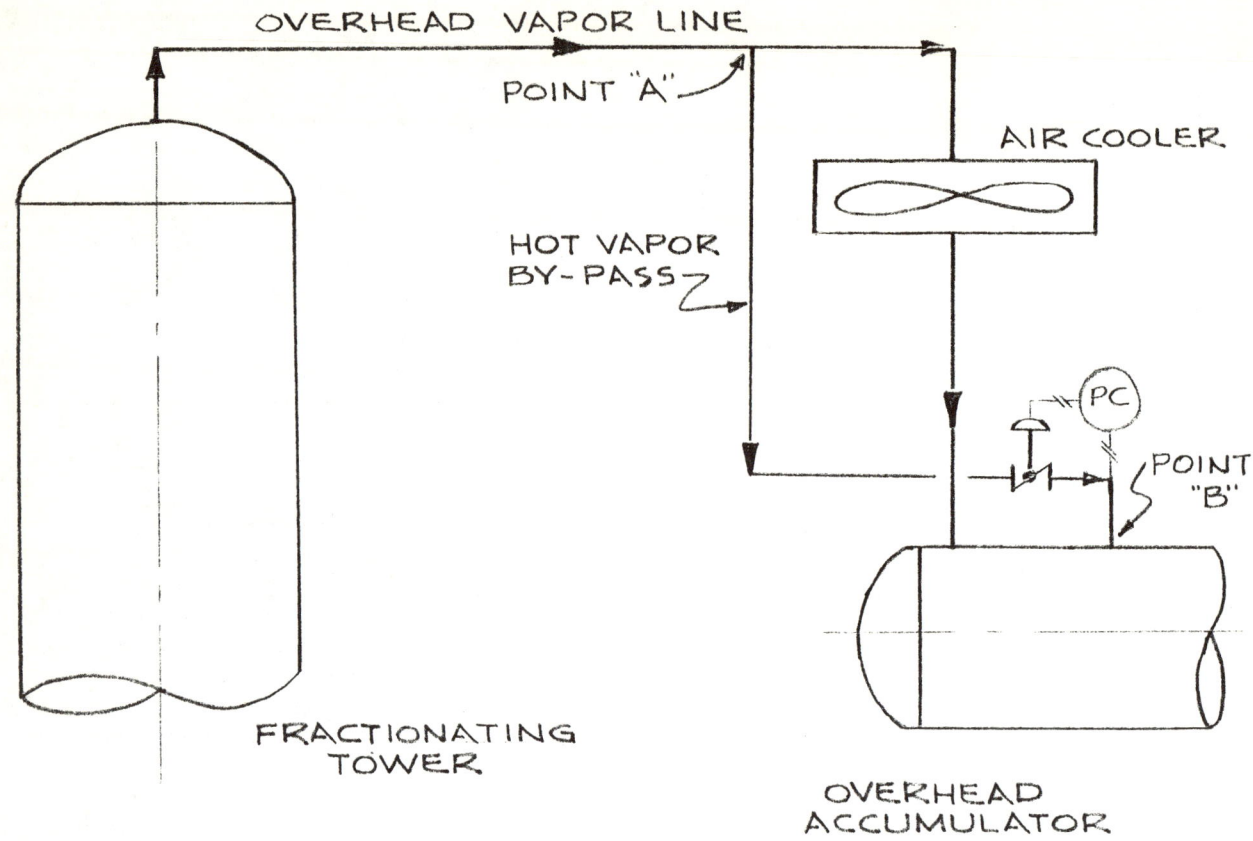

Figure 2-6. Hot vapor by-pass.

In the proper illustration, the two-phase flow enters the horizontal pipe midway between the two exchangers. The pressure drop is the same to either one so there is no path of least resistance and the flow of both liquid and vapor will be equal to both exchangers.

Figure 2-5 is a flow sketch of piping at a four-pass fired heater. This is a very common piping installation where the heater inlet is all liquid flow and the outlet is two-phase flow. And in nine out of ten cases this installation is completely unnecessary and excessively expensive!

To analyze this statement, the liquid heater feed, item 1, is divided into four streams to match the four passes of heater tubes. Each stream has a globe valve, item 2, for throttling and a flow indicator (orifice or meter run), item 3. Each heater pass outlet has a temperature indicator. To operate the heater, flow is regulated with the globe valves to ensure that each pass has the same flow, while the temperature indicator shows the desired outlet temperature. The piping is correct to this point.

The incorrect design occurs downstream of the TI located in each pass outlet. Someone has coined the magic term *two-phase flow* and has decided that this means symmetrical piping. And symmetrical piping means money and piping problems. This can be doubly expensive if the heater outlet is alloy material.

Symmetrical piping is necessary for two-phase flow if there is no method of control, and distribution must be made. But in Figure 2-5 distribution is made while the fluid is all liquid, controlled by the globe valve and metered by the flow indicator to insure that streams *A, B, C* and *D* are all equal flow. If the flow is equal going into the heater it must be equal coming out. Only in cases of very low pressure, such as a crude charge heater service, should symmetrical piping be considered. In higher pressure installations the outlets should be com-

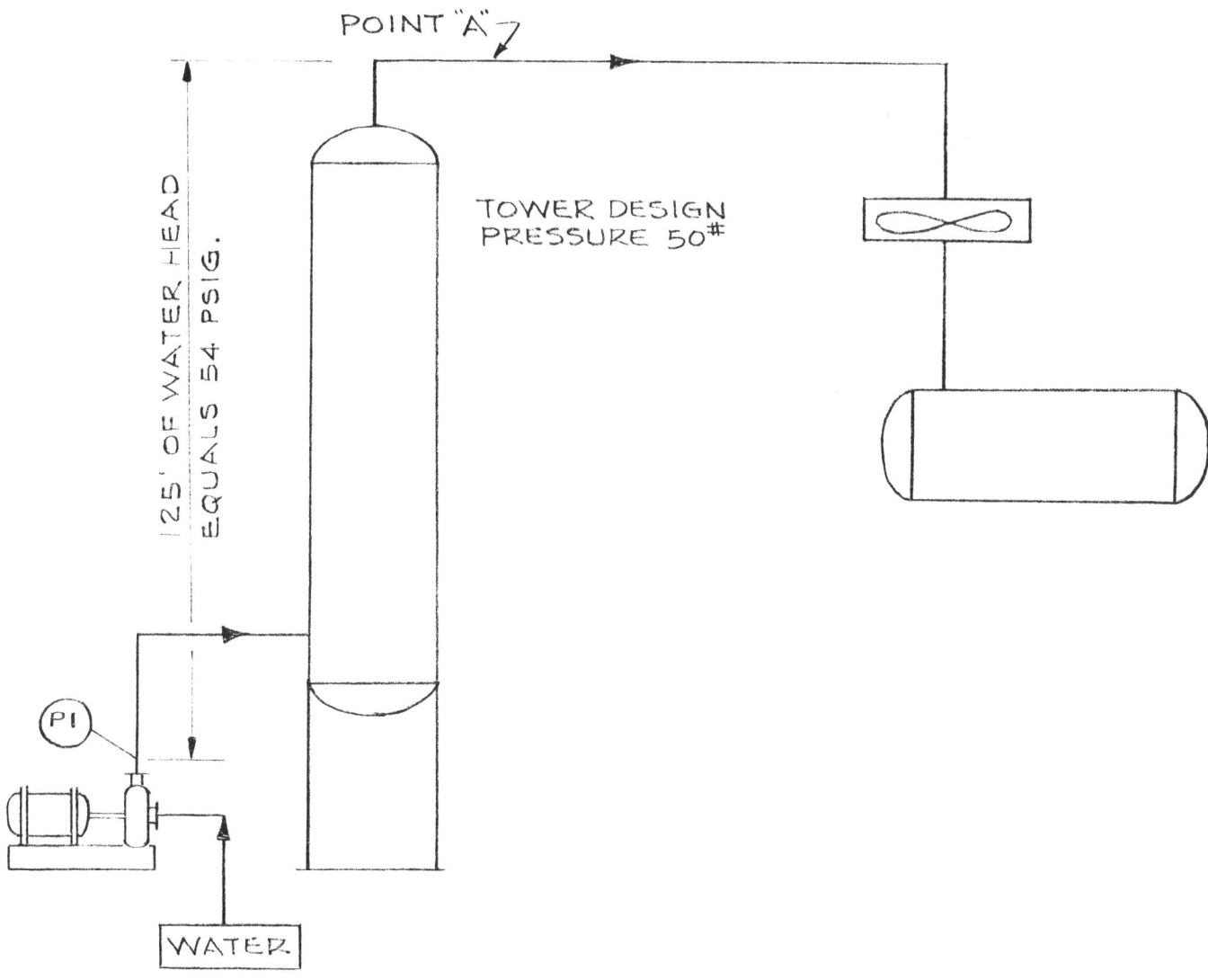

Figure 2-7. Impact of water head.

bined in the most economical manner and routed on its way. The outlet pass differential pressure drop is so minor that it is of no actual consequence.

Any time a competent piping designer sees symmetrical piping called for he should seriously question its need, especially if it is on a fired heater outlet.

Hot Vapor By-Pass

Figure 2-6 is a hot vapor by-pass schematic. The tower overhead vapor line is routed to an air cooler (fin-fan). This could be a shell and tube exchanger. The condensed vapor is then directed to the overhead accumulator. To maintain pressure on the accumulator, a hot vapor by-pass is installed. Hot overhead vapor is by-passed around the cooler and is routed to the pressure control valve, which allows pressure to enter the accumulator as required.

Hot vapor by-passes should never be pocketed. The pressure control valve should be installed above the top of the accumulator and the by-pass piping should continuously drain from point A to point B.

As this by-passed vapor cools, due to rainfall on the line or cool air cooling the line, condensate

Table 2-2
Water Static Head Pressures

Feet of Water	Pressure psig	Feet of Water	Pressure psig	Feet of Water	Pressure psig	Feet of Water	Pressure psig
1	0.43	26	11.26	51	22.09	76	32.92
2	0.86	27	11.69	52	22.52	77	33.35
3	1.30	28	12.12	53	22.95	78	33.78
4	1.73	29	12.55	54	23.39	79	34.21
5	2.16	30	12.99	55	23.82	80	34.65
6	2.59	31	13.42	56	24.26	81	35.08
7	3.03	32	13.86	57	24.69	82	35.52
8	3.46	33	14.29	58	25.12	83	35.95
9	3.89	34	14.72	59	25.55	84	36.39
10	4.33	35	15.16	60	25.99	85	36.82
11	4.76	36	15.59	61	26.42	86	37.25
12	5.20	37	16.02	62	26.85	87	37.68
13	5.63	38	16.45	63	27.29	88	38.12
14	6.06	39	16.89	64	27.72	89	38.55
15	6.49	40	17.32	65	28.15	90	38.98
16	6.93	41	17.75	66	28.58	91	39.42
17	7.36	42	18.19	67	29.02	92	39.85
18	7.79	43	18.62	68	29.45	93	40.28
19	8.22	44	19.05	69	29.88	94	40.72
20	8.66	45	19.49	70	30.32	95	41.15
21	9.09	46	19.92	71	30.75	96	41.58
22	9.53	47	20.35	72	31.18	97	42.01
23	9.96	48	20.79	73	31.62	98	42.45
24	10.39	49	21.22	74	32.05	99	42.88
25	10.82	50	21.65	75	32.48	100	43.31

is formed. With the small amount of differential pressure between points A and B, a pocket of liquid head might not be overcome and the hot vapor by-pass would not work. There is generally less than 10 psig differential between these two points. The pressure control valve is usually specified as a butterfly type to keep pressure drop to a minimum but this will consume 2-3 psig. Line loss due to friction may consume another 2-3 psig. This only leaves about 4 psig and that would not overcome much liquid head.

Static Head

Hydraulics is the term used for the action of liquids in motion or at rest. Unconfined liquids seek the lowest possible level and a horizontal position.

Liquids at rest cause a pressure equal in all directions and perpendicular to any surfaces in contact with the liquid. This pressure is caused by the weight of the liquid above the point in question plus the pressure at the top level of the liquid. The liquid height is called *static head, pressure head* or quite often just *head*.

To calculate the head pressure one must multiply the height of liquid by the specific gravity and divide by 2.309.

Specific gravity is the item's weight compared to water, which has the assigned specific gravity of one. Consequently, a column of water 2.309' high would exert 1 psig of pressure at its base. This would be called 1 pound of head. To this head pressure one must add the pressure above the liquid to get the actual pressure at the base of the liquid.

Table 2-3
Saturated Steam Data

Steam psig	Temp °F	Steam psig	Temp °F	Steam psig	Temp °F
14.7	212	140	353	290	414
15	213	145	356	300	417
20	228	150	358	320	424
25	240	155	361	340	429
30	250	160	364	360	434
35	259	165	366	380	440
40	267	170	368	400	445
45	274	175	371	420	449
50	281	180	373	440	454
55	287	185	375	460	459
60	293	190	378	480	463
65	298	195	380	500	467
70	303	200	382	550	477
75	308	205	384	600	486
80	312	210	386	650	495
85	316	215	388	700	503
90	320	220	390	750	511
95	324	225	392	800	518
100	328	230	394	850	525
105	331	235	396	900	532
110	335	240	397	950	538
115	338	245	399	1000	545
120	341	250	401		
125	344	260	404		
130	347	270	408		
135	350	280	411		

This becomes very important when calculating the hydrostatic pressure of vessels and piping systems. For instance, in Figure 2-7 there is a tower system designed for 50 psig. This requires a hydrostatic test pressure of 75 psig. This test pressure must be measured at the highest point in the system, point A.

To attain this pressure at point A the field brings in a hydrotest pump and fills the system full of water. The piping designer knows this pump is located at grade. He must calculate the height from the pump's pressure gage to point A, divide by 2.309 and add this to the required test pressure. In this case 54 psig must be added to 75 psig, totaling 129 psig, which is the pressure that must be obtained at the pressure gage to satisfy the 75 psig hydrostatic conditions.

The vessel designer must also consider the static water head when calculating the tower head and shell thickness. This added weight is considered in the design of the support for the tower. It becomes very critical when vessels are located high in a steel or concrete structure.

In most organizations it is the piping designer's responsibility to furnish the field hydrostatic test flow diagrams, showing the required hydrotest pressures for each line in the system. It is imperative that static head calculations be made for each system and that proper test pressures are assigned.

Table 2-2 gives the pressures caused by water head from 1' to 100'. For greater heights the figures shown may be added, i.e., for 125' add the figure for 100' and 25'.

Steam Data

Steam is generated by heating water to such a temperature that it vaporizes and this vapor is steam. Steam is invisible.

The boiling point of water is 212°F, or 100°C, at sea level atmospheric pressure, 14.7 psig. *Boiling point* is the temperature at which a liquid boils, changing into a vapor with the application of heat. The pressure applied to a liquid such as water greatly changes its boiling point. Water will boil at 32°F with a pressure of 0.0885 psig. Place water under 1000 psig pressure and the temperature must reach 544°F to make steam.

Saturated steam is steam which is at its minimum temperature for its pressure. No additional heat (called superheat) has been added. Saturated steam data is given in Table 2-3.

Superheated steam is saturated steam to which additional "degrees of superheat" have been added. Its condition is usually expressed by the degrees of superheat above the saturation point and by its pressure. Superheated steam has gas characteristics; when compressed the pressure increases, when heated its volume increases at constant pressure and its pressure rises if the volume remains constant. Adding superheat to steam and maintaining the same pressure then gives more steam for each pound of water.

One cubic foot of water at 14.7 psig and 212°F becomes 1606 cubic feet of saturated steam. One pound of this steam will occupy 26.8 cubic feet. Water weighs 62.4 pounds per cubic foot.

Chapter 2
Review Test

1. Define "mercaptan." _____

2. What is a "sweet" stream? _____

3. What is the lightest hydrocarbon? _____

4. What is a saturated hydrocarbon? _____

5. Define "fractionation." _____

6. Define "equilibrium liquid." _____

7. What is two-phase flow? _____

8. Define "flashing." _____

9. What is the purpose of a reboiler? _____
_____

10. What is symmetrical piping? _____
_____

11. Hot vapor by-passes shall never be _____.

12. Define "hydraulics." _____

13. What is "head?" _____

14. Define "specific gravity." _____

15. Superheated steam is usually expressed by _____
_____.

3 Plant Arrangement and Storage Tanks

This chapter will deal with general plant arrangement and will probably be the most controversial chapter in the book. For each "grass roots" plant being designed a different approach can be justified. There is not enough space in this book to explore all possibilities and their justifications. One basic conception is presented with some alternate methods of plant layout.

Site Data

Assuming the plot area is ample, the first thing a layout designer must know is the topography. If the site is relatively flat, many problems will not be encountered. But for a flat site, the most important item to know is the soil condition. Some flat sites have been filled by some means over the years. How good is the load-bearing characteristic? Generally, the natural soil will have good allowables while the filled areas may not, requiring all foundations with heavy loads to be pile supported.

The plant is basically divided into two areas, offsite and onsite. The offsite portion consists of storage tanks, loading or unloading facilities, cooling tower, steam generation, electrical power generation when not purchased, electrical transformer substation(s), flare(s), waste disposal facilities, buildings, plant roads and other items as needed. The onsite part of a plant is limited to the process units. This is the area of heaviest concentration of loads and should be located in the area of best soil. The offsite will need from 75-80% of the total plot area and, depending on storage capacities, will be about the same percentage of the total plant cost.

Hilly Terrain

When the plant is to be built on a hilly site, the soil is usually pretty good and is not a major consideration for locating onsite vs. offsite. The author advocates knocking off the top of the hill to construct a relatively flat site for the process units, with the offsites located below. This has several advantages which outweigh the disadvantages. First, and most important, it offers maximum safety to the greatest number of personnel working in the process units and offices. The higher elevation produces head for natural flow into product or rerun storage tanks.

By locating the loading facilities still lower, the designer can make the loading pumps smaller if not eliminate them entirely. The entire sewer system is greatly simplified with waste treatment located at the lowest possible elevation, eliminating costly waste lift pumps.

Soil removed from the hilltop is utilized as tank dike material or for leveling sublevels.

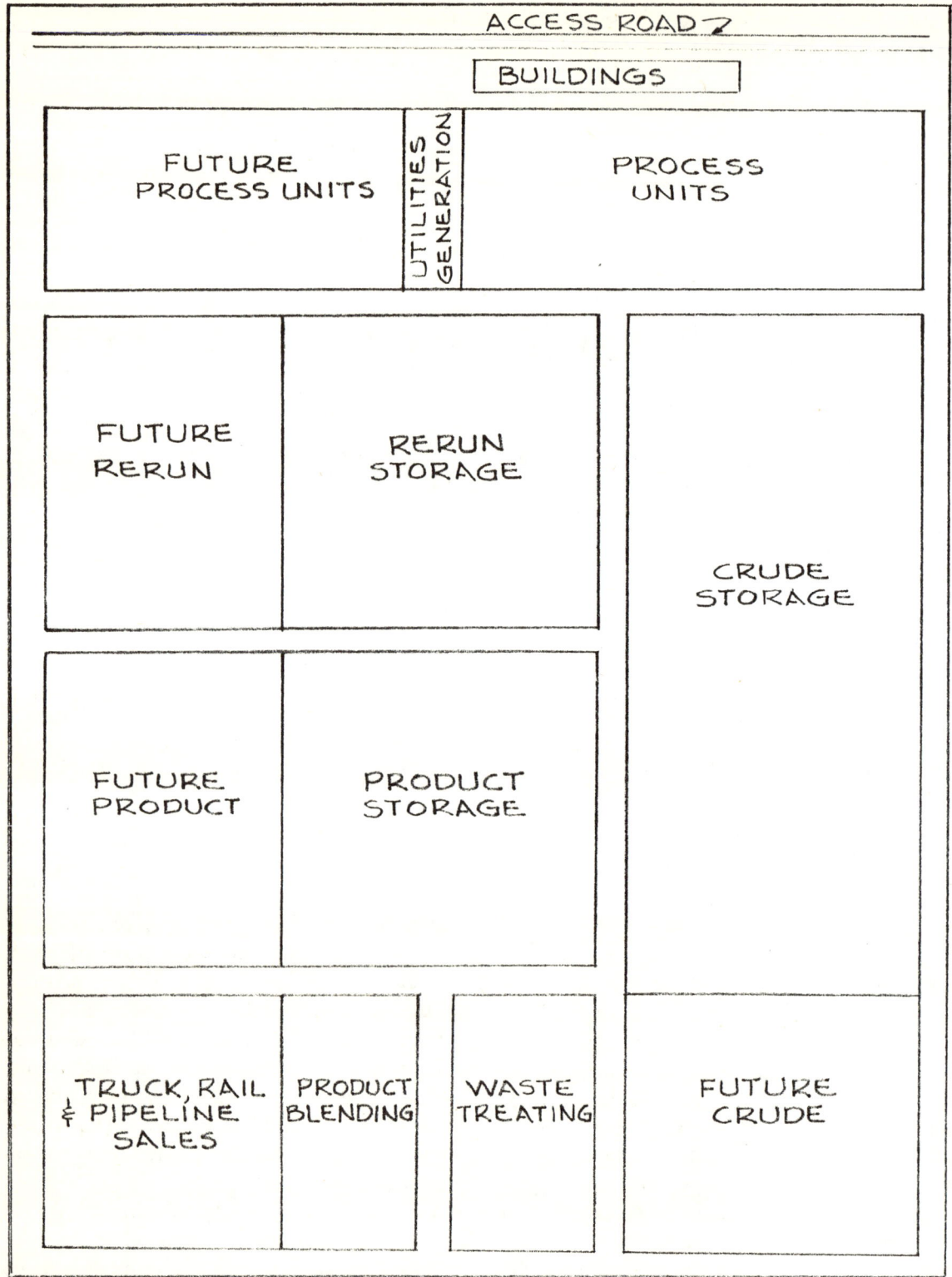

Figure 3-1. Refinery block plot plan.

Block Plot Plan

After the site topography has been established and the preliminary soil report has been studied, the plot designer is ready to make a block plot plan. When the plant site is flat and soil conditions are good, the designer needs to know the quantity and size of the tanks, location of access highways, railroads, barge or ship docks, prevailing wind direction and the area's climatic conditions. If there are incoming utilities such as purchased electrical power or potable water, these locations may effect the plot layout.

Figure 3-1 is a block plot plan for a refinery. To develop this, the designer must:

1. Establish plot limits
2. Define process units and area required
3. Estimate storage dike areas
4. Know and allow for plant expansion
5. Locate waste treating at lowest elevation
6. Locate process units on best soil
7. Locate rerun storage near process units
8. Locate truck, rail and pipe line sales near roads, railroads or existing pipe line tie-in point
9. Locate product blending near sales area
10. Locate product tanks near product blending
11. Keep plant road system in mind
12. Locate utilities generation near process units
13. Keep in mind the cooling tower and electrical substation location

These blocked areas are estimated and located in relation to each other within the plot limits. When this is approved the designer must establish the actual space requirements for each block.

The Process Block

To determine the space required for the process units one must know what units are to be built, the general capacity of each unit and the flow between units. This data can be obtained from the process engineer.

Figure 3-2 is the block plot plan for the process area of Figure 3-1. The crude oil is pumped from the crude storage tanks to the crude unit so this process unit is to be located near the crude storage area. The interconnecting pipeway routes both process and utility lines to the different units. By establishing this blocked area, the designer is ready to set the sizes of each process unit, the unit pipeway sizes, the interconnecting pipeway size and finally the required plot size for all the process units. Tentatively set are the junctions to the interconnecting pipeway from crude storage, other storage and utilities generation.

Storage Tanks

To establish plot areas for storage tanks the designer must have a fairly good knowledge of tanks, how they are designed and when to use what type as well as operation and maintenance procedures.

Tanks for storing liquids at atmospheric pressure are built in two basic styles, floating roof design where the roof floats on top of the liquid, rising and falling with the level, and the cone roof design where the roof is fixed.

Cone Roof Design

Figure 3-3 is a cone roof tank of 200' in diameter and 48' in shell height, storing 368,000 barrels. The size of large tanks is limited by the plate thickness of the bottom course, which should not exceed 1½". By using special high-strength steels, it is possible to utilize very large tanks without exceeding this maximum plate thickness.

Soil conditions and cost of land are economic factors in the selection of tank dimensions. For very poor soil, tanks with low shell height but larger diameters usually are more economical than high shells and smaller diameters because the higher the shell, the greater the head pressure on the soil. If the pressure becomes greater than the soil allowable bearing pressure, pile-supported foundations will be necessary and this is very expensive.

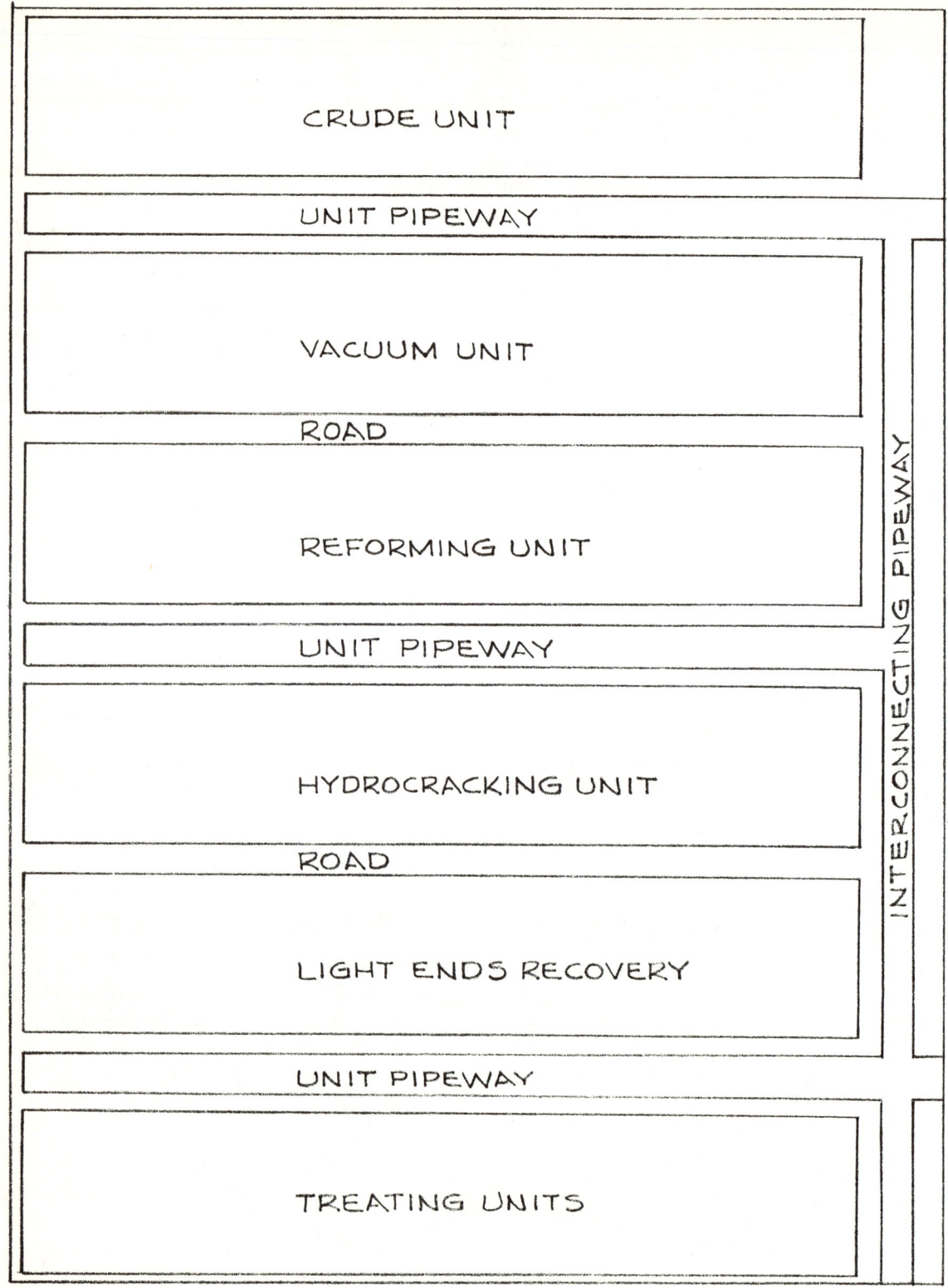

Figure 3-2. Process area block plot plan. (Rotated 90° counterclockwise)

Figure 3-3. Cone roof tank. Courtesy of Chicago Bridge and Iron Co.

Piling is to be avoided but there are times when there is no choice. If land is very costly or just is not available, it is justifiable to construct high shells and utilize the expensive piling.

Figure 3-4 shows a "pancake" tank design, 200' in diameter but only 14' for the shell height, providing 78,000 barrels of storage on "marshy" soil unable to support conventional tank height without elaborate pile foundations.

Large cone roof tanks usually have columns to support the roofs with a roof slope of ¾" to 1'-0". Since these roofs are fixed, there is always a vapor space above the liquid. These tanks must be vented to the atmosphere and quite often the loss of this vapor is so extensive that it warrants specifying the more expensive floating roof tank design.

Floating Roof Design

For storing volatile liquids, floating roof storage tanks have three distinct advantages. They reduce evaporation loss, greatly increase safety from fire and minimize air pollution. Figure 3-5 shows how the floating roof (a) eliminates filling loss, (b) reduces breathing loss and (c) provides effective fire protection and lightning safety.

42   Process Piping Design

Figure 3-4. Cone roof "pancake" tank. Courtesy of Chicago Bridge and Iron Co.

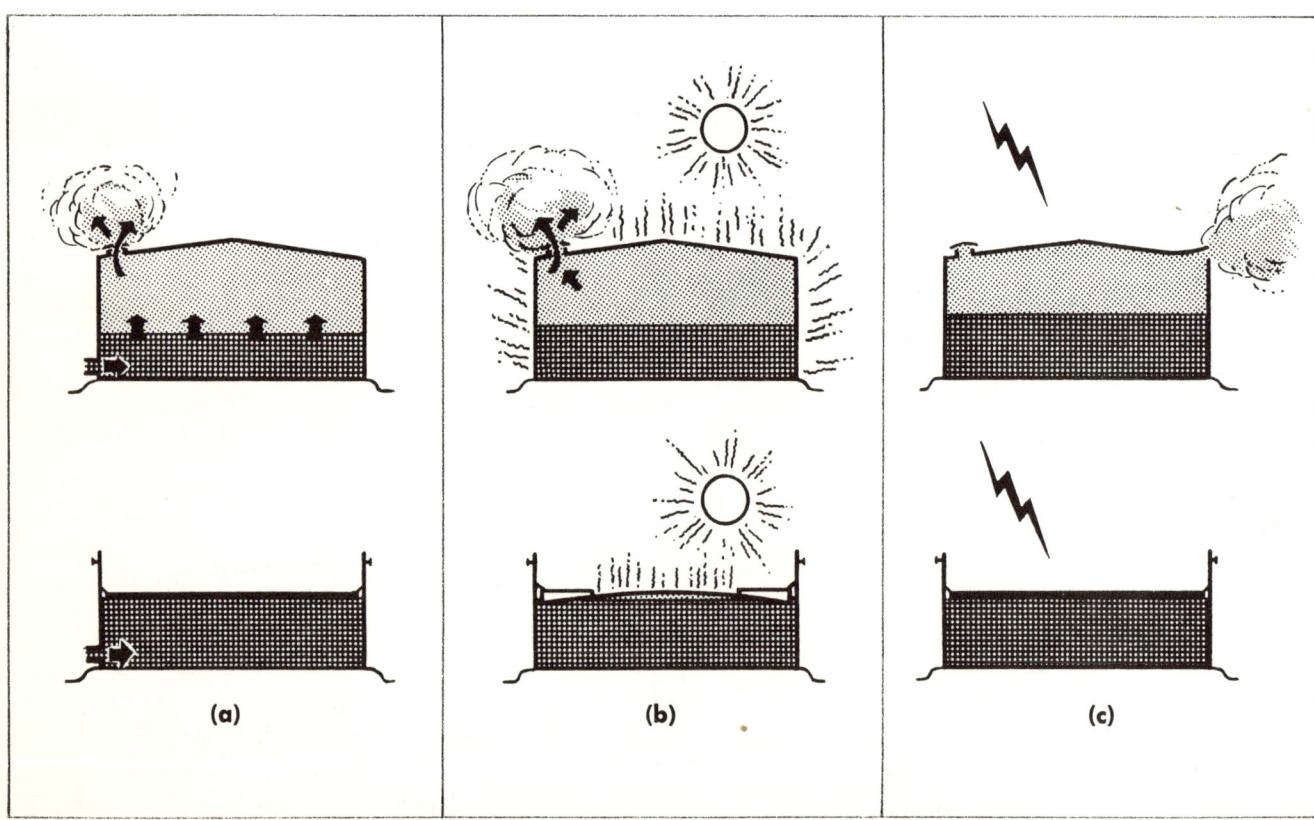

Figure 3-5. Floating versus cone roof design. The floating roof, by minimizing vapor space, eliminates filling loss (a), reduces breathing loss (b), and provides effective fire protection and lightning safety (c). It also inhibits roof and shell corrosion. Courtesy of Chicago Bridge and Iron Co..

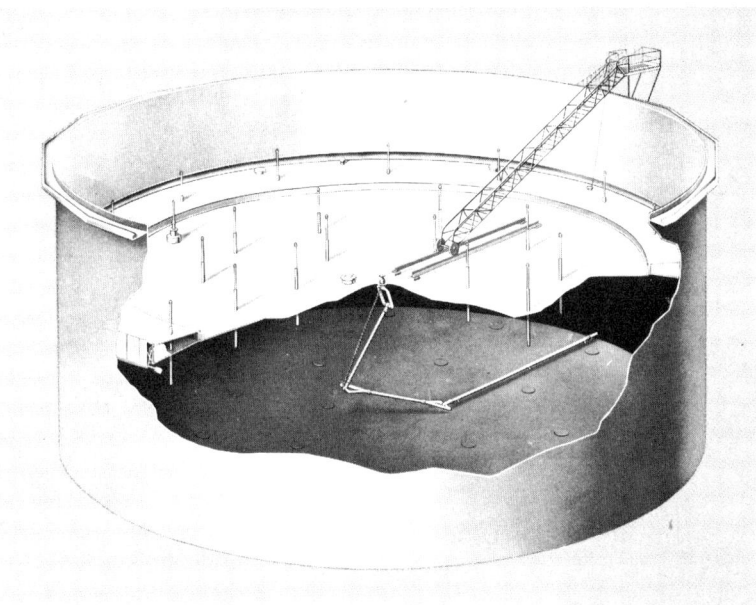

Figure 3-6. Pontoon roof design tank. The pontoon roof has a compartmented annualar ring of pontoons and a single deck. Courtesy of Chicago Bridge and Iron.

Figure 3-6 depicts the pontoon roof design. This has a compartmented annular ring of pontoons and a single - deck center. The most expensive design is the double-deck roof which has two complete decks with an insulating air space between, which makes it the most efficient of all floating roofs for handling the more volatile liquids.

Why do people specify the more expensive floating roof tank when the cone roof is more economical? In terms of economy only, consider an 80,000 barrel cone roof tank in average service. In normal operation it will lose 2700 barrels of gasoline per year by evaporation. At 42 gallons per barrel that loss would be over 100,000 gallons a year, a very substantial loss. And over 90% of this loss is preventable with a floating roof design.

Table 3-1 gives data on storage tanks designed of standard steel in accordance with API Standard 650 for welded tanks.

Table 3-2 gives tank data for typical API Standard 650 welded oil storage tanks for Appendix D and high - stress design (Appendix G).

Tank Spacing

The National Fire Protection Association has published several volumes of recommended practice concerning fire protection. Volume 1, Flammable Liquids, pertains to storing liquids and supplys guidelines for storage tank location, spacing and dike requirements. The following text and tables comply with the NFPA recommendations.

Liquids are divided into three basic classes: flammable liquids with flash points below 100°F, flammable liquids with flash points over 100°F but below 140°F and combustible liquids with flash points over 140°F but below 200°F. Table 3-3 defines these classes.

Table 3-4 shows the further restrictions based on the actual storage conditions. Tables 3-5 through 3-9 are used to set the actual tank spacing in feet, from property lines, public ways or important buildings.

Shell to shell tank spacing is restricted by paragraph 2120 of the NFPA as follows:

Table 3-1

API Standard 650 Welded Oil Storage Tanks

Capacity in Barrels Exact	Tank Dimensions in Feet and Inches		Weight in Pounds With 3/16" Cone Roof 1/4" Bottom	Based on 1.0 Product Specific Gravity Shell Plate Thickness in Inches						Top Angle Size in Inches
	Diameter	Height		1	2	3	4	5	6	
505	15-0	16-0	11,400	3/16	3/16					2½ x 2½ x ¼
1,010	21-3	16-0	17,400	3/16	3/16					2½ x 2½ x ¼
1,515	21-3	24-0	22,000	3/16	3/16	3/16				2½ x 2½ x ¼
1,512	26-0	16-0	23,100	3/16	3/16					2½ x 2½ x ¼
2,020	21-3	32-0	26,800	3/16	3/16	3/16	3/16			2½ x 2½ x ¼
2,100	25-0	24-0	27,100	3/16	3/16	3/16				2½ x 2½ x ¼
3,025	26-0	32-0	34,400	3/16	3/16	3/16	3/16			2½ x 2½ x ¼
3,020	30-0	24-0	34,800	3/16	3/16	3/16				2½ x 2½ x ¼
3,765	33-6	24-0	40,500	3/16	3/16	3/16				2½ x 2½ x ¼
4,030	30-0	32-0	41,300	3/16	3/16	3/16	3/16			2½ x 2½ x ¼
5,040	30-0	40-0	47,800	3/16	3/16	3/16	3/16	3/16		2½ x 2½ x ¼
5,020	33-6	32-0	47,700	3/16	3/16	3/16	3/16			2½ x 2½ x ¼
5,485	35-0	32-0	50,900	3/16	3/16	3/16	3/16			2½ x 2½ x ¼
6,040	30-0	48-0	55,300	.205	3/16	3/16	3/16	3/16	3/16	2½ x 2½ x ¼
6,855	35-0	40-0	58,900	.199	3/16	3/16	3/16	3/16		2½ x 2½ x ¼
6,010	36-8	32-0	54,200	3/16	3/16	3/16	3/16			2½ x 2½ x 5/16
7,160	40-0	32-0	61,300	3/16	3/16	3/16	3/16			2½ x 2½ x 5/16
7,515	36-8	40-0	62,900	.208	3/16	3/16	3/16	3/16		2½ x 2½ x 5/16
8,950	40-0	40-0	71,500	.227	3/16	3/16	3/16	3/16		2½ x 2½ x 5/16
10,100	42-6	40-0	79,100	.241	.192	3/16	3/16	3/16		2½ x 2½ x 5/16
10,315	48-0	32-0	81,600	.217	3/16	3/16	3/16			2½ x 2½ x 5/16
11,330	45-0	40-0	86,700	.256	.203	3/16	3/16	3/16		2½ x 2½ x 5/16
12,100	42-6	48-0	93,300	.291	.241	.192	3/16	3/16	3/16	2½ x 2½ x 5/16
12,100	52-0	32-0	104,900	¼	¼	¼	¼			2½ x 2½ x 5/16
12,890	48-0	40-0	96,200	.273	.217	3/16	3/16	3/16		2½ x 2½ x 5/16
13,595	45-0	48-0	102,500	.308	.256	.203	3/16	3/16	3/16	2½ x 2½ x 5/16
13,985	50-0	40-0	114,100	.284	¼	¼	¼	¼		2½ x 2½ x 5/16
15,470	48-0	48-0	114,200	.328	.273	.217	3/16	3/16	3/16	2½ x 2½ x 5/16
15,130	52-0	40-0	122,100	.295	¼	¼	¼	¼		2½ x 2½ x 5/16
15,060	58-0	32-0	123,600	.262	¼	¼	¼			2½ x 2½ x 5/16
16,785	50-0	48-0	133,600	.342	.284	¼	¼	¼	¼	2½ x 2½ x 5/16
20,140	60-0	40-0	154,600	.341	.271	¼	¼	¼		2½ x 2½ x 5/16
24,170	60-0	48-0	181,800	.411	.341	.271	¼	¼	¼	2½ x 2½ x 5/16
25,120	67-0	40-0	185,400	.380	.302	¼	¼	¼		3 x 3 x 3/8
27,415	70-0	40-0	201,600	.397	.316	¼	¼	¼		3 x 3 x 3/8
30,140	67-0	48-0	219,200	.458	.380	.302	¼	¼	¼	3 x 3 x 3/8
30,100	73-4	40-0	218,700	.416	.331	¼	¼	¼		3 x 3 x 3/8
32,905	70-0	48-0	238,400	.479	.397	.316	¼	¼	¼	3 x 3 x 3/8
35,810	80-0	40-0	258,100	.454	.361	.268	¼	¼		3 x 3 x 3/8
40,425	85-0	40-0	290,300	.483	.384	.285	¼	¼		3 x 3 x 3/8
42,970	80-0	48-0	305,700	.547	.454	.361	.268	¼	¼	3 x 3 x 3/8
45,320	90-0	40-0	313,700	.511	.406	.301	¼	¼		3 x 3 x 3/8
44,760	100-0	32-0	316,600	.451	.335	¼	¼			3 x 3 x 3/8
54,390	90-0	48-0	379,100	.616	.511	.406	.301	¼	¼	3 x 3 x 3/8
54,165	110-0	32-0	380,600	.496	.368	¼	¼			3 x 3 x 3/8
55,950	100-0	40-0	382,600	.568	.451	.335	¼	¼		3 x 3 x 3/8
67,140	100-0	48-0	460,700	.684	.568	.451	.335	¼	¼	3 x 3 x 3/8
67,705	110-0	40-0	456,800	.625	.496	.368	¼	¼		3 x 3 x 3/8
81,245	110-0	48-0	552,800	.753	.625	.496	.368	¼	¼	3 x 3 x 3/8
80,580	120-0	40-0	554,800	.681	.542	.402	5/16	5/16		3 x 3 x 3/8
96,690	120-0	48-0	666,800	.821	.681	.542	.402	5/16	5/16	3 x 3 x 3/8
100,470	134-0	40-0	682,900	.761	.605	.449	5/16	5/16		3 x 3 x 3/8
109,700	140-0	40-0	744,500	.795	.632	.469	5/16	5/16		3 x 3 x 3/8
120,565	134-0	48-0	821,500	.917	.761	.605	.449	5/16	5/16	3 x 3 x 3/8
125,895	150-0	40-0	853,000	.852	.677	.502	.328	5/16		3 x 3 x 3/8
131,600	140-0	48-0	894,600	.958	.795	.632	.469	5/16	5/16	3 x 3 x 3/8
143,200	160-0	40-0	971,900	.909	.722	.537	.349	5/16		3 x 3 x 3/8
150,995	150-0	48-0	1,025,700	1.026	.852	.677	.502	.328	5/16	3 x 3 x 3/8
171,900	160-0	48-0	1,168,500	1.095	.909	.722	.537	.349	5/16	3 x 3 x 3/8
181,300	180-0	40-0	1,225,600	1.022	.812	.603	.393	5/16		3 x 3 x 3/8
217,500	180-0	48-0	1,476,900	1.232	1.022	.812	.603	.393	5/16	3 x 3 x 3/8
223,800	200-0	40-0	1,487,000	1.136	.903	.670	.437	5/16		3 x 3 x 3/8
268,600	200-0	48-0	1,815,300	1.369	1.136	.903	.670	.437	5/16	3 x 3 x 3/8

Source: Chicago Bridge and Iron Co.

Weights include stairway, roof and shell manholes, connection in roof for vent and usual inlet and outlet connections in shell. Plate thicknesses shown are for eight feet wide rings, but ring widths and plate thicknesses may vary to suit available stock. Standard tanks are furnished with shell plates conforming to the latest edition of ASTM Specification A283, grade C.

Plant Arrangement, Storage Tanks

Table 3-2
API Standard 650 Tanks for High Stress Design
For Appendix D and Appendix G

Capacity in Barrels Exact	Tank Dimensions in Feet		Weight in Pounds for 0.85 Product Specific Gravity—Weight Includes ¼″ Bottom, 3/16″ Roof, and Cantilevered Roof Framing			
			Appendix D Design		High Stress Design (App. G)	
	Diameter	Height	No Corrosion Allowance (See Note 2)	With 1/16″ Corrosion Allowance	No Corrosion Allowance (See Note 2)	With 1/16″ Corrosion Allowance
42,970	80-0	48-0	278,200	288,600	—	—
50,130	80-0	56-0	321,000	334,000	—	—
67,140	100-0	48-0	412,200	427,200	—	—
78,330	100-0	56-0	479,000	496,400	—	—
96,690	120-0	48-0	596,800	612,200	—	—
112,800	120-0	56-0	693,200	710,000	—	—
131,600	140-0	48-0	800,200	814,800	—	—
153,500	140-0	56-0	931,500	946,400	—	—
171,900	160-0	48-0	1,037,000	1,051,000	—	—
200,500	160-0	56-0	1,209,000	1,222,000	—	—
217,500	180-0	48-0	1,304,000	1,321,000	—	—
253,800	180-0	56-0	1,520,000	1,538,000	—	—
268,600	200-0	48-0	1,599,000	1,617,000	—	—
313,300	200-0	56-0	1,867,000	1,885,000	—	—
325,000	220-0	48-0	1,942,000	1,955,000	1,762,000	1,778,000
379,100	220-0	56-0	2,266,000	2,279,000	2,013,000	2,029,300
387,000	240-0	48-0	2,301,000	2,316,000	2,086,000	2,101,000
451,200	240-0	56-0	2,687,000	2,702,000	2,385,000	2,401,000
453,500	260-0	48-0	2,691,000	2,705,000	2,438,000	2,452,000
526,000	280-0	48-0	3,112,000	3,124,000	2,792,000	2,802,000
529,500	260-0	56-0	See Note 1	See Note 1	2,789,000	2,803,000
604,000	300-0	48-0	—	—	3,204,000	3,216,000
614,000	280-0	56-0	—	—	3,199,000	3,209,000
672,500	293-0	56-0	—	—	3,505,000	3,517,000
687,500	320-0	48-0	—	—	3,634,000	3,645,000
705,000	300-0	56-0	—	—	See Note 1	See Note 1
776,000	340-0	48-0	—	—	4,092,000	4,102,000
789,800	343-0	48-0	—	—	4,163,000 See Note 1	4,172,000 See Note 1

Source: Chicago Bridge and Iron Co.

1) Maximum tank size limited by 1½″ maximum plate thickness.
2) The design of tank shells storing products with a specific gravity of 0.91 or less is governed by the hydrostatic test (provided no corrosion allowance is specified).
3) Type of steel supplied for App. D. design varies with design metal temperature specified; Either ASTM A283 Grade C; ASTM A131 Grade A, B, or C; or ASTM A131 Grade C. Normalized. In cold climates, the higher cost of the quality steel required may offset any anticipated saving from weight reduction when compared with an API Standard 650 shell design.
4) For certain tank sizes, circumferential stiffening rings are not mandatory by Appendix D design but are recommended by CB&I.
5) The high stress design uses high strength steels having a minimum tensile strength of 70,000 psi, a minimum yield of 50,000 psi, and an improved notch-toughness. The use of high strength steels makes it possible to build larger tanks without exceeding 1½″ plate thickness. The weight of circumferential windgirders is included in the Appendix G design. (Basis of 100 mph wind).

Table 3-3
Tank Liquids Classification

Class I	Flammable Liquid	Flash point below 100°F and vapor pressure not exceeding 40 psia at 100°F
Class IA	Flammable Liquid	Flash point below 73°F, boiling point below 100°F
Class IB	Flammable Liquid	Flash point below 73°F, boiling point 100°F or above
Class IC	Flammable Liquid	Flash point 73°F or above but below 100°F
Class II	Flammable Liquid	Flash point below 140°F but at or above 100°F
Class III	Combustible Liquid	Flash point above 140°F but below 200°F
No Class	Unstable Liquid	Will polymerize, decompose, condense or become self-reactive under shock, temperature or pressure
No Class		Liquids with boilover characteristics such as crude oil

Table 3-4
Restrictions for Conditions of Storage

Condition 1	Class I through III liquids stored below 2.5 psig or equipped with emergency venting which will not permit pressure to exceed 2.5 psig. Use table 3-5.
Condition 2	Class I through III liquids stored above 2.5 psig or equipped with emergency venting which will permit pressure to exceed 2.5 psig. Use Table 3-6.
Condition 3	Unstable liquids. Use Table 3-7.
Condition 4	Liquids with boilover characteristics. Use Table 3-8.

1. Three feet minimum distance between any two flammable or combustible liquid storage tanks.
2. Minimum distance between adjacent tanks (all services except unstable liquids and crude in production areas), 1/6 the sum of their diameters; unless one tank is less than ½ the diameter of the other tank, spacing shall be ½ the diameter of the smaller tank.
3. Crude Petroleum in production areas-minimum spacing; capacity under 3000 barrels–3'; capacity over 3000 barrels–diameter of smaller tank.
4. Unstable flammable or combustible liquid-½ the sum of tank diameters.
5. Local authorities (fire protection, insurance, etc.) may require greater spacing for grouped tankage (3 or more rows or irregular pattern) to allow access for fire fighting.
6. LPG Containers (Liquified Petroleum Gas)
   6.1 20' minimum spacing to flammable or combustible liquid tank
   6.2 When adjacent flammable or combustible liquid tanks are diked, LPG containers shall be a minimum of 10' outside centerline of dike. Suitable means shall be taken to prevent accumulation of flammable liquid under LPG tanks (dikes, curbs or grading is acceptable).

Tank Dike Regulations

The NFPA establishes minimum dike regulations in Volume 1, paragraph 2170. Earthen dikes

Plant Arrangement, Storage Tanks

Table 3-5
Tank Spacing Table—Condition 1

Type of Tank	Protection	Minimum Distance (Feet) from Property Line Which May Be Built On, Including the Opposite Side of a Public Way	Minimum Distance (Feet) from Nearest Side of Any Public Way
Floating roof	Protection for exposures*	½ times diameter of tank but need not exceed 90'	1/6 times diameter of tank but need not exceed 30'
	No Protection	Diameter of tank but need not exceed 175'	1/6 times diameter of tank but need not exceed 30'
Vertical with weak roof to shell seam	Approved foam or inerting system on the tank	½ times diameter of tank but need not exceed 90' and not less than 5'	1/6 times diameter of tank but need not exceed 30' and not less than 5'
	Protection for exposures*	Diameter of tank but need not exceed 175'	1/3 times diameter of tank but need not exceed 60'
	No protection	Twice diameter of tank but need not exceed 350'	1/3 times diameter or tank but need not exceed 60'
Horizontal and vertical, with emergency relief venting to limit pressures to 2.5 psig	Approved inerting system on the tank or approved foam system on vertical tanks	½ times Table 3-9 but not less than 5'	½ times Table 3-9 but not less than 5'
	Protection for exposures*	Table 3-9	Table 3-9
	No protection	Twice Table 3-9	Table 3-9

*Protection for exposures shall mean fire protection for structures on property adjacent to tanks. When acceptable to the authority having jurisdiction, such structures located (1) within the jurisdiction of any public fire department or (2) within or adjacent to plants having private fire brigades shall be considered as having adequate protection for exposures.

Table 3-6
Tank Spacing Table—Condition 2

Type of Tank	Protection	Minimum Distance (Feet) from Property Line Which May Be Built On, Including the Opposite Side of a Public Way	Minimum Distance (Feet) from Nearest Side of Any Public Way
Any type	Protection for exposures	1½ times Table 3-9 but shall less than 25'	1½ times Table 3-9 but not less than 25'
	No protection	3 times Table 3-9 but not less than 50'	1½ times Table 3-9 but not less than 25'

**Table 3-7
Tank Spacing Table—Condition 3**

Type of Tank	Protection	Minimum Distance (Feet) from Property Line Which May Be Built On, Including the Opposite Side of a Public Way	Minimum Distance (Feet) from Nearest Side of Any Public Way
Horizontal and vertical tanks with emergency relief venting to permit pressure not in excess of 2.5 psig	Tank protected with any one of the following: Approved water spray Approved inerting Approved insulation and refrigeration Approved barricade	Table 3-9 but not less than 25′	Not less than 25′
	Protection for exposures	2½ times, Table 3-9 but not less than 50′	Not less than 50′
	No Protection	5 times Table 3-9 but not less than 100′	Not less than 100′
Horizontal and vertical tanks with emergency relief venting to permit pressure over 2.5 psig	Tank protected with any one of the following: Approved water spray Approved inerting Approved insulation and refrigeration Approved barricade	Twice Table 3-9 but not less than 50′	Not less than 50′
	Protection for exposures	4 times Table 3-9 but not less than 100′	Not less than 100′
	No protection	8 times Table 3-9 but not less than 150′	Not less than 150′

are usually specified by the designer; however, concrete wall dikes are sometimes used where plot costs are high or where earth fill material is either expensive or just not available. Dike requirements are generally specified by the customer, local regulatory bodies or, if neither of them have established specifications, the minimums noted below may be used:

1. The area surrounding a tank or group of tanks shall be provided with either drainage or dikes to protect adjacent property and waterways (from a rupture and/or spill) unless the jurisdictional authority has waived this requirement (which never happens).
2. Drainage system shall comply with the following:

2.1. Slope not less than 1% away from the tank toward the drainage system.
2.2. Drainage system shall terminate in vacant land or impounding basin having minimum capacity of the largest tank served.
2.3. The terminating area and the route of the drainage system shall be so located that burning liquids in the system will not seriously expose tanks or adjacent property.
2.4. The drainage system, including drainage pumps, shall not discharge to adjoining property, natural water courses, public sewers or public drains unless no hazard is constituted or its design does not per-

Plant Arrangement, Storage Tanks

Table 3-8
Tank Spacing Table—Condition 4

Type of Tank	Protection	Minimum Distance (Feet) from Property Line Which May Be Built On, Including the Opposite Side of a Public Way	Minimum Distance (Feet) from Nearest Side of Any Public Way
Floating roof	Protection for exposures	Diameter of tank but need not exceed 175'	1/3 times diameter of tank but need not exceed 60'
	No protection	Twice diameter of tank but need not exceed 350'	1/3 times diameter of tank but need not exceed 60'
Fixed roof	Approved foam or inerting system	Diameter of tank but need not exceed 175'	1/3 times diameter of tank but need not exceed 60'
	Protection for exposures	Twice diameter of tank but need not exceed 350'	2/3 times diameter of tank but need not exceed 120'
	No protection	4 times diameter of tank but need not exceed 350'	2/3 times diameter of tank but need not exceed 120'

Table 3-9
Tank Spacing Table Reference
from Tables 3-5 through 3-8

Capacity Tank (Gallons)	Minimum Distance (Feet) from Property Line Which May Be Built On, Including the Opposite Side of a Public Way	Minimum Distance (Feet) from Nearest Side of Any Public Way
275 or less	5	5
276 to 750	10	5
751 to 12,000	15	5
12,001 to 30,000	20	5
30,001 to 50,000	30	10
50,001 to 100,000	50	15
100,001 to 500,000	80	25
500,001 to 1,000,000	100	35
1,000,001 to 2,000,000	135	45
2,000,001 to 3,000,000	165	55
3,000,001 or more	175	60

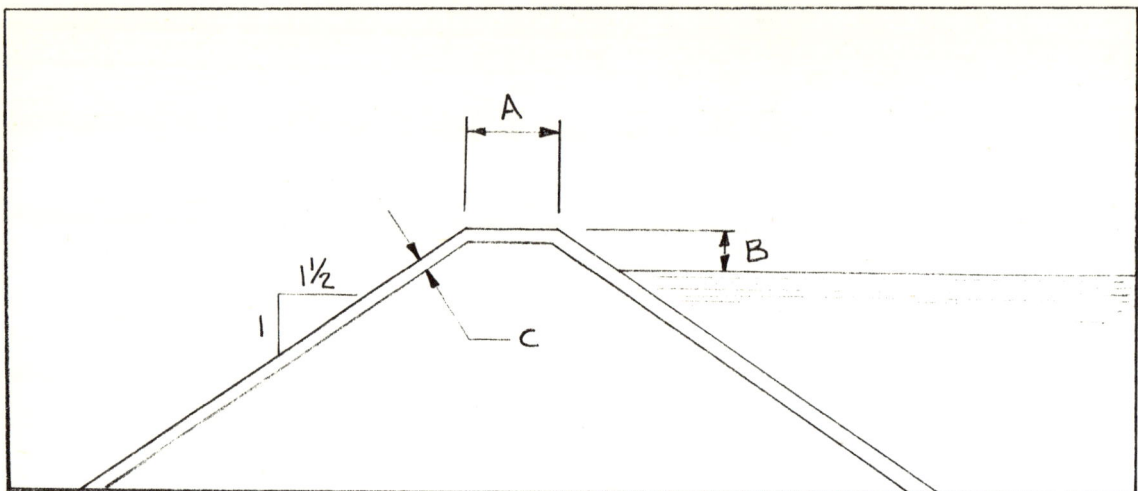

Figure 3-7. Typical dike design.

mit flammable or combustible liquids to be released. (Note: this implies the requirement of separator facilities in all but the most remote locations)
3. Diked areas shall comply with the following when protection of adjoining property is accomplished by retaining the liquid around the tank with dikes:
3.1. For other than crude petroleum tanks with fixed roofs, the minimum capacity of the diked area shall be not less than can be released from the largest full tank. (Diked area capacity not to include capacity of enclosed tanks [other than the largest tank] below height of the dike.)
3.2. For crude petroleum tank or group of tanks *with fixed roofs*, minimum diked area capacity shall be the full capacity of the tank or tanks enclosed. (Diked area capacity not to include capacity of enclosed tanks below the height of the dike.)
3.3 Dike construction requirements:
3.3.1. Maximum average height of 6' above interior grade.
3.3.2. Earthen walls over 3' high shall have 2' wide flat section at the top.
3.3.3. Slope consistent with the angle of repose of the material.
3.3.4. Drains from diked areas to be located at most remote location from tank, controlled to prevent flammable or combustible materials from entering adjacent property, natural waterways or public drains and sewers.
3.4 Diked areas containing two or more tanks shall be subdivided by drainage channels or curbs to prevent spills from endangering adjacent tanks within the diked area as follows:
3.4.1. Normally stable liquids in:
  a. Cone roof tank with weak roof to shell seam.
  b. Floating roof tanks
  c. Crude petroleum in production areas-any type tanks:
  1 subdivision for each tank in excess of 10,000 barrel capacity
  1 subdivision for each group of tanks (no tank exceeding 10,000 barrel) having aggregate capacity not exceeding 15,000 barrel capacity
3.4.2. Normally stable liquid in tanks not covered in 3.4.1:
  1 subdivision for each tank in excess of 100,000 gallons capacity (2500 barrels)
  1 subdivision for each group of tanks (no tank exceeding 100,000 gallons) having aggregate capacity not exceeding 150,000 gallons (3570 barrels)

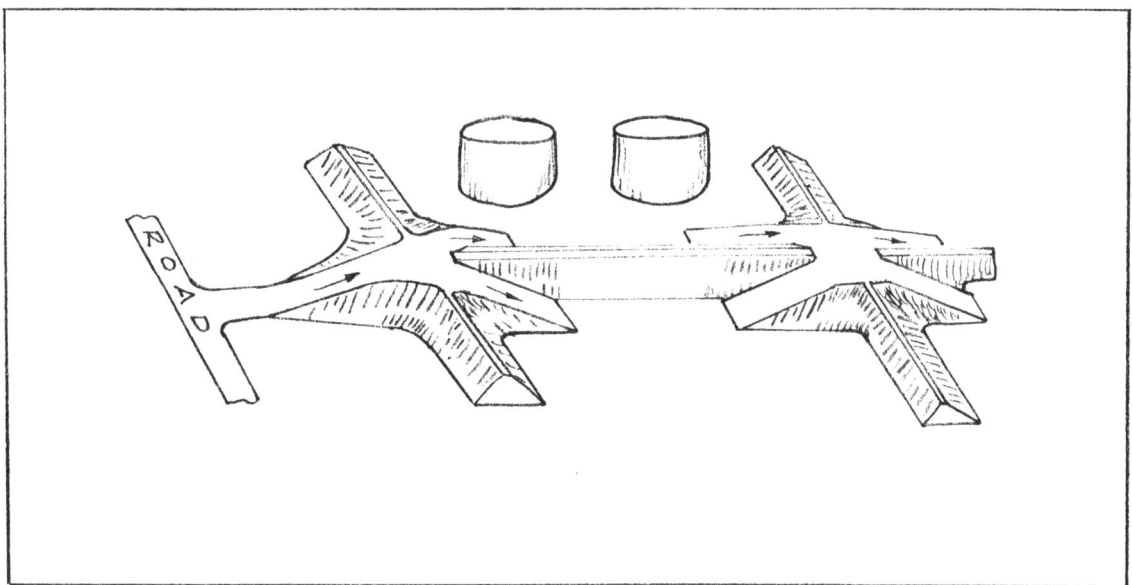

Figure 3-8. Truck access into diked area.

3.4.3. Unstable liquid-any type tank-1 subdivision each tank (exception-tanks protected by water spray system per NFPA No. 15 require no additional subdivision) (Subdivision by drainage channels is preferred).

3.4.4. Curbs shall be not less than 18" high.

Figure 3-7 shows a typical earthern dike design. Item A should be 2'-0" minimum with 3'-0" preferred for tall dikes. Item B is the freeboard and should be 1'-0" minimum for crude oil for boilover wave. Tanks holding other commodities require no freeboard. C is a 1'-0" layer of clay that prevents penetration of liquid through the bank. Top soil and grass should be applied above this clay blanket to prevent erosion.

The slope of the dike wall should match the angle of repose for the area's soil. A slope of 1½ to 1 is depicted as a design guide and may be used until actual soil data is known.

Many authorities will demand truck access into the tank's diked areas for fire fighting. When large pumps are located within the diked area, truck access also is necessary. Some customers require truck access for general maintenance. Figure 3-8 shows how to design the truck access. Ramp slope is 10% maximum, or 1' to 10'.

Storage Tank Design

In many companies the design of atmospheric storage tanks rests with the piping designer since it is a nonpressure vessel. It then becomes his task not only to orient the nozzles but to specify the tank materials and sizes. To do this he must have more than a passing knowledge of tank design and material specifications.

Tank materials for hydrocarbon service are divided into three basic types: intermediate strength steels such as ASTM A285-C for general service (also ASTM A516 for atmospheric and low-temperature service and ASTM A515 for intermediate and higher temperature service), high-yield strength steels used for larger and taller tanks to keep shell thicknesses to a minimum (ASTM A514 and A517) and the low-temperature steels used for a pressure containing tank (ASTM A537). Water tanks are specified usually as A283-C, a low grade steel.

Before specifying the material, the designer must consider that the higher strength steels will cost more per pound. Freight will be less but how much less will depend on the plant location. By utilizing the higher allowable stresses, the shell wall thickness is reduced. Base plate, roof and framing are not affected. Any corrosion allowance specified

Figure 3-9. Aerial view of British Petroleum's refinery in Germany. Courtesy of Chicago Bridge and Iron. Co.

must come out of the wall of the more expensive steel. Some of these materials require special welding and quality control procedures which add to the total cost. Large thin shells may require special bracing during construction and permanent stiffening due to wind loading.

Some designers may proudly say they saved 10,000 pounds by using the high-strength steel for their tank, but this may have made the installed cost more expensive. The installed cost is what is important and a careful summation of the many factors is necessary before final material selection.

The advantage of the high-strength steel is utilized in designing tanks to larger diameters and heights than would be permitted by the allowable stresses of the lower strength metal. Usually larger and fewer tanks are more economical than more tanks of a smaller size. The height is limited by the soil bearing allowable. The diameter is limited by the shell material thickness.

Economy cannot be the only consideration in sizing tanks. Flexibility of operation, plant loss in case of a tank failure or fire and the quantity of energy in one tank must be considered. A tank farm fire is a powerful menace and the larger the tank, the bigger the fire.

Each tank selection is a difficult problem and must be based on the items listed plus factors peculiar to the application. Selection can be made after a careful summation of all conditions.

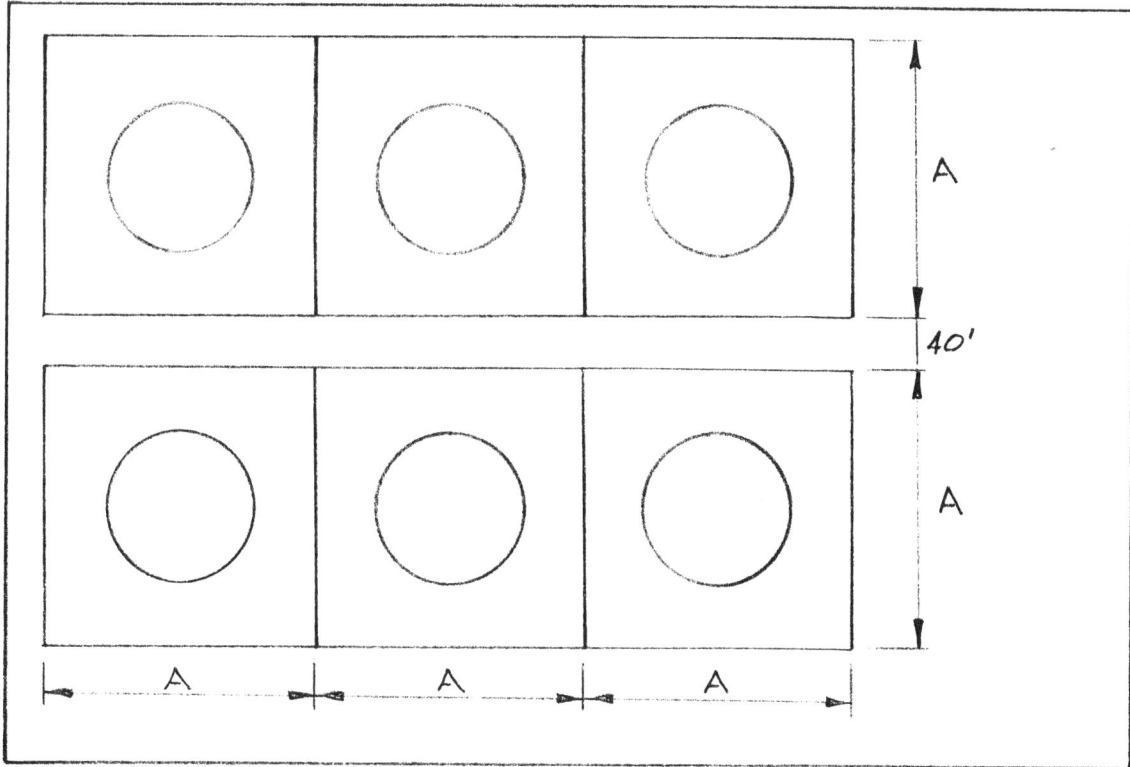

Figure 3-10. Crude tank dikes.

A Refinery

Figure 3-9 is an aerial view of British Petroleum's Vohburg refinery in West Germany where Chicago Bridge and Iron's German division built 161 tanks and vessels with an aggregate capacity of over 6.5 million barrels of storage. The refinery has a capacity of 100,000 barrels per day.

These tanks range in size from a small 10' diameter corrosion inhibitor tank to the large 168' in diameter Horton Floating Roof tanks used for crude oil storage. In this installation, 41 tanks are of the floating roof design, 112 are dome roof (or cone roof) and eight are spheres. Spheres are used to store propane, butane, propylene and butylene under pressure.

In the foreground, the large crude tanks are located four to a unit with road access around the unit for the fire-fighting equipment. Large crude tanks are diked separately. It is usual practice to limit a diked area to 250,000 barrels.

In the center of the tank farm are the rerun and transfer storage tanks, unitized with like or similar commodities bunched together in one diked area. Farther in the background are the product storage tanks, again unitized, with the LPG spheres in the far background, located near the product loading area, the railroad.

The process units, right center, utilize two common stacks for all the fired heaters in the plant with the waste heat exhausting almost 500' high. Huge ducts, large enough to walk through, connect each heater to these stacks. (In most of Germany this type of design is required to try to keep the air as pollution free as possible.)

The Vohburg installation is considered a large refinery; however, there are many over twice its size.

Exercise

A crude storage area has six floating roof storage tanks with a total storage capacity of 3,000,000 barrels. The soil bearing will allow a tank height of 56'-0". What tank diameter is needed? (Tanks are the same diameter.)

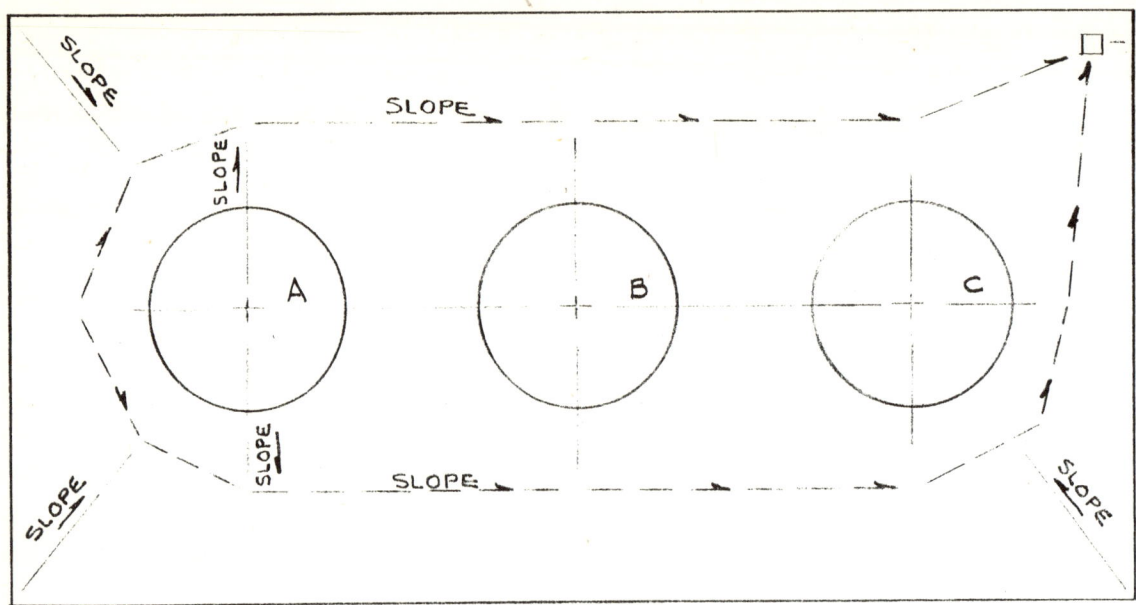

Figure 3-11. Diked area drainage plan.

Refer to Figure 3-10 and supply the following:

1. Dimension A with 6'-0" dike height.
2. Cubic yards of earth needed for dikes.
3. Dimension A with 5'-0" dike height.
4. Cubic yards of earth needed for dikes.
5. Square feet of plot area saved with 6'-0" dike.
6. If this oil has a specific gravity of 0.85, what is the head pressure on the tank's base plate with the tank full. Express in pounds per square foot.

Diked Area Drainage

Each diked area must be drained to rid the area of rainwater. Figure 3-11 shows how to design this drainage system. All gravity flow is directed toward the catch basin. Note that surface water is sloped away from the tanks (see tank A, which is typical). This is done to keep the tank supports as dry as possible. For very large diked areas, two catch basins may be required.

Coming from the catch basin a line is routed through the dike to a gate valve which is normally closed. This valve discharges to the storm water drain system. The valve is kept closed to contain the oil within the diked area in case of a rupture. After a rain, the valve is opened to drain off the surface water and is then closed again.

Storage Tank Piping

Each storage tank will have most of the following piping systems: inlet, outlet, overflow, manual drain, vent, steam if tank heating is necessary and possibly a foam system for fire fighting. The inlet and outlet nozzles are both located near the bottom of the tank.

The outlet goes to a "booster pump", a low-head, high-velocity pump that boosts the pressure enough to get it from the tank to the unit's "charge pump." If possible, the booster pump is to be located outside of the diked area. To keep this pump suction as short as possible, locate tanks closer to the dike near the pump. It is not necessary to have the tank in the middle of the diked area. Also, diked areas may be rectangular instead of square.

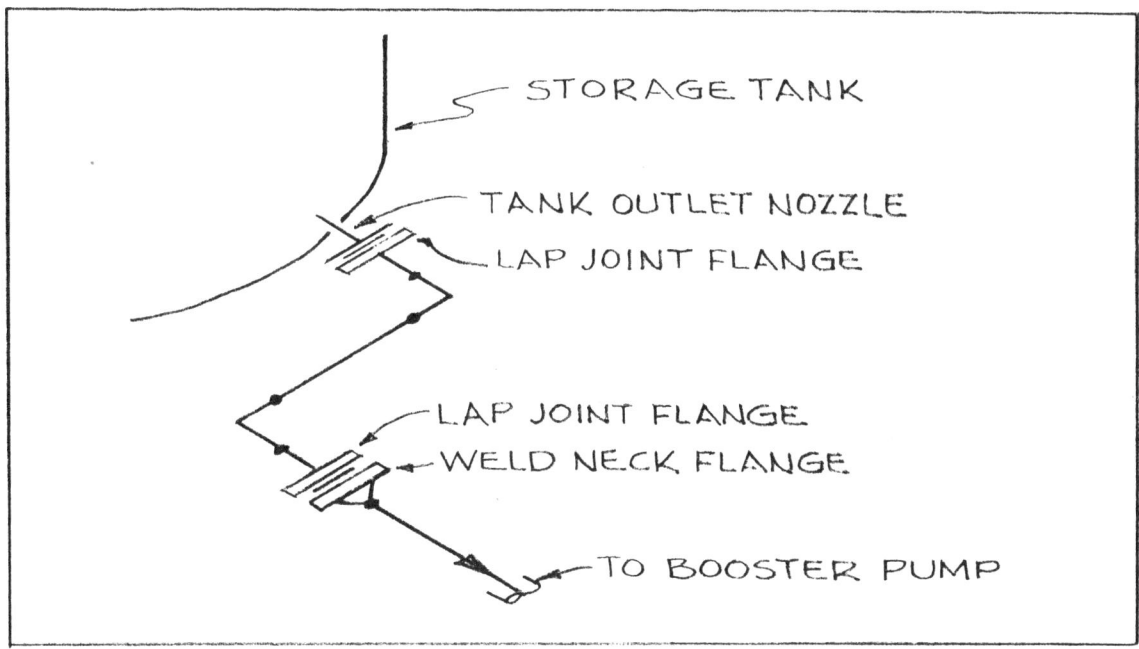

Figure 3-12. Lap joint flange detail for tank settlement.

Since large diameter tanks have a tendency to settle on their foundations, provision must be made in the suction piping to take care of tank settlement. This may require the use of expansion joints, victaulic couplings or a lap joint flange installation as shown in Figure 3-12. Here the piping to the tank outlet nozzle has a horozontal offset with two lap joint flanges installed. The offset would be several feet, depending on the calculated tank settlement. (The soil report will usually supply good data on anticipated tank settlement.)

With the design of lap joint flanges, as the tank gradually settles, the lap joint stub end can rotate within the flange, maintaining its gasketed seal. It is good design to set the outlet nozzle higher than the line to the booster pump for a new installation. This dimension should be equal to the anticipated settlement so that the piping is horizontal after settlement occurs.

The API code for storage tanks gives detailed design data for nozzles such as projection, reinforcement, minimum dimension above base ring and manhole requirements. Location is the piping designer's responsibility but must comply with the code. Nozzles on the top head of cone roof tanks should be grouped together for ease of access. Handrails are to be provided around the tank edge for the protection of personnel who service these nozzles but shall be located only in the nozzles' vicinity. Shell nozzles are to be oriented to keep piping runs to a minimum. Manways at grade can be located almost any place. The target gage type level indicator must be located so it may be read from the operating aisle or roadway if in a diked area.

Foam Protection System

Many safety-conscious companies have a firefighting system called foam blanketing. This is usually specified only for cone roof tanks. Foam is a powder which, when mixed with the proper ratio of water, forms a foam resembling thick shampoo suds. It attaches itself to the walls of the tank and also floats, preventing air from getting to the flame and suffocates the fire.

Foam chambers are located inside the tanks, with foam piped to them from a central foam-mixing unit. When activated, a blanket of foam about 6" thick will cover the tank liquid in a very few minutes and snuff the flame.

Foam connections are located at the very top of cone roof tanks. With tall tanks, many times an internal foam trough is specified which slopes from the nozzle to the bottom to keep the foam from a long fall which might damage it.

Chapter 3
Review Test

1. Define "onsite" and "offsite." _____

   _____

2. Which of the above should be located on the best soil and why? _____

   _____

3. Give the two basic storage tank styles. _____

4. Name two items that determine maximum storage tank height. _____

5. Name three advantages of floating roof storage tanks. _____

   _____

6. Define LPG. _____

7. Maximum height of tank dikes is limited to _____

8. High strength tank steel saves weight and shell thickness. However, _____, _____,

   and _____ are not affected. _____

9. Spheres are used to store _____

10. Define "angle of repose." _____

    _____

4 Process Unit Plot Plans

The purpose of any drawing is to communicate information to the construction personnel, office people, customers and other interested parties. The equipment plot plan of a process unit is the most carefully scrutinized drawing the piping department will make.

Unit plot plans are made for (*a*) equipment location, (*b*) foundation location, (*c*) excavation drawings, (*d*) paving plan, (*e*) flow diagram transposition and (*f*) for the model shop to fix components to the model plot board if a model is utilized.

Equipment Plot Plan

Equipment plot plans are drawn to scale, usually 1″ = 20′-0″ or if possible on one sheet of paper, 1″ = 10′-0″. Process unit plot plans are to be drawn to as large a scale as feasible and each item depicted is to be drawn to scale. These plot plans should *not* have any dimensions, as all equipment will be located on the foundation location plan. Many costly errors are committed by dimensioning the plot plan and duplicating equipment locations on the foundation location plan.

Plot plans should show all equipment, main pipeways, buildings, major structures, housed electrical gear and starter rack location, roads, accessways and any other item of importance. True north and plant north arrows are to be shown. A bar graph scale should be drawn in above the title block to help visulize distances if the reader is viewing a reduced copy or microfilm.

Pipeways are shown by centerlines only. Indicate simple outlines for all equipment, especially for pumps and compressors. Indicate all equipment by item number but do not try to give equipment titles as this will clutter the drawing. Figure 4-1 is an example of a plot plan of new equipment in an existing area. Here a bar scale was not drawn and relative distances cannot be determined.

To establish a preliminary plot plan the piping designer must first know basic maintenance and layout rules. These are usually designated in the general piping specifications or the customer specifications and are listed below for guidance:

1.0 Clearances beneath main pipeway shall be 12′-0″ minimum. To maintain this, the bottom of pipe in the lower rack is to be about 15′-0″ above the high point of the finished surface.
2.0 Clearance of 7′-6″ will be maintained for all lines inside buildings, miscellaneous lines in the process units and lines running over aisles and platforms.
3.0 Roadway clearance shall be 17′-6″ for main roads and 15′-0″ for secondary roads.

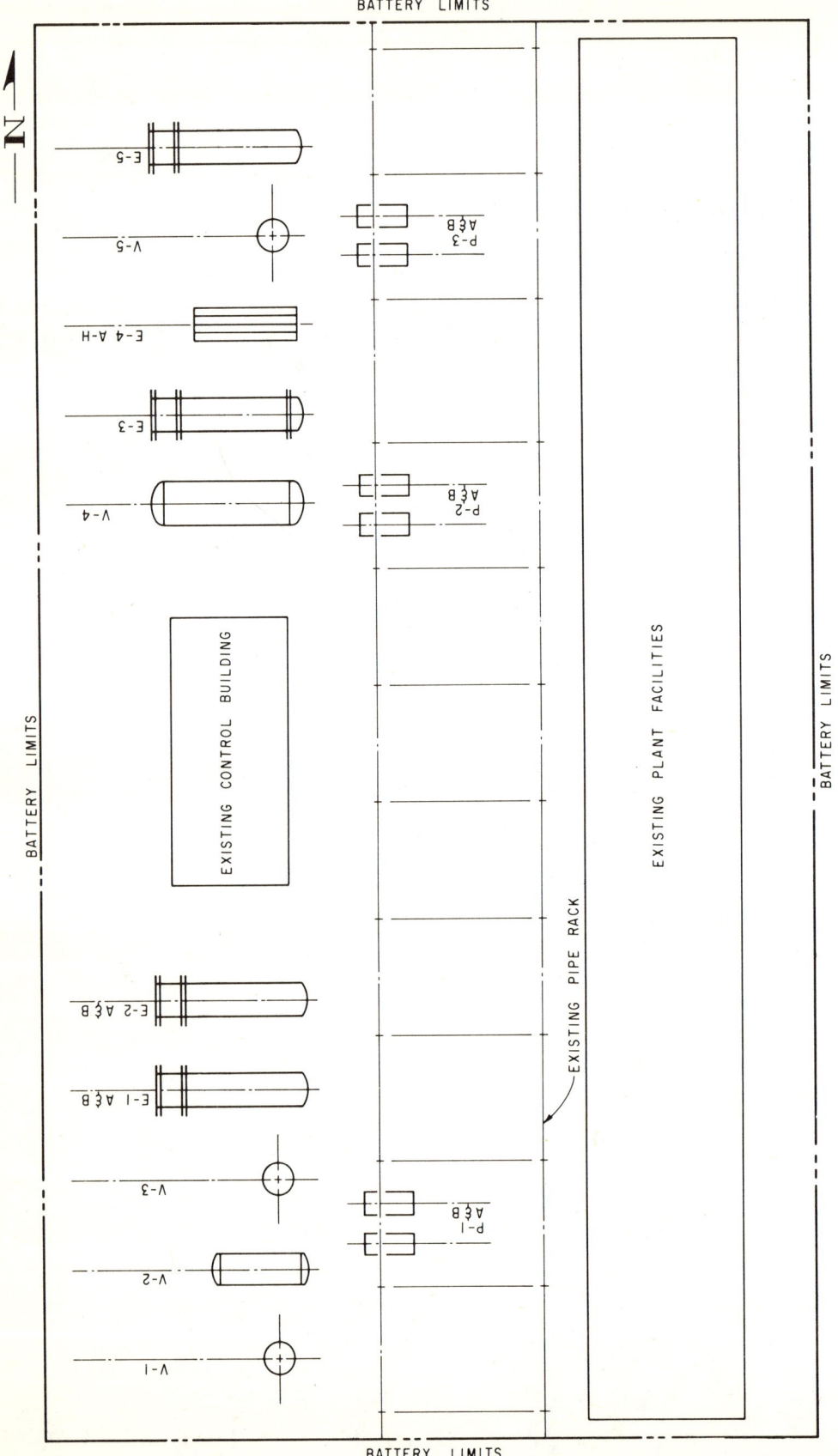

Figure 4-1. Plot plan for process unit.

4.0 Minimum horizontal clearance between equipment and/or piping shall be 2'-6". Exception: clearance between exchanger flanges may be 1'-6".

5.0 Main pipeway supports will be spaced evenly and at 20'-0" maximum. Pickup supports shall be supplied for lines 3" and below, supporting them from the larger lines.

6.0 For pipe racks supporting air coolers (fin-fans), 30'-0" width is desirable. Air cooler tube length will be 32'-0".

7.0 Locate pumps under the pipeway with the centerline of discharge 2'-0" out from under the pipe rack (2'-0" from centerline of rack columns).

8.0 There are two operating aisles, one under the pipe rack and one between the pumps and the equipment line. Control valve stations located 2'-0" from the rack column centerline shall have their handwheels turned in under the rack.

9.0 Locate exchangers for adequate tube removal space at the channel end (referred to as the front end) and locate the back head on a line 8'-0" from the pipe rack column centerline. This is called the equipment line.

10.0 Locate vertical vessels by locating the largest diameter vessel's OD on a line 2'-0" away from the equipment line or 10'-0" from the rack column centerline. Line up all other vertical vessels' centerlines with this one.

   10.1 Horizontal vessel heads will line up with the equipment line.

Laying Out the Plot Plan

Most plot plans are first roughly sketched on the back of an old print to establish relative locations. Many designers use paper cut-outs, taping them to a scaled drawing of the plot area. The easiest method is to use a 1/8" = 1'-0" scale model with a magnetized plot board. Once the plot plan has been completed it is photographed for record purposes and for customer approval.

Since a book cannot utilize a magnetized model or paper cut-outs, the rough sketch procedure will be explained. The following data is needed but does not have to be complete:

1. Process flow diagram for unit.
2. Plot size.
3. Rough equipment sizes.
4. Control building location (off-plot or in unit); if in unit show the approximate size.
5. Location of off-plot main pipeway (for unit connections).
6. Which lines are alloy, large carbon steel or other, which must be kept as short as possible.

For item 1, refer to Figure 4-2. This rough process flow diagram has been sketched in a hurry by a process engineer to give the piping designer some idea of the equipment and the major piping involved. It is incomplete and subject to careful study by the designer laying out the plot plan. Some items the piping designer will consider are:

1. Pumps are shown as single units for P-1–P-6, but each unit is actually two pumps, the operating one and the spare. Common spares may be utilized later.
2. FF-1–FF-5, air coolers, should be located on the pipe rack. Each unit may be several cells, but the process engineer has given a hint by showing FF-4 as the largest one. This indicates that it is larger than the others but not necessarily just two cells.
3. Vessel sizes are not known at this time. This is not important for relative locations of equipment but must be known to finalize equipment locations. A good rule of thumb is to consider all vessels as 96" diameter for units with large throughput and 72" for smaller units until sizes are determined.
4. Exchanger sizes are also unknown. The reboiler E-2 is guessed as 72" OD while other shell and tube units are considered as 36" diameter. Estimate tube bundles of 20'-0" length for E-1, E-3, and E-4. Figure 16'-0" tube bundle length for reboilers.
5. Since F-1 and F-2 are pump-through type reboiler furnaces, it can be assumed that they are vertical type heaters about 20'-0"

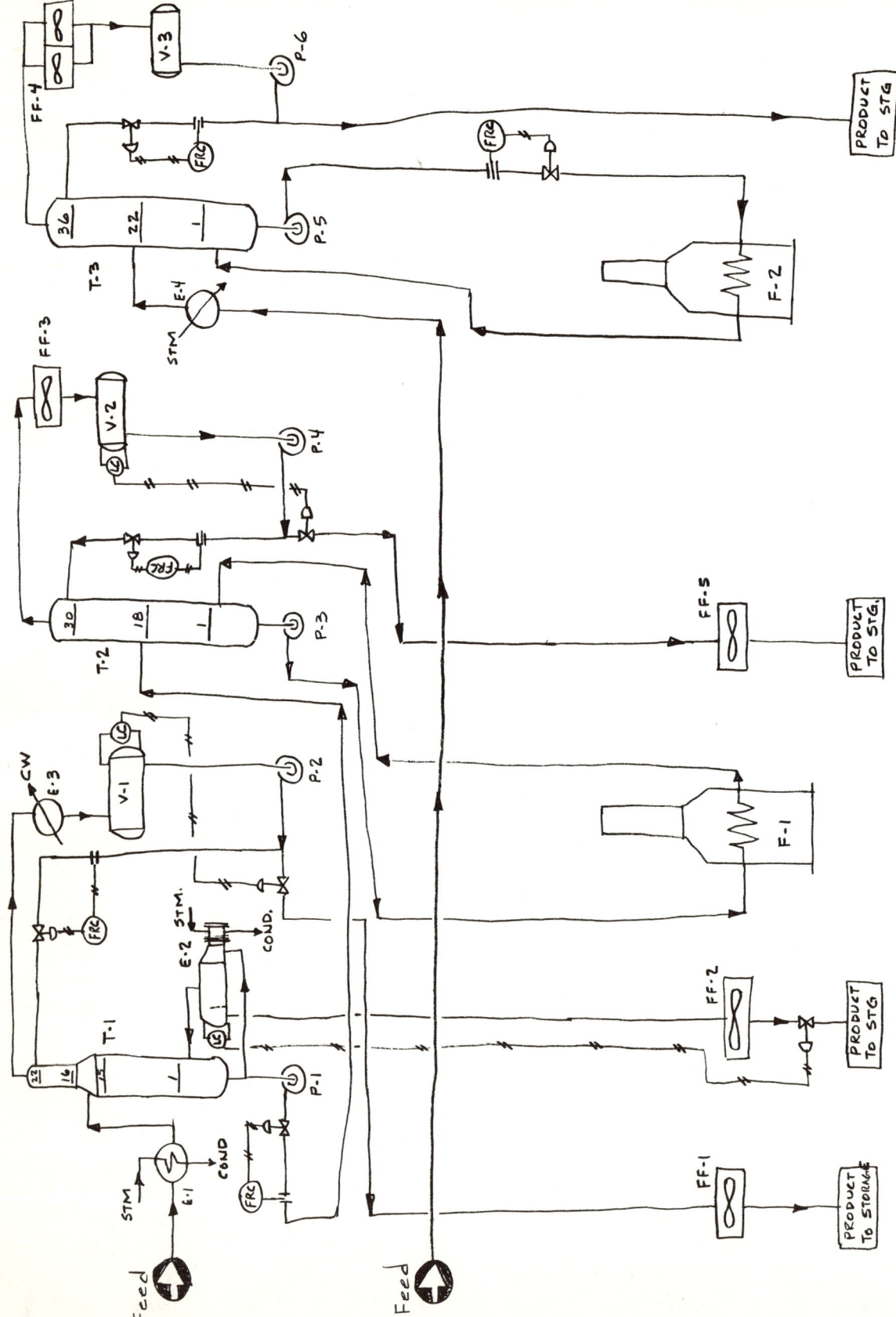

Figure 4-2. Rough process flow diagram.

diameter. Piping at reboiler heaters is usually all carbon steel so assume no alloy piping in the unit.

The plot plan designer knows to keep the fired heaters at a minimum of 50'-0" from any piece of equipment containing hydrocarbon. The only exception to this is reactors, which are located about 20'-0" from heaters to keep the alloy lines shorter. This unit has no reactors so this is not a consideration here.

Figure 4-3 is a plot plan for multiple process units. The dotted spaces indicate the future light ends unit, the equipment shown in Figure 4-2. The plot is sized at 60' x 220' with the fired heater area immediately south of the unit.

This process area is well laid out with the fired heaters to the outside away from the units but with ample access. The units themselves have a common control building with easy access to all areas for the operators. The main pipeway will contain all utility, feed and product headers.

Figure 4-4 is a rough layout for the light ends unit plot plan. It is made freehand and to no scale. The existing rack is drawn and equipment is sketched in according to flow of the process streams. Note in Figure 4-2 that the first feed goes to E-1 to preheat the T-1 feed liquid. So E-1 is located as the first piece of equipment on the plot. The flow goes from E-1 to T-1, the fractionating tower, so it is placed next to E-1, leaving some room between them for operator walkway and, anticipating a large spread footing octagon for T-1, ample room to keep the two foundations from touching. The reboiler, E-2, is located next to T-1. Since the liquid line from T-1 goes to E-2, usually low and blocking walkthrough access, reboilers are located close to towers and their back foundation usually rests on the tower octagon.

The overhead condenser E-3, is located next to E-2, since the overhead line will be oriented on the east side of T-1 and will be able to span across E-2 to E-3 shellside connection.

The next piece is the accumulator V-1, which is placed next to the condenser. This keeps the complete system from E-1 to V-1 flowing in sequence.

The same philosophy is used for locating the other pieces of equipment. The only items out of sequence are the combined air coolers, FF-1, 2, 3 and 5. The largest duty will occur at FF-3, the overhead condenser; consequently, its piping will be the most expensive, which governs its placement between T-2 and V-2. The air coolers (fin-fans) were not located on top of the pipe rack because there was no space left on the existing rack.

A small rack has been added to route process and utility lines to the two reboiler fired heaters. This rack will have the two liquid lines to the reboiler and the return lines to the two towers T-2 and T-3, plus fuel oil, fuel gas, atomizing steam, snuffing steam, instrument air, utility station air, water and steam, and electrical conduits plus possibly the instrument tray, which carrys the pneumatic signals to the control building.

The two heaters are located close together so that one ladder or stairway from grade can serve a platform common to both heaters. The operator can then check both heaters without having to go up and down and back up again. This design is often overlooked but is an initial savings and an operator convenience.

Now the designer is ready to get the supervisor's approval on the sequence layout. With a critical eye the supervisor scans the plot plan and the flow diagram. To him it looks good except for one piece of equipment. Can the student find the error without reading on? It is actually a costly error in judgment.

Large insulated lines are costly to install. Every insulated elbow is expensive. The experienced supervisor's eye tells him the most expensive lines in this unit are the hot lines to and from the fired heaters. Consequently, these lines should be routed as straight as flexibility will allow and with a minimum number of elbows. The error is in locating T-2. By changing places with T-2 and V-2 a great savings is assured. The liquid line from the bottom pump is shortened and about eight elbows are saved. The reboiler return line has an even greater saving.

T-3 is not exchanged with F-4 as no fittings are involved. Some straight pipe could be saved but this would be offset by adding more pipe to the feed to E-4 coming from the west.

Pump locations are not considered during this preliminary layout. It is planned that they will be located under the rack but their final position will not effect the length of the process string.

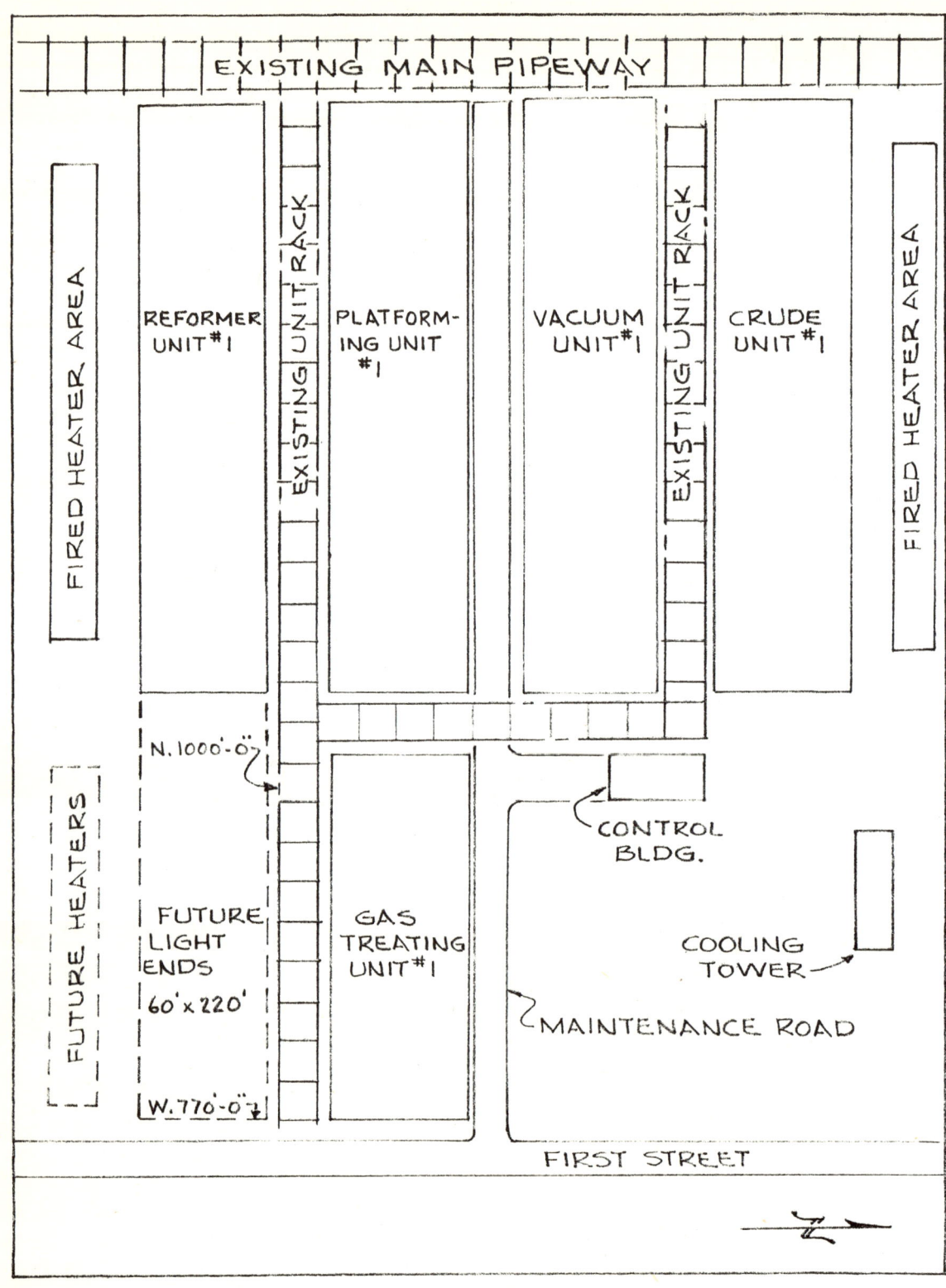

Figure 4-3. Plot plan for multiple process units.

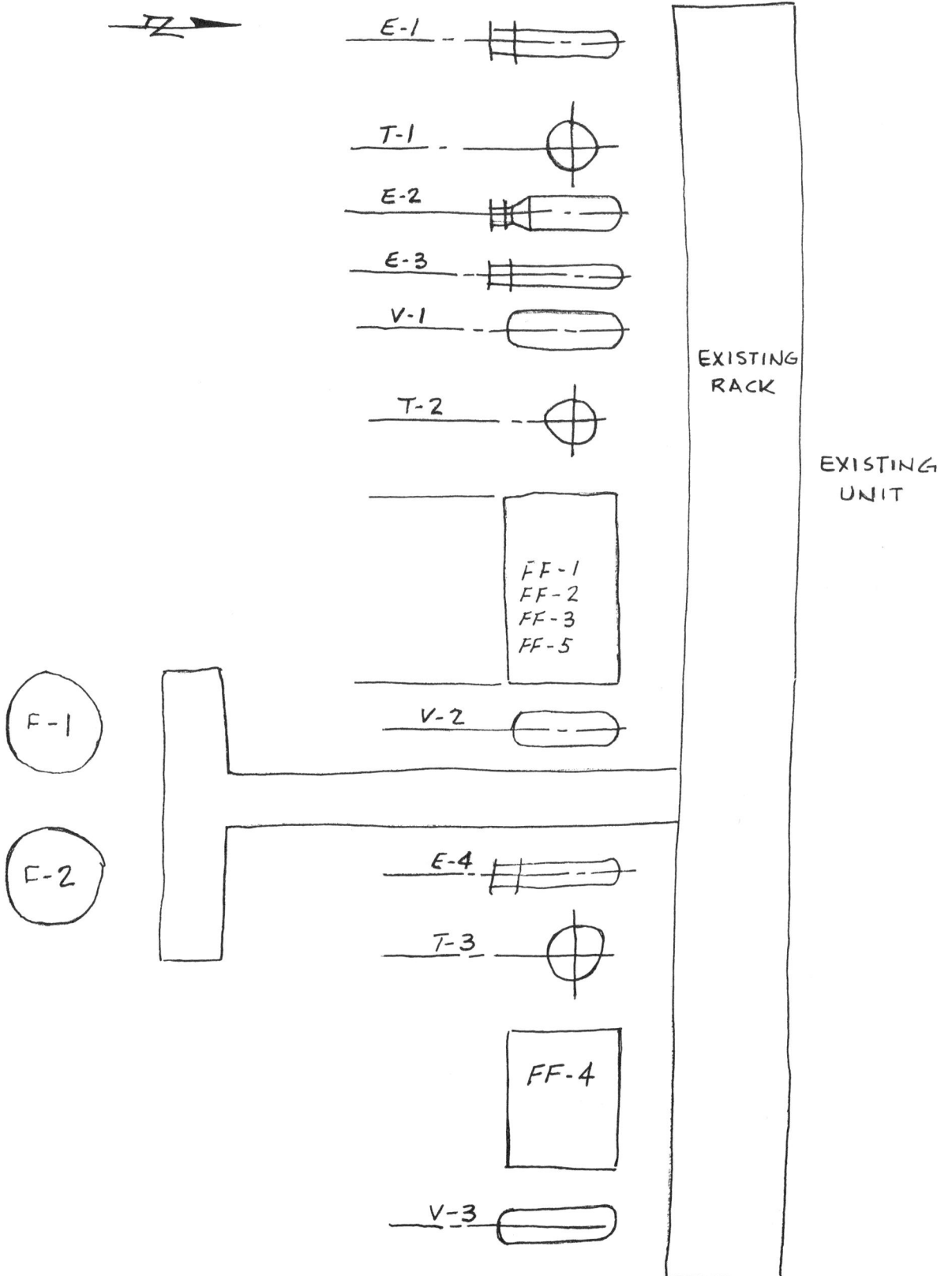

Figure 4-4. Rough layout for light ends plot plan.

Now that the designer has an approved sequence plot plan, he is ready to establish preliminary dimensions. Again, this can be done with no equipment sizes. A lot of progress is lost on design because people think they need all the answers before they can do anything.

Preliminary Plot Plan Dimemsions

Figure 4-5 is a freehand sketch made to establish preliminary plot plan dimensions. Final dimensioning is done when equipment sizing is completed and piping layout is in the final stage. To establish preliminary dimensions the designer makes certain assumptions based on his experience and training. Assumptions and conclusions are listed below.

Dimension 1. Locate E-1 from the plot limit line; 5'-0" clearance is maintained. Assuming a maximum of 48" exchanger channel flanges, set this dimension at 5'-0" plus 2'-0", or 7'-0".

Dimension 2. Locate T-1 from E-1. T-1 is assumed to be 72" diameter. Since there are pumps taking suction, the skirt will be about 20' high. Below tray 1 there is a surge space of about 14'. Assume trays are spaced at 2'-6". Between tray 1 and tray 15 there are 14 spaces at 2'-6", or 35'-0". Between trays 15 and 16 there is a cone section of about 4'-0". From trays 16 to 22, there are six spaces at 2'-6", or 15'-0". The vapor space above tray 22 is about 4'-0". Adding all of these, the designer learns that the tower is about 82'-0" tall. With normal soil bearing, a tower of this size will require a foundation with a spread footing about 18' wide. These are octagon shaped. E-1 footing is estimated to be 4'-0" wide. To keep spread footings clear and to allow ample walkway between, dimension 2 is established as 9'-0" (one-half of T-1 octagon) plus 2'-0" (one-half of E-1 footing) plus 2'-0" for clearance to total 13'-0".

Dimension 3. Locate E-2 from T-1. E-2 is a kettle type reboiler and will be elevated a little above grade. The liquid line from T-1 will probably be low enough to block walking between these two pieces of equipment, so E-2 is located so the back support will rest on the T-1 octagon. This makes the octagon a "combined footing" therefore, dimension 3 should be less than 9'-0" (one-half the octagon). One-half of T-1 OD is 3'-0". One-half of E-2 OD is assumed to be 3'-0". Allowing 1'-0" clearance, dimension 3 becomes 3'-0" plus 3'-0" plus 1'-0" or 7'-0".

Dimension 4. Locate E-3 from E-2. There are three things to consider in estimating this dimension. The spread footings must clear (about 6'-0" will accomplish this). The equipment must clear (6'-0" will clear this). Since there was no walkway west of E-2, access for maintenance and operation should be provided on the east side. The 2'-6" access path added to one-half of E-2 OD of 3'-0" plus one-half of E-3 diameter of 1'-6" makes dimension 4 total 7'-0".

Dimension 5. Locate V-1 from E-3. V-1 has pumps taking suction from it so it will be elevated about 14'-0" or higher. The main consideration is to clear the spread footings. The V-1 footing is assumed to be 6'-0" wide. E-3 footing is about 4'-0" wide so 5'-0" plus 1'-0" clearance would suffice. Dimension 5 then becomes 6'-0".

Dimension 6. Locate V-2 from V-1. V-2 also has pumps taking suction from it so it will be elevated about the same as V-1. Assuming 72" diameter for both vessels and a common platform between them with 6'-0" for working space, dimension 6 becomes 12'-0".

Dimension 7. Locate combined air cooler from V-2. Equipment clearance is all that is necessary since foundations will be relatively small. Dimension 7 becomes 8'-0".

Dimension 8. Estimate width of air cooler. Four separate cooling cells are located in one cooling frame. The product coolers, FF-1, 2 and 5, are generally small-duty units and will require one or two cells per unit. FF-3 is an overhead condenser and has larger duty which may have three or four cells. Assume a total of eight cells 5'-0" wide or 40'-0" plus frame of 2'-0" to total 42'-0" for dimension 8.

Dimension 9. Locate T-2 from air cooler. T-2 has 30 trays so it will be taller than T-1. The spread footing (octagon) is assumed to be wider than the one required for T-1 by 2'-0", which makes it a 20'-0" octagon with an allowance of 2'-0" for one-half of the air cooler footing and 2'-0" for clearance, dimension 9 totals 10'-0" plus 4'-0", or 14'-0".

Dimension 10. Locate pipe rack from T-2. Here the footings must clear so the 14'-0" dimension is used.

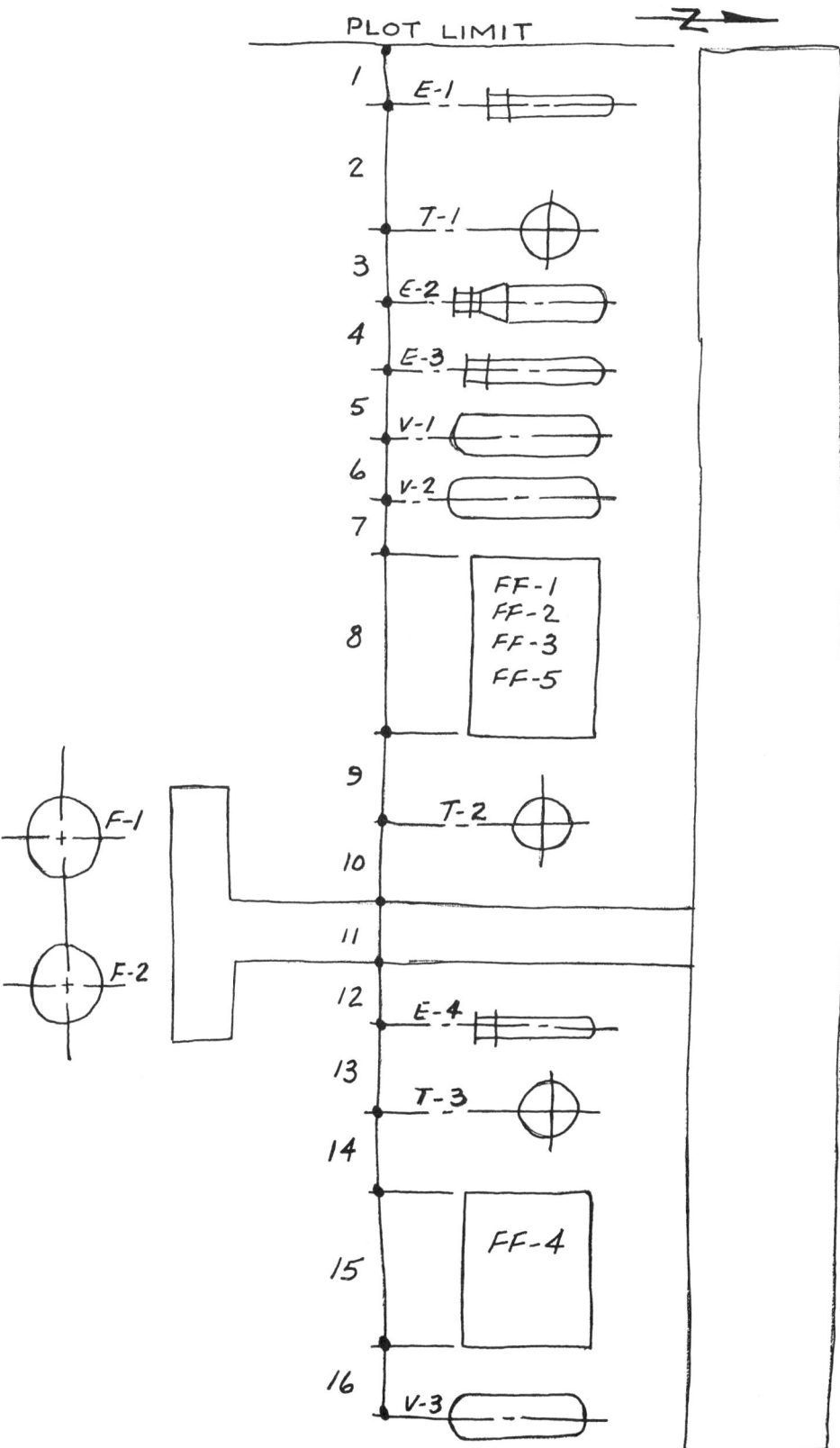

Figure 4-5. Preliminary plot plan dimensions.

Dimension 11. Establish pipe rack width. To do this accurately the pipe sizes must be known. For this purpose assume a 14'-0" width.

Dimension 12. Locate *E*-4 from pipe rack. Footings must clear and some accessway between rack columns and exchanger should be provided. Assume 6'-0".

Dimension 13. Locate T-3 from E-4. T-3 has 36 trays so it will be taller than T-2, but we can assume that the 20'-0" octagon will be large enough. By using the same procedure used for determining dimension 9, dimension 13 also is 14'-0".

Dimension 14. Locate FF-4 from T-3. Use the same 14'-0".

Dimension 15. Estimate width of air cooler. FF-4 is in overhead condensing service and the process flow diagram indicates it is the largest of all the air coolers. Assume five cells, 5' wide plus 2'-0" frame, or 27'-0".

Dimension 16. Locate V-3 from FF-4. Use same procedure as dimension 7. Dimension is 8'-0".

Adding all these estimated dimensions, the designer finds he has used 213'-0" of plot length. Since the total plot length is 220'-0" (Figure 4-3), V-3 is located 7'-0" from the east battery limits. As more firm information is developed some of the above dimensions may vary slightly, but if so there is no more plot length available and the designer will have to adjust other dimensions to suit, perhaps combining some foundations.

Plot Plan Exercise

The designer is now ready to draw the plot plan to scale. The student is to do this, assuming that the existing rack is 30'-0" wide and rack bents are spaced at 20'-0" with the last bent at the east plot limit, 7'-0" from V-3 centerline. Assuming no common spare pumps, locate pumps under the rack with the centerline of discharge 2'-0" out from the rack column centerline. Each pumping service shown in Figure 4-2 has two pumps, one operating and one spare. Locate pumps to keep suction lines short. Use equipment sizes and dimensions as estimated in this chapter. Fin-fans are 26'-6" long. Draw all pumps 2'-6" wide by 6'-6" long. Use 1" = 10'-0" scale. Remember, plot plans do *not* show dimensions.

Foundation Location Plan

Figure 4-6 is the foundation location plan for the plot plan shown in Figure 4-1. This shows the location of all underground concrete by coordinate and to scale. The underground portion of foundation is shown dotted while the concrete portion projecting above grade is shown with a solid line.

Using Figure 4-6 as a guide, the student is to prepare a foundation location plan for the equipment shown in Figure 4-5. Figure 4-3 supplies the two basic coordinates of the existing unit, called the "bench mark." Bench marks also supply existing elevation base.

Excavation Plan

When the field construction crew moves onto the job site the first thing they want to do is excavate. To do this they need a plot plan marked with the bottom of concrete elevations or an excavation plan, usually a transparency of the plot plan showing the excavation depth and periphery.

Flow Diagram Transposition

The flow diagram transposition is the first effort of laying out the major piping systems. It is made on a print or sepia of the plot plan. Piping is sketched in from equipment to equipment, showing all meter runs and control valve locations. No effort is made to orient tower nozzles at this stage but the piping is shown going to the towers. This transposition is sometimes called the flow diagram overlay.

It is best to develop this drawing after mechanical and utility flow diagrams are prepared showing line sizes and full instrumentation. Then one transposition is made for the process lines and one for the utility lines.

Piping Drawing Index

A piping drawing index is made for each process unit to show the extent of area covered by each piping drawing and the drawing numbers showing the piping details for that area. Figure 4-7 is a piping drawing index for the plot plan shown in Figure 4-1. The student is to prepare a piping drawing

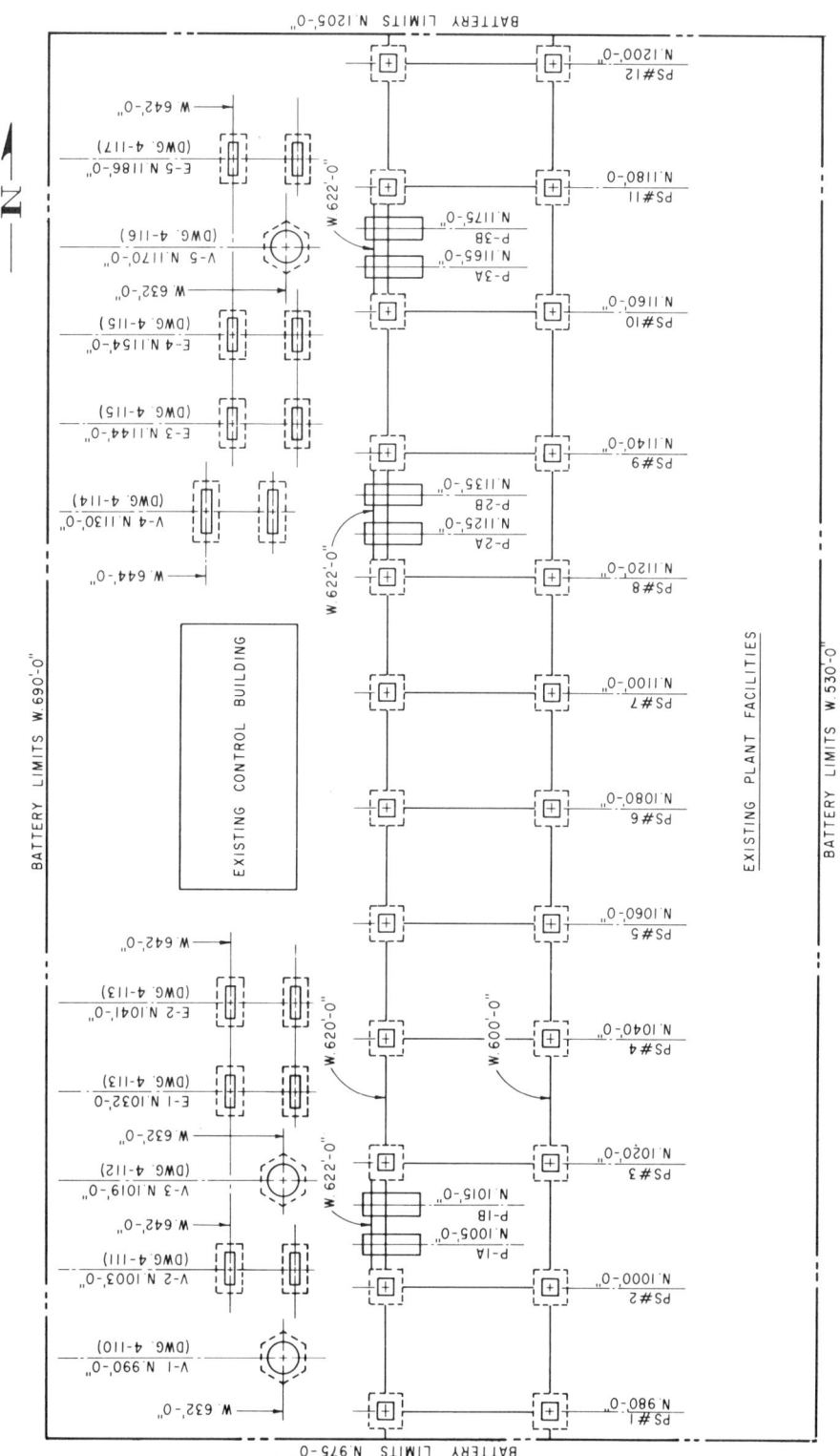

Figure 4-6. Foundation location plan for Figure 4-1.

index for his plot plan of the light ends unit. Try to keep the piping areas no wider than 60-65'.

Equipment Setting

The plot plan designer will come in contact with many different kinds of equipment, depending on the type of unit he must lay out. Each subsequent chapter in this book will go into detail about most of these items but, in general, the designer must know how to locate equipment for plot plan purposes and the following is offered as a guide to this.

Fractionating Towers

Fractionating towers are located on a common centerline about 12-14' from the rack columns. The first ladder from grade should be located on the pipe rack side for easy access by the operators. Towers over 50' high are to have davits for handling vessel trays. A clear drop area is to be provided on the side away from the rack.

Exchangers

Exchangers set the "equipment line." This is the location of the back head, usually set at 8' from the pipe rack column. Shell and tube type exchangers may have a removable shell cover, a flanged head. Access must be provided for equipment to handle this cover, usually from under the rack. Tube pulling or rod cleaning area must be allowed at the channel end. This should be the tube length plus 5' from the tube sheet. Double pipe exchangers are located with the front end toward the rack. Tube removal space should be allowed but, for one to four units, is not mandatory if grade mounted, as mobil maintenance equipment can pick up the entire unit and transport it to the repair shop.

Control Buildings

Control buildings are to be centrally located because they are home base for all the unit operators. This also keeps instrument leads shorter. Control buildings should have road access. Keep process equipment 25' away. In the original layout keep hydrocarbon equipment 50' away as control buildings usually get larger than originally planned. To size control buildings, get the instrument engineer to size the control board and space needed for future board. Then consider requirements for offices, toilet facilities, lockers, tables and chairs for a lunch room, stoves, refrigerators, any electrical switchgear (consult the electrical engineer) or possibly an air compressor and related equipment for the plant instrument air system. Allow space for the heating and/or air-conditioning unit. Some customers want a small lab area in the control building to test samples taken in the unit.

Fired Heaters

Fired heaters are located a minimum of 50' from hydrocarbon-containing equipment; however, reactors may be closer. The fired heaters must have road access for equipment needed for tube repair or replacement. Vertical heater tubes (tubes are mounted vertically) are pulled up from the top with a crane. Horizontal box type heaters must have tube removal space allocated behind the heater equal to the tube length plus 10'. A lightly traveled road can be utilized as part of this maintenance area.

Cooling Towers

Cooling towers are to be located where the prevailing wind is directed to the small side. This allows both long sides to intake an equal amount of circulating fresh air. Many people locate cooling towers exactly opposite, directing the prevailing wing to the slatted long side. This allows one-half the tower to intake fresh air but the downwind side is starved. Locate cooling towers away from fired heaters, flare stack or any heat producing item. Air to the cooling tower must be as cool as possible to do its duty. Supply road access to cooling towers for maintenance of pumps, chemical additive equipment and for handling screens. Pump pits may be located anywhere around the tower. It is not necessary to center them along the long side. Many installations utilize the short side. Locate pump pits to keep piping runs to a minimum.

Piping

Piping determines most equipment locations. Alloy piping costs much more than carbon steel, sometimes a thousand dollars a foot, so special at-

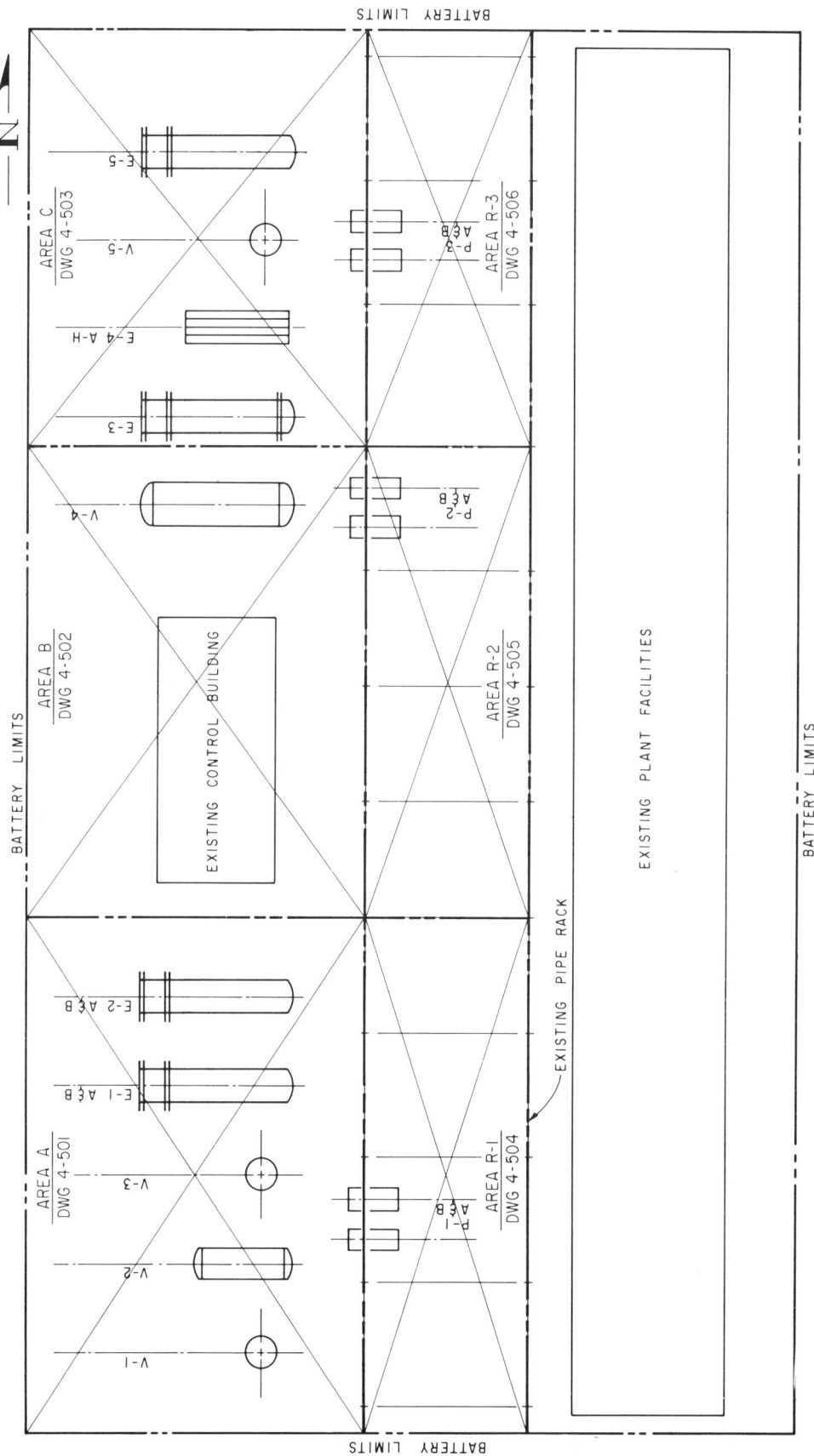

Figure 4-7. Piping drawing index for Figure 4-1.

tention is needed to keep these lines short. Large insulated lines are also kept short. Many times the insulation is almost as costly as the pipe and is an expense often overlooked by the designer.

Reboilers

Reboilers are to be located next to the tower they serve except for pump-through type fired heaters. Kettle type reboiler elevation is determined by the tower liquid. Thermosiphon reboilers are regular shell and tube type exchangers. Horizontal thermosiphon exchangers are located at a minimum elevation. Vertical thermosiphon types are usually supported by the tower and are located on the back side to be accessible to the maintenance equipment. Large vertical types may require a supporting structure. Consult the vessel engineer to see if the tower can support the reboiler or reboilers. There may be more than one.

Compressors

Compressors have two basic categories, centrifugal and reciprocating. Drivers may be electric motor, gas engine, gas fired turbine or steam turbine. Locate compressors to keep suction lines as short as possible.

Reciprocating compressors are usually housed in a tropical type building with a drop-curtain wall extending to within 8' of the floor. In areas of severe winter, such as Canada, they are fully housed. Keep the building far enough away from the pipe rack to allow for the suction drum and the suction and discharge headers on the sleepers. Large motor-driven compressors may require room behind the building for transformers. Always allow room at one end of the building for parts drop and pick-up area. Compressor buildings must have truck access.

Centrifugal compressors should be mounted outside unless the customer requires a shelter. Allow a large area for the lube and seal oil console, usually about 15 x 20', with truck access. Locate the console outside the building, if a building is required, and at grade, behind the building, away from the pipe rack. Motor-driven compressors may require a large transformer. If the compressor is turbine driven, a large surface condenser may be needed. If it is necessary, locate the surface condenser above the compressor.

Electrical Starter Racks

Electrical starter racks and switchgear must be located as early as possible. Consult the electrical department to get the size and, with them, determine the location.

Maintenance

Maintenance requirements must be considered during early plot plan development. Shell and tube exchangers should not be stacked over three units high and then only if the centerline of the top unit is not over 15' from grade. For exchangers requiring four shells, stack them two high by two wide.

Between two major processing units allow for a road, if space is available, and clearance for a 25-ton crane. Provide accessways from main plant roads into unit roads.

Erection

Erection of the new equipment must be considered. For large diameter and long towers, consider how the field crew can get them in the plant and where they can lay them for erection. For field-constructed equipment such as very large vessels and reactors space must be allowed for field construction.

Future Expansion

Future expansion must always be considered. Future pumps, vessels and exchangers are often overlooked. Consult the process engineer for these requirements.

Chapter 4
Review Test

This is a composite review of the first four chapters. The student should be able to answer 20 of the questions in his own words, without referring to the text.

1. Ferrous metals differ from nonferrous because they _____.

2. Define ANSI. _____

3. A joint efficiency is applied to a _____ joint.

   Seamless piping has a _____% joint efficiency.

4. Pipe made from plate has a mill tolerance of _____.

5. Seamless pipe has a mill tolerance of _____%.

6. Random length carbon steel pipe is ±___ feet long.

7. What is the difference between pipe and tubing? _____

8. Name the three basic types of valves, _____, _____, _____.

9. Which valve is for throttling? _____

10. What type of check valve is specified for pulsating flow? _____

11. Define hydrocarbon. _____

12. Define mercaptan. _____

13. Which is the lightest hydrocarbon? _____

14. Define fractionation. _____

15. What is flashing? _____

16. Define equilibrium. _____

17. Define head. _____

18. Define two-phase flow. _____

19. What is hydrotesting and why is it done? _____

20. Define topography. _____

21. Define cone roof vs. floating roof tank. _____

22. What determines tank height? _____

23. What are the advantages of the floating roof tank? _____

24. There are _____ gallons per barrel of oil.

25. What is the purpose of "foam" and how does it work? _____
_____

5 Piping Systems and Details

Every piping installation is resplendent with systems, each having something in common with—yet different from—the last unit the designer finished. The fractionation system (see Chapter 2) is the most common system since it is the basis of all process units. All offsite and onsite units have drainage systems to handle surface or rain water, oily drips and drains or special liquids such as caustics, acids, etc. Most process units have one or two flare systems. Steam tracing systems are similar from job to job. This chapter will discuss these and other systems.

Details are also similar from unit to unit. Many details should be standardized in the industry, but because they are not, countless millions of dollars are spent and man-hours are wasted developing a way to do something that has already been done 50 times before by other designers. This chapter will present details showing ways to solve problems which are not standardized but do the job.

Piping fabrication is done on all piping. Most process pipe is fabricated in a pipe shop and shipped to the job site in "shop spools" to be assembled by the erection workers. Most piping designers know very little of a pipe fabricator's problems or how he gets his work done. While his costs are high, the shop's streamlined efficiency and assembly line techniques make shop fabrication cost less than field fabrication.

Underground Piping

Underground piping is broken down into two main categories: process systems and utility systems. Underground process lines should be avoided; however, there are times where this is the best installation and this will be discussed in other chapters. Underground utility systems are classified into two divisions: gravity flow and pressurized systems.

Gravity Flow Systems

Gravity flow systems depend on the pull of gravity for flow. Consequently these lines must have a constant slope from liquid origin to terminus. Recommended minimum slope is 1/8" per foot, or 1' per 100' of line.

Piping is supplied for the following systems:

1. *Storm water* or clean water includes rain water, wash water and fire water run-off. This is usually collected from paved areas by a catch basin and piped to a separation pit or API separator to separate any oils that may get into the system and will then go to a creek, river or possibly a large evaporation pond.

2. *Process sewer,* sometimes called oily sewer or dirty water sewer, includes drips and drains from pumps, vessels, sample connection funnels and other dirty drains. This system is routed through a separator and the hydrocarbons are usually recovered.
3. *Combined sewer* is the sewer that collects both storm and process sewers, utilizing only one piping system. This must be routed through a large API separator capable of handling the combined flow to separate hydrocarbons from water. While the combined system saves money on piping costs, the savings are usually offset by larger separation facilities costs. The combined system is rarely used today.
4. *Sanitary sewers* are the systems that carry human waste. These are routed to a large sanitary system or, if handled locally, to a septic tank and its related field.
5. *Corrosive sewers* are designed as a separate sewer system within the unit. This includes acids, amines, carbonates and others. Some of these circulate within the unit, with all funnels draining to a separate header which is routed to a local sump. A sump pump either returns sewage to the pressurized system or sends it out of the unit for disposal, possibly to a neutralizing pit. Some acids, such as sulphuric acid, are pumped to storage, then shipped to a reclaiming plant where the acid is recovered.

As these systems vary widely, the materials of construction will be somewhat different than normal and each selection must be thoroughly investigated. If vitrified clay or cast iron soil pipe is selected, careful attention must be given to the material selected to be used at the joints.

Sewer terms are different and often confused with terms applied to overhead systems. *Inverts* are used as a dimensional reference point on all but carbon steel piping. Inverts are the elevations of the *inside bottom* of the sewer line. In the larger sizes of vitrified clay pipe, the thickness is very large and must be considered when calculating clearances from other lines or underground concrete. The use of a concentric reducer or increaser will change the invert elevation and must be calculated.

When carbon steel pipe is selected for gravity sewer systems, dimensioning is to the bottom of pipe (BOP), expressed as an elevation. Inverts are also expressed as an elevation.

The liquid flow is determined by the slope of a gravity system which in turn is governed by the inverts' set. The designer must be concerned with the elevation of the internal surface, the invert, regardless of wall thickness. A joint's internal surface must line up even though the two wall thicknesses differ. The gravity systems' flow must be as smooth as possible, without projections which would form weirs or dams, providing a point for solids to accumulate and block flow.

Reducer Selection

When reductions in line sizes for headers are required, special care must be taken in selecting the type of reducer. Because of inconsistent terminology, there is a great deal of confusion when defining a line reduction in either clay or cast iron material.

Proper terminology is defined below. All terms are specified in the direction of flow.

1. Vitrified clay: Concentric is all that is available.
    a. For increasing line size use *increaser* with bell on *small* end.
    b. For reducing line size use *reducer* with bell on *large* end.
2. Cast iron soil pipe: Concentric only is available.
    a. For increasing line size use *reducer* with bell on *small* end.
    b. For reducing line size use *increaser* with bell on *large* end.
3. Cast iron pressure pipe: Concentric and eccentric are available.
    a. For increasing or reducing line size use reducer. End types must be called out for both ends to match adjoining fitting or pipe.
    b. Eccentric reducers are available in limited sizes and patterns but their use should be avoided. Before specifying, the designer must be sure proper sizes are commercially available.

Sewer System Terms

To discuss any sewer system, verbally or on paper, the designer must know the terms that make up the system. These are:

Mains: Sewers for collection from two or more laterals; usually located in roadway easements. Mains shall be sealed at regular intervals, with manholes, to prevent the spread of fire or gas backup.

Laterals: Sewer lines collecting from two or more sublaterals and discharging to mains through a sealed manhole.

Sublaterals: Sewer lines connecting branches and catch basins to laterals.

Branches: Collect from various drain funnels or catch basins and tie into sublaterals.

Funnel: Liquids collection point, usually projecting 2" above the finished surface. For carbon steel systems, a 6"x4" concentric swage is an economical funnel. Lines from funnels should not be smaller than 4" to prevent clogging. For other materials a 6"x4" reducer could be used; 6" should be the minimum size for the funnel collection end. The swage is specified over the reducer because the swage is longer and will have a deeper bowl area for splashing.

Catch Basin: Used to collect surface drainage. Paving or other surface is sloped to the catch basin. Catch basin is usually about 2' square by 1' to 1½' deep and covered with grating.

Manhole: A central collection box of a size that a man can enter to clean the sewer laterals. Incoming lines are usually sealed to prevent backflow of fire or gases.

Sewer Materials

Selection of sewer material depends on pressure, temperature, durability, cost (of material and labor), availability and the fluid. Carbon steel pipe, properly coated, is quite often used. Vitrified clay and cast iron soil pipe are widely used. It is very difficult to establish hard and fast rules for material selection. Experience with the particular application is the best asset. The plant sites soil corrosiveness must also be considered.

1. *Vitrified clay pipe* is economical. It is used for gravity systems handling surface drainage and sanitary sewers. Since this piping has joints mechanically assembled, it should not be used under buildings or 6" or thicker concrete paving. If the commodity being handled is 150°F or hotter, special joint material may be necessary.

2. *Cast iron soil pipe* is used for gravity systems. When vitrified clay is specified, use cast iron soil pipe under buildings and thick paving. If the fluid is too hot for vitrified clay, use cast iron soil pipe for the necessary length to cool the fluid.

3. *Carbon steel piping* is preferred by many designers because it is easily installed. It is subject to external corrosion and must be coated to ensure long life. The old standard coating is tar and felt paper, usually referred to as "tar and feathers." This is the cheapest coating but has many disadvantages. It is easily damaged during handling and damages are not always properly repaired. The author recommends coating with plastic at the mill, such as "Scotchkote" or "Plicoflex." This costs more initially but is almost impossible to damage and lasts for many years in the ground.

4. *Cast iron water pipe* is used for gravity sewers where the use of 12' or longer sections makes a more economical installation than the 4' to 6' lengths of vitrified clay or cast iron soil pipe. Also specified for pressurized water systems.

5. *Concrete pipe* is generally used for surface drainage headers for sizes 24" and larger. Investigate cost, freight and availability before specifying it.

6. *Concrete-lined steel pipe* is used for pressurized corrosive service where pressures are greater than allowed for cast iron piping.

7. *Duriron pipe* is sometimes specified for highly corrosive special drains. Because of its high silicon content it is very brittle—like glass—and has high breakage from handling, which adds to its already expensive cost.

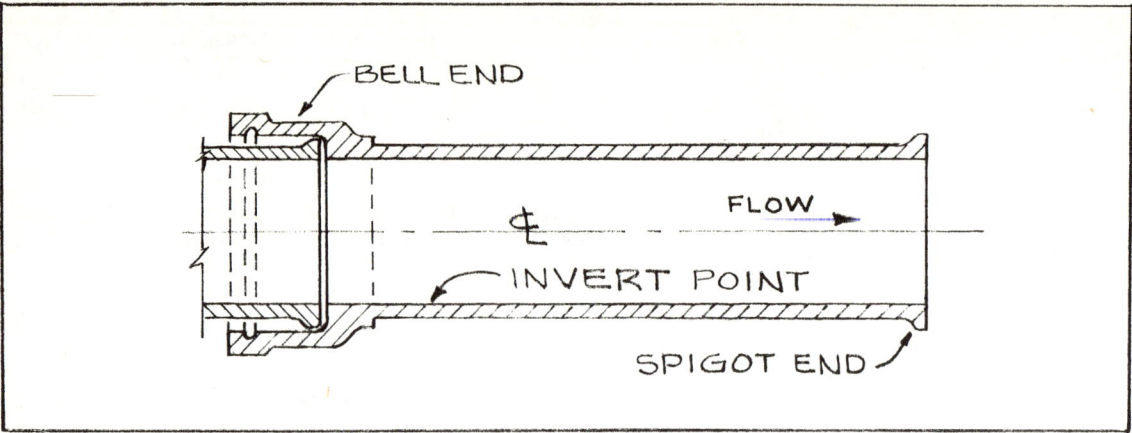

Figure 5-1. Cast iron bell and spigot pipe detail.

Joints for Cast Iron Piping

Cast iron pipe and fittings are equipped with bell and spigot ends. Figure 5-1 pictures the pipe detail. The joint can be sealed with lead and oakum, cement or a rubber or neoprene ring fitting into the groove shown. Lead and jute or oakum is standard joint material for surface water systems. To ensure operational tightness, the joint material must be kept moist. The conveyed liquid accomplishes this.

Bell and spigot joints are rarely used in gas pressurized systems. Pressures of up to 200 psig can be held by the joint for liquid service. For high pressure water service, such as fire water systems, bell and spigot pipe and fittings are available with bolting lugs to prevent joint separation.

For pressurized systems, thrust blocks are applied at turns or points of thrust to restrict joint separation. Thrust blocks are generally a mass of concrete.

The mechanical joint, a modification of the bell and spigot joint, is used in low pressure gas distribution systems. However, the main use of the mechanical joint is for higher pressure fire water systems operating above 150 psig. The author restricts all cast iron piping to water and sewer systems. For any hydrocarbon gas service, welded carbon steel is recommended.

The mechanical lock joint groove which is shown in Figure 5-1 is located in the bell end. The groove accommodates a rubber or neoprene gasket. For installations where sagging is likely to occur the mechanical lock joint is recommended. This might be river crossings running above the water, supported by the bridge, or underground where soil heaving is expected. The Gulf Coast soil calls for this joint detail rather often.

The mechanical roll-on joint is a low cost mechanical joint with a round rubber gasket over the spigot end. When the spigot is pulled into the bell, the ring is seated in the bottom of the bell. Outside the rubber gasket, braided jute wedged behind a projected ridge confines the gasket to retain pressure in the piping system. A bituminous compound seals the mouth of the bell, retaining the jute and gasket. Both bell and spigot and mechanical joint fittings can be used with this pipe.

Table 5-1
Sewer Flow Diagram Symbols

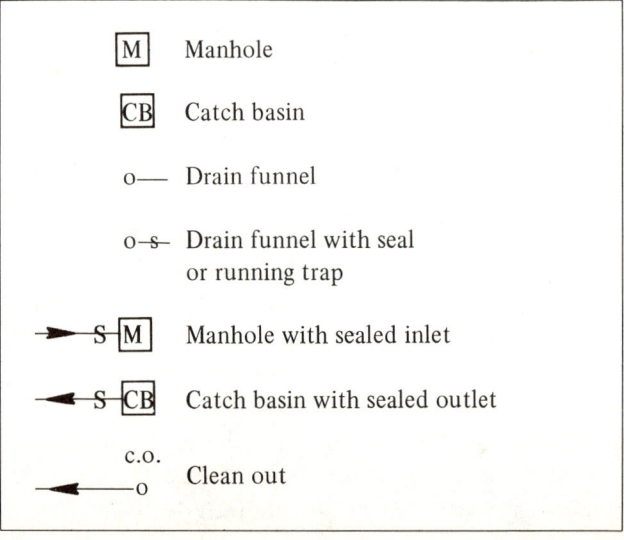

Piping Systems

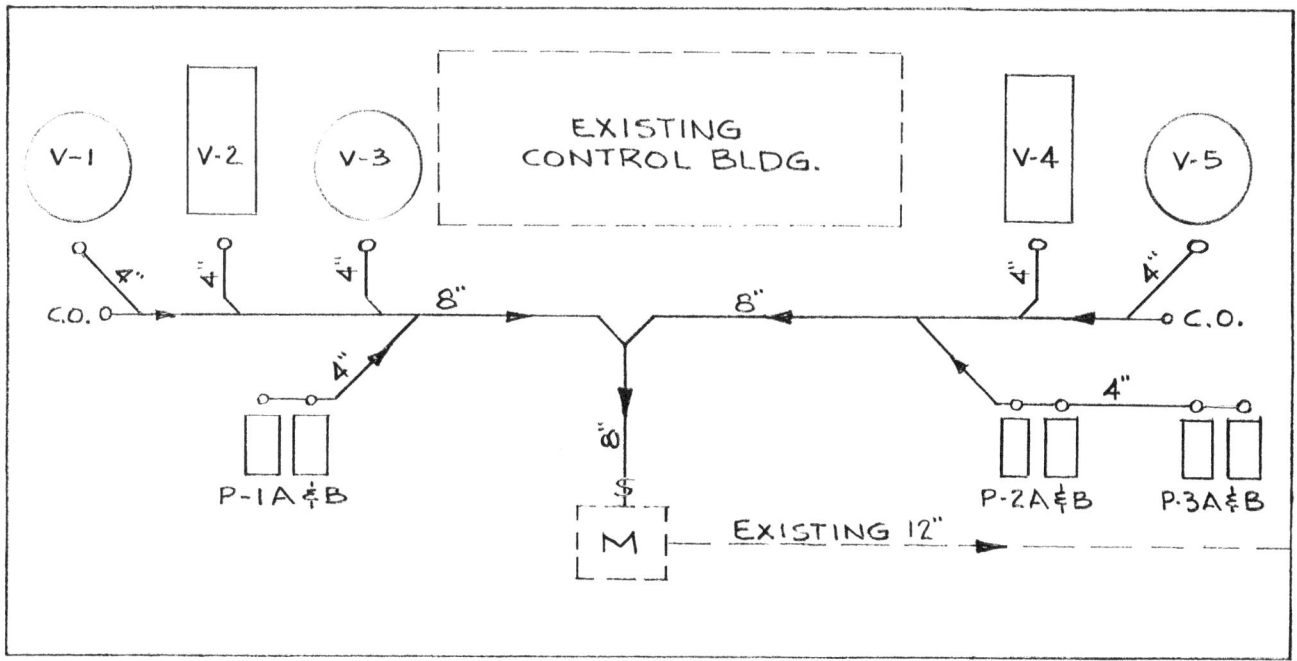

Figure 5-2. Typical process sewer flow diagram.

For special service conditions, the piping designer might employ the mechanical screw-gland joint which will contain oil, gas and water at greater pressures, the ball-and-socket joint which provides great flexibility at the joints, or the universal pipe joint which is a rather stiff joint.

The Dresser coupling is a compression-sleeve coupling. It is used with plain end cut pipe of either cast iron or steel in both underground and aboveground installations for air, gas, water and oil. This type of joint is quite often installed without proper anchors or bracing and when the line is pressured it will come apart. Properly installed, this joint will have anchors or braces at turns on both sides of the joint. These joints are a fine economical design for lines needing a small amount of expansion. Improperly installed they result in total chaos.

The Victaulic coupling is a split-coupling joint used with cast iron and carbon steel piping. This coupling is joined to the pipe or fitting via grooves cut near the ends. It is acceptable for use in liquid or vapor service, will take some expansion and allows considerable angular displacement. This joint is easily installed and/or broken apart.

Sewer Flow Diagram

The sewer flow diagram is the easiest way to communicate information about the drainage system to the layout designer and the customer. Finalizing locations of drain funnels, where manholes are in the system, locations of seals, etc., are much easier settled on a flow diagram than on the finished sewer drawing. Although most companies do not prepare a sewer flow diagram, many man-hours would be saved if they did. If proper design information is given to the layout designer and his checker, a better job can be done in about half the expended man-hours.

Because these flow diagrams are not commonly made, the symbols used are different in each company that does make them. Table 5-1 depicts symbols which are used and could be standardized by the industry. Figure 5-2 is a typical process sewer flow diagram.

Cast Iron Soil Pipe and Fittings

Cast iron soil pipe and fitting standards are established by the Cast Iron Soil Pipe Institute, 2029

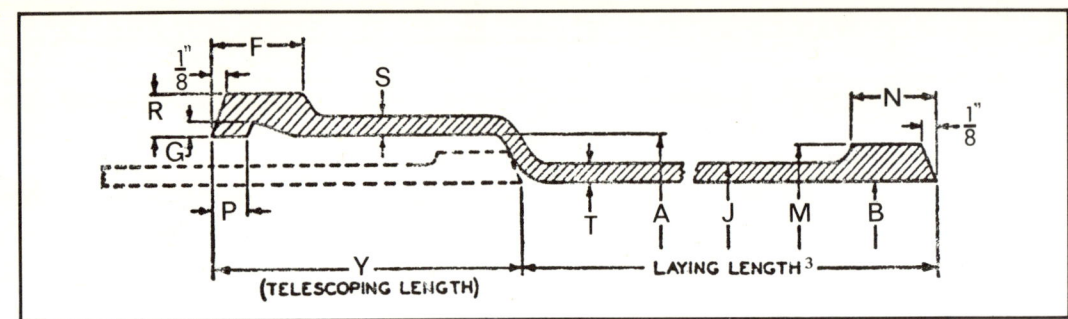

Figure 5-3. Dimensions of service cast iron soil pipe and fittings. Courtesy of Cast Iron Soil Pipe Institute.

Table 5-2
Dimensions of Hubs, Spigots, and Barrels for
Service Cast Iron Soil Pipe and Fittings

Size[1]	Inside diameter of hub[2]	Outside diameter of spigot[2,4] Bead	Outside Diameter of barrel[2]	Telescoping lengths[2]	Inside diameter of barrel[2]	Thickness of barrel[2]
	A	M	J	Y	B	T
inches	inches	inches	inches	inches	inches	inches
2	2.94	2.62	2.30	2.50	1.96	0.17
3	3.94	3.62	3.30	2.75	2.96	.17
4	4.94	4.62	4.30	3.00	3.94	.18
5	5.94	5.62	5.30	3.00	4.94	.18
6	6.94	6.62	6.30	3.00	5.94	.18
8	9.25	8.75	8.38	3.50	7.94	.23
10	11.38	10.88	10.50	3.50	9.94	.28
12	13.50	12.88	12.50	4.25	11.94	.28
15	16.75	16.00	15.62	4.25	15.00	.31

Size[1]	Thickness of hub		Width of hub head[2,5]	Width of Spigot bead[2,5]	Distance from lead groove to end, pipe and fittings[2]	Depth of lead groove	
	Hub body	Over bead					
	S (min.)	R (min.)	F	N	P	G (min.)	G (max.)
inches	inches	inches	inches	inches	inches	inches	inches
2	0.13	0.34	0.75	0.69	0.28	0.10	0.13
3	.16	.37	.81	.75	.28	.10	.13
4	.16	.37	.88	.81	.28	.10	.13
5	.16	.37	.88	.81	.28	.10	.13
6	.18	.37	.88	.81	.28	.10	.13
8	.19	.44	1.19	1.12	.38	.15	.19
10	.27	.53	1.19	1.12	.38	.15	.19
12	.27	.53	1.44	1.38	.47	.15	.19
15	.30	.58	1.44	1.38	.47	.15	.19

Source: Cast Iron Soil Pipe Institute, *Cast Iron Soil Pipe and Fittings Handbook.*
[1] Nominal inside diameter.
[2] For tolerances see Table 3, *Cast Iron Soil Pipe and Fittings Handbook.*
[3] Laying length, all sizes—Single hub 5'0" less Y, for 5-foot lengths; single hub 10'0"; Double hub 10'0" less Y, for 10-foot lengths.
[4] If a bead is provided on the spigot end, M may be any diameter between J and M.
[5] Hub ends and spigot ends can be made with or without draft, and spigot ends can be made with or without draft, and spigot ends can be made with or without spigot bead.

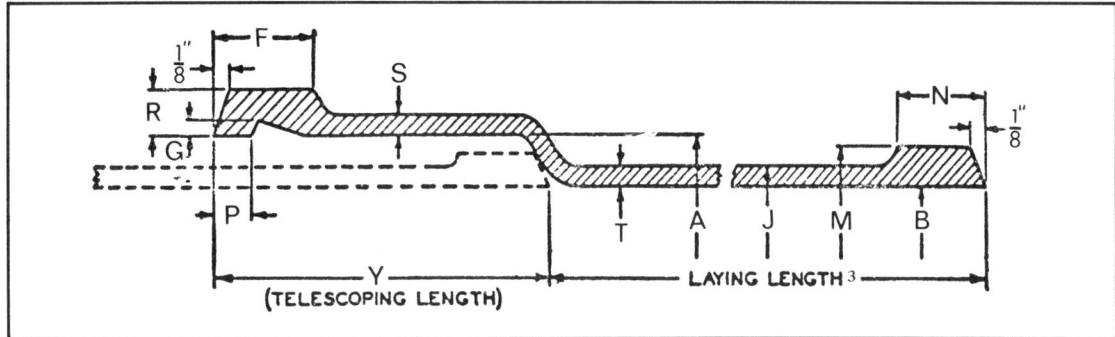

Figure 5.4. Dimensions of extra heavy cast iron soil pipe and fittings. Courtesy of Cast Iron Soil Pipe Institute.

Table 5-3
Dimensions of Hubs, Spigots, and Barrels for
Extra Heavy Cast Iron Soil Pipe and Fittings

Size[1]	Inside diameter of hub[2]	Outside diameter of spigot[2,4] Bead	Outside diameter of barrel[2]	Telescoping length[2]	Inside diameter of barrel[2]	Thickness of barrel[2]
	A	M	J	Y	B	T
inches	inches	inches	inches	inches	inches	inches
2	3.06	2.75	2.38	2.50	2.00	.19
3	4.19	3.88	3.50	2.75	3.00	.25
4	5.19	4.88	4.50	3.00	4.00	.25
5	6.19	5.88	5.50	3.00	5.00	.25
6	7.19	6.88	6.50	3.00	6.00	.25
8	9.50	9.00	8.62	3.50	8.00	.31
10	11.62	11.13	10.75	3.50	10.00	.37
12	13.75	13.13	12.75	4.25	12.00	.37
15	17.00	16.25	15.88	4.25	15.00	.44

Size[1]	Thickness of hub		Width of hub head[2,5]	Width of Spigot bead[2,5]	Distance from lead groove to end, pipe and fittings[2]	Depth of lead groove	
	Hub body	Over bead					
	S (min.)	R (min.)	F	N	P	G (min.)	G (max.)
inches	inches	inches	inches	inches	inches	inches	inches
2	0.18	0.37	0.75	0.69	0.28	0.10	0.13
3	.25	.43	.81	.75	.28	.10	.13
4	.25	.43	.88	.81	.28	.10	.13
5	.25	.43	.88	.81	.28	.10	.13
6	.25	.43	.88	.81	.28	.10	.13
8	.34	.59	1.19	1.12	.38	.15	.19
10	.40	.65	1.19	1.12	.38	.15	.19
12	.40	.65	1.44	1.38	.47	.15	.19
15	.46	.71	1.44	1.38	.47	.15	.19

Source: Cast Iron Soil Pipe Institute, *Cast Iron Soil Pipe and Fittings Handbook*.
[1] Nominal inside diameter.
[2] For tolerances see Table 3, *Cast Iron Soil Pipe and Fittings Handbook*.
[3] Laying length, all sizes—Single hub 5'0" less Y, for 10-foot lengths; single hub 10'0"; Double hub 10'0" less Y, for 10-foot lengths.
[4] If a bead is provided on the spigot end, M may be any diameter, between J and M.
[5] Hub ends and spigot ends can be made with or without draft, and spigot ends can be made with or without draft, and spigot ends can be made with or without spigot bead.

K Street, N.W., Washington, D.C. 20006. They have published an excellent book on all cast iron piping called *Cast Iron Soil Pipe and Fittings Handbook,* which should be in every designer's library.

Cast iron soil pipe is available in sizes 2-15″. Sizes are the nominal *inside* diameter of the pipe. The wall thickness varies with the class specified, which changes the outside diameter. Laying lengths vary, but in general will be either 5 or 10′. See Tables 5-2 and 5-3 and Figures 5-3 and 5-4 for details.

Fittings are manufactured in sizes to match the pipe. Terms are different than the ones for carbon steel and the designer must specify the correct term. For instance, the 90° ell is called a 1/4 bend. The 45° ell is a 1/8 bend. Cast iron fittings also employ bends of 1/5, 1/6 and 1/16. These terms are devised according to what fraction of 360° they turn.

All junctions are made with fittings. Welding is not employed. Fittings are available for line intersections, clean-outs, etc. Flow is always directed into the barrel of the pipe or fitting (see Figure 5-1).

Joints are usually made with twisted jute (or oakum) and calked with lead. Jute is a vegetable fibre. Cotton and hemp are also sometimes used. These materials are ordered by the pound. Lead joint requirements are calculated to be 12 ounces per inch of diameter. Thus an 8″ diameter pipe would require 6 pounds of lead. Jute is estimated at 10% of the lead requirement by weight. The 8″ pipe would need 0.6 pounds of jute. To order joint material, the piping designer must count all joints by sizes and calculate the total weight of joint materials. Ten precent is added to this total to allow for waste.

Designing Systems

To design a sewer system the designer must know the depth of cover required. The highest point in his horizontal piping must be below the frost line or the liquids could freeze in the line. With a slope of 1′ per 100′, the lowest point could be very deep. Lines are also located below frost lines to protect them from "heaving," rising and falling caused by frozen earth thawing and freezing. Frost lines of each area are designated in the general specifications. In any event, underground lines should have a minimum of 3′ of cover in unpaved areas for protection against truck traffic. A cover of 1′-6″ is the minimum under paved areas.

Student Exercise

The plot plan, Figure 4-1, and the foundation location plan, Figure 4-6, comprise a unit for which the designer must prepare an underground piping plan and elevation. The student will make these drawings as a class exercise. Figure 5-2 is the process sewer. Assume a frost line of 3′. Use carbon steel material and supply dimensions or coordinates necessary for fabrication and installation. Locate funnels and lines by coordinates in plan.

To design the storm sewer, the designer must divide the process area into drainage areas to facilitate the removal of liquids as quickly as possible. If possible, drainage areas should be square. Paved drainage areas should have a maximum of 2500 square feet of surface per catch basin. The student is to designate the surface drainage areas, locate the catch basins and route the total flow to the existing manhole. The centerline coordinates of the existing manhole are W.610′-0″ and N.1090′-0″. High points of paving extremities are W.610′-0″ and W.655′-0″, N. 975′-0″ and N.1205′-0″.

Drainage area slopes shall be limited to a minimum of 0.01″ per foot and a maximum of 0.04″ per foot. The maximum total drop allowed is 6″ from the high point of paving to the lowest point at the catch basin.

After completing the drawing, prepare a bill of material listing all material necessary to install the two sewer systems.

Refer to Figures 5-5, drain funnel installation, and 5-6, typical clean-out detail, for design data.

Upon completion, students may exchange drawings, checking one anothers work.

Design Guidelines

Manholes should be located at all major intersections and in line turns of major headers, which are 90°. For sewer line sizes up to 24″, provide manholes every 300′ and for line sizes over 24″, provide them every 500′ for cleaning the system. Manholes may be square or round.

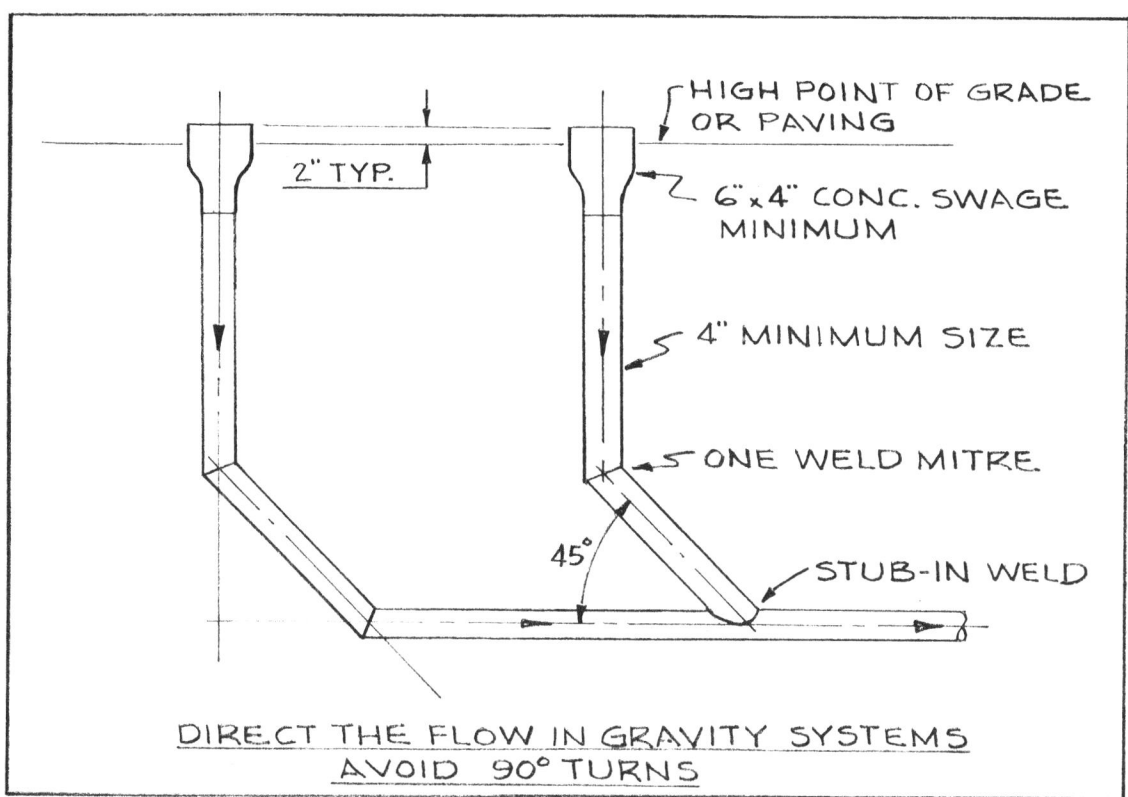

Figure 5-5. Drain funnel installation.

Clean-outs shall be installed at dead ends of process sewers, or where the line turns a total of 90° or at a 45° turn preceded by 50' of straight-run pipe. Locate clean-outs so that cleaning may be accomplished in the flow's direction. Drain funnels may be considered as clean-outs if the line from the funnel is short and total turns do not exceed 135°.

When routing underground lines always check:

1. Location of underground electrical envelopes, their sizes and elevations.
2. Lines entering or leaving buildings routed by architectural or structural groups.
3. Location and elevation of all foundations, whether they are spread footings or grade beams. Sewer piping must clear.
4. Angle of repose of the soil. Foundations must not be underreamed by excavation for sewer piping. Any time a pipe is run deeper than the foundation (if not on piling) and within a few feet of it, check the structural group about possible underreaming. If there is no other route for the pipe, the structural group may have to lower their footing.

Underground Pressurized Systems

Fire water, cooling water, closed process drains and pump-out systems are the more common pressurized underground lines. Carbon steel is the most commonly used material for these systems.

Transite piping is often used in pressurized brine systems and some people specify it for fire-water service. Transite is an asbestos material, very light in weight and, since it does not rust, it needs no external protection when laid underground. It is also used in some aboveground special service.

The Johns-Manville "Ring-tite" joint is the author's selection of transite joints. It is easily joined by the field people and keeps installed costs at a minimum while furnishing an excellent pressure-containing closure. Johns-Manville will be happy to supply catalogs, at request, showing full details of their transite pipe and Ring-tite coupling joint.

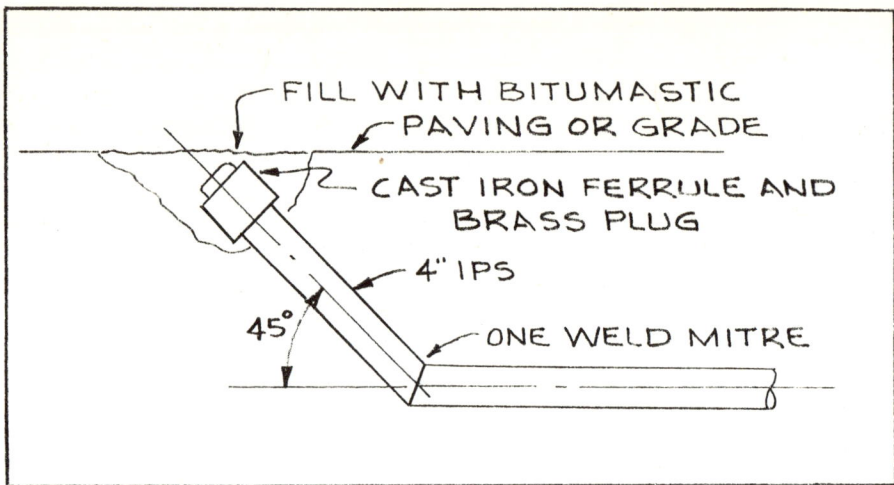

Figure 5-6. Typical clean-out detail.

Fittings of transite material are very different and limited in types. Corrosion Proof Fitting Company manufactures most fittings and Johns-Manville sells their fittings. Tyler Pipe Industries of Tyler, Texas, makes a cast iron fitting for use with transite pipe.

Mechanical joint cast iron is also very popular for pressurized systems. The Clow Company is one of the larger manufacturers of this. When specifying mechanical joint piping, either tie rods or thrust blocks must be used. Consult the Clow catalog for details.

Fire Water

Every day people see fire plugs but not many consider the design of the underground fire system. Fire plugs and monitors are furnished in plant protection systems.

Monitors are used to direct water to protect specific risks which may not be accessible with a portable hose stream or where the fire hazard is such that prompt application of water is necessary. They are also used where the area would be extremely hazardous to occupy during an emergency. Monitors may be fixed on one target, but usually they are left free to rotate by the operator. They emit as much water as a 2½" hose which requires three men to handle.

The Elkhart Brass Mfg. Co., of Elkhart, Indiana, is a large producer of monitors. Figure 5-7 shows three types of their monitors.

Fire plugs are located around the perimeter of each process unit, 5' from the road's edge, at all road intersections and one between road intersections if this distance exceeds 250'. Locate them about 50' from an area or building where special protection is desired. Do *not* locate them where falling buildings or walls would injure the fire fighters or possibly knock over the fire plug, resulting in a loss of system pressure when it is needed the most.

Fire water headers are usually 8", while branches to plugs are 6". Valves should be installed at intervals in the main header to make it possible to shut off small sections for repair or a new line tie-in without having to shut down the system. Provide valves so that no break or repair will shut down more than 1000' of the system.

Fire water mains should be looped, allowing hydrants and monitors to be fed from two directions, greatly increasing the possible delivery of water without excessive friction loss. Flushing connections, 4" size, should be located at the ends or far corners of the main. Fire water will freeze and must be located under the area frost line. For safety, locate it 1' below frost line. Minimum cover should be 3' except under railroads where this is increased to 4'. If the main is routed under a railroad the author suggests a culvert or pipe sleeve for load protection.

Underground Cooling Water

For most installations, underground cooling water supply and return lines are utilized for two reasons. First, they eliminate possible freezing, and second, they are cheaper. Some companies insist

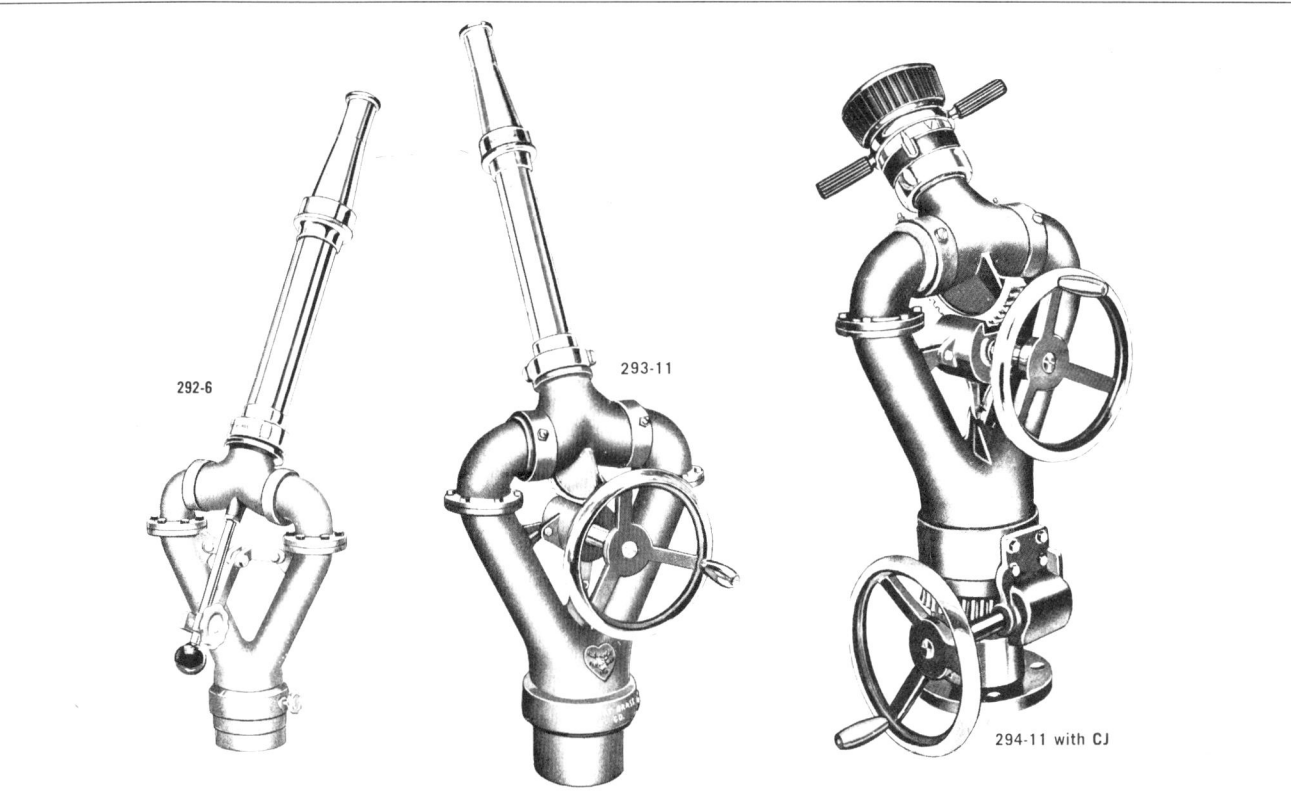

Elkhart Monitor Nozzles

Elkhart manufactures a complete line of Deck Pipes or Monitor Nozzles. There are eight different styles and three different sizes to choose from. Elkhart monitor nozzles are the only ones manufactured with double row, full diameter nylon bearings. These "King-sized" bearings make the nozzles extremely easy to operate at high pressures and never need to be lubricated. Even at high pressures, one man can easily control any of these nozzles. Nozzles are normally supplied with a discharge tube and one tip. 600 G.P.M., 1100 G.P.M. and hydrant Monitor nozzles are supplied with a 2½" x 2½" No. 282 discharge tube and No. 181 tip. 2000 G.P.M. Monitor nozzles are supplied with a 3½" x 3½" No. 284 discharge tube and No. 181-3 tip. All 293 and 294 Monitors are available with a gauge and gauge guard at extra cost. Frequently it is desirable to use a combination straight stream and fog nozzle on these monitors such as the CF (500 G.P.M.), CJN (2000 G.P.M.), J (500 G.P.M.), JN (1000 G.P.M.) or the remote controlled Sky Chief nozzle. When these are used, tubes and tips are not necessary and price of these can be applied against the price of the fog nozzle. Monitors are available with either female TIPT or ASA flat-faced flanged bases (see list of available bases). All discharge tubes are equipped with double stream shapers. Brass body painted red — specify either polished brass or chromium plated trim.

292 Handle Operated Monitor

Full $360°$ rotation with lock — Maximum elevation with handle control — Has positive elevation lock.

> 292-6 for discharges from 200 to 600 G.P.M.
> 292-11 for discharges from 600 to 1100 G.P.M.
> 292-20 for discharges from 1100 to 2000 G.P.M.

293 Single Wheel Operated Monitor

Full $360°$ rotation with lock — Maximum elevation by wheel operated worm gears which lock positively unless the wheel is turned.

> 293-6 for discharges from 200 to 600 G.P.M.
> 293-11 for discharges from 600 to 1100 G.P.M.
> 293-20 for discharges from 1100 to 2000 G.P.M.

294 Double Wheel Operated Monitor

Wheel-operated worm gears control both vertical and horizontal operation and lock automatically — full $360°$ rotation.

> 294-11 for discharges from 600 to 1100 G.P.M.
> 294-20 for discharges from 1100 to 2000 G.P.M.

Figure 5-7. Fire monitor nozzles. Courtesy of Elkhart Brass Mfg. Co., Inc.

on cooling water in overhead pipeways, but this means that their racks will be larger and laterals to exchangers will be longer and consume more fittings. The most economical system will employ underground headers routed outside of the channel end of exchangers, causing very short lateral lines. If possible during plot plan development, locate all exchangers using cooling water on one side of the unit pipeway. This would eliminate having two sets of supply and return headers. If one or two small water users must be located on the opposite side, a subheader could be run across the unit from the main header to feed them.

Locating the cooling water headers at the channel end of the exchangers leaves the area under the unit pipe rack clear for underground drains and electrical distribution envelopes. It simplifies maintenance when needed and keeps these big headers free and clear of foundations.

Space underground cooling water supply and return headers at least 5' apart (clearance between steel) to keep heat from being conducted through the earth from the hotter return header to the cooler supply header. Branches are to have 1'-6" clear for shovel room only.

Pump-Out System

The closed process pump-out system is usually located underground. It is a pressurized carbon steel pipe connecting to the unit's vessels, exchangers and some pumps, routed to a pump manifolded so it can be used as a pump-out pump, which pumps the unit down discharging to a storage tank in the tank farm. This system is run with 90° elbows; clean-outs and manholes are not provided.

Glycol Lines

Glycol lines in cold process gasoline plants should always be run underground. These are routed to inlet gas chillers and are discharged from a positive displacement pump. This pump raises the pressure from 50 psig to over 1000 psig and due to the pump design the small glycol line is subject to intense pulsation from each thrust of the pump's plunger. When this line is located overhead it has a good chance of shaking, rattling in the rack, and eventually fatiguing and leaking at joints.

Steam Tracing

Tracing is the term used to describe the transfer of heat from a foreign source to piping systems. Tracing is usually accomplished by a steam tracer line, but sometimes this gives way to electric tracing, where an electrical wire transmits heat to the pipe.

Figure 5-8 shows how steam tracers are installed inside insulated lines. Steam traced lines may have one or more tracers depending on the amount of heat transfer required for the fluid. Insulation for steam traced lines must be sized one size larger than line size to accommodate the tracer. For example, 8" insulation would be ordered for 6" lines, etc. Since electrical tracing employs a flat tape, oversized insulation is not necessary.

Steam tracing is used to keep liquids from freezing in cold climates, to keep a viscous liquid from setting up or solidifying and to vaporize small amounts of hydrocarbon condensate that may form in a vapor service.

One-half inch copper tubing, with flared fittings, is the most common and least expensive method of steam tracing. Properly installed, all flared joints will be located outside the insulation where any leak can be observed and repaired. This also will prevent damage to insulation should a leak occur. Steam of 100 to 150 psig is usually economically available for tracing.

Steam Tracing Design Practice

The "Low January Average" is available from the U.S. Weather Bureau office for any particular area. The Weather Bureau calculates this by averaging the low temperatures for the entire month. This Low January Average is used as the design low temperature for determining extent of tracing or "winterizing."

For a Low January Average of 30°F or higher, winterizing is specified only where a sustained reading below 30°F is recorded several days of the month, lasting 24 hours or longer.

For a Low January Average of 0°F to 29°F, water must be protected from freezing. Trace water to maintain a temperature of approximately 75°F. Hydrocarbons containing water must have tracing applied to "dead leg" areas, where water (which is heavier) can settle out and become a freezing hazard.

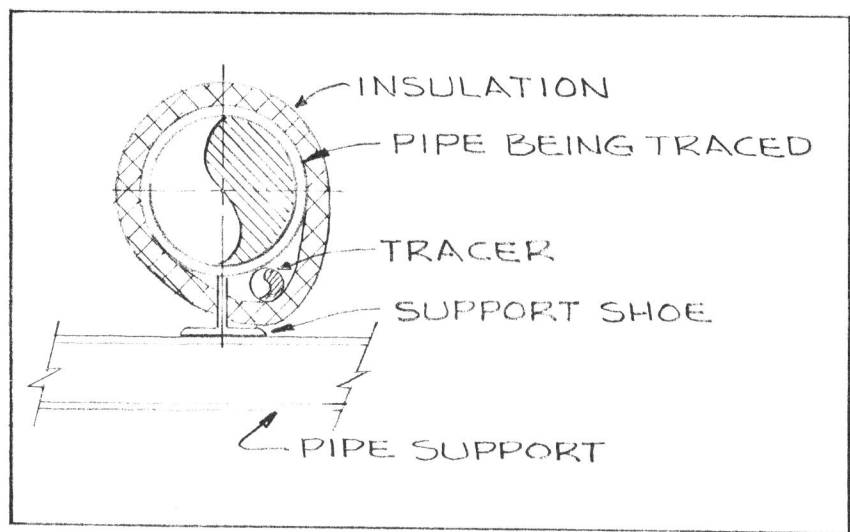

Figure 5-8. Section of steam tracer.

All fluids with pour points at or above the low ambient design temperature should be traced to maintain a temperature approximately 100°F above their pour point. Pour point temperatures of all fluids are available from chemical engineering handbooks.

For a Low January Average below 0°F:

Winterizing is required for water and aqueous solutions. Trace to maintain temperature of approximately 75°F.

Fluids with pour points at or above the low ambient design temperature should be traced to maintain a temperature approximately 100° above their pour point.

Special consideration is given to the following:

Buried lines, because of the deep frost line.

Air intakes to boilers and other heating equipment.

Cooling towers and air coolers.

Equipment Winterizing:

Pumps are preferably winterized with companion piping. Compressors, blowers and other mechanical equipment are specified for operation at low ambient design temperature.

Storage tanks and vessels have winterized type drains.

Water draw off sections of vessels and bottoms of Air or Gas receivers where water can collect are heat traced.

Materials for Steam Tracing

Tracers shall be O.D. tubing. Soft annealed copper tubing shall be used where temperature of the product line or tracing steam does not exceed 400°F. Above this temperature dead soft annealed hydraulic quality, low carbon seamless steel tubing shall be used.

For stainless steel lines, the tracer material shall be low carbon steel. Stainless steel instrument leads shall be traced with copper tubing.

For aluminum pipe lines, the tracer material shall be stainless steel.

For conditions where the tracer could overheat lines containing acid, caustic, amine, phenolic water, or other chemicals insulation spacer blocks shall be installed between tracer and pipe.

Each tracer shall have its own trap. Tracer traps shall discharge to sewer. If condensate must be collected, minimum usable pressure is 25 psig.

Tracer Size and Length

a. Required tracer size is determined by piping heat loss and tracer steam pressure found in the "Heat Loss Chart" (Figure 5-9).

Process Piping Design

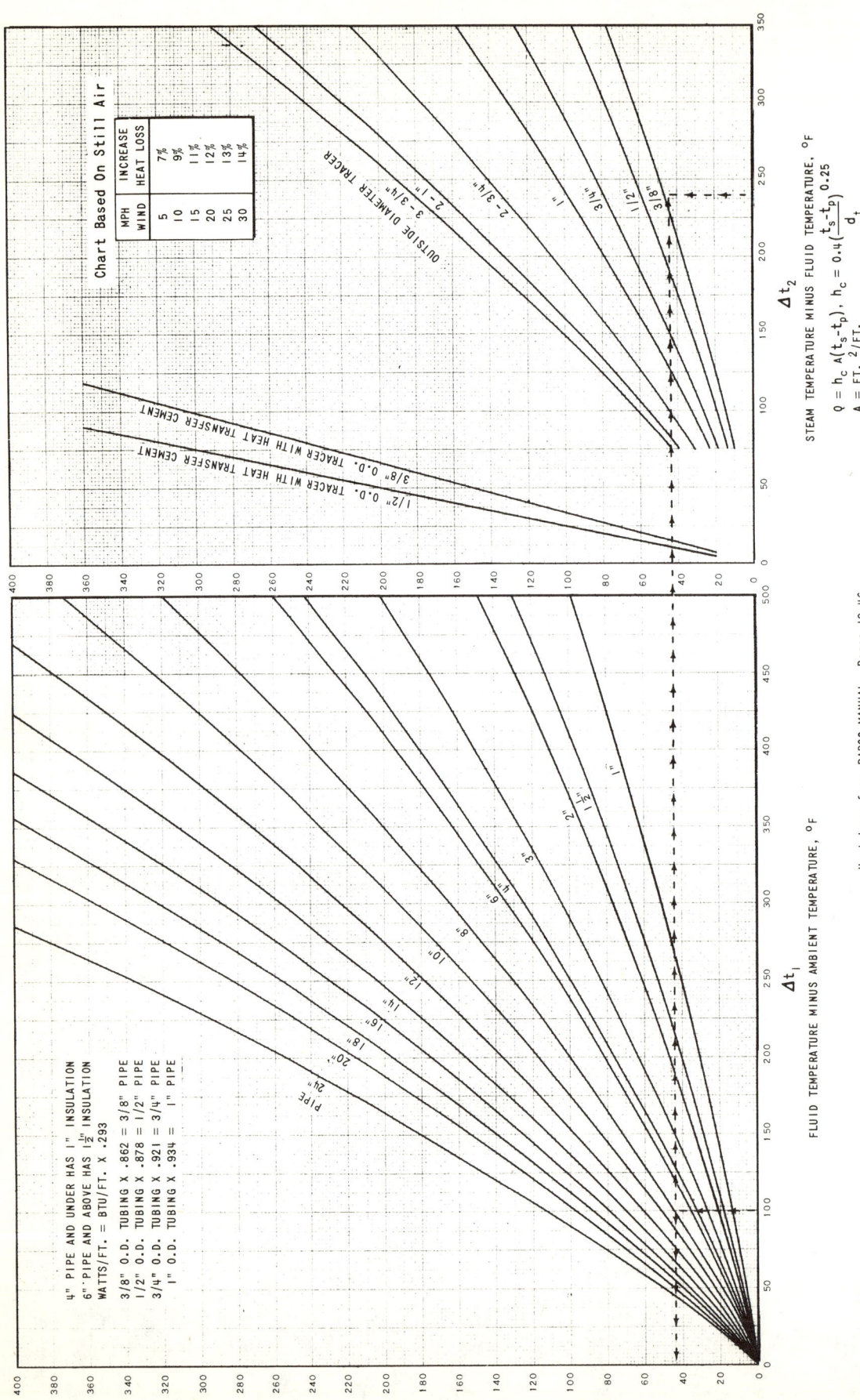

Figure 5-9. Heat loss chart for steam tracing. Courtesy of Fluor Corp.

Piping Systems

b. Minimum tracer size is 3/8°" O.D. tubing; maximum size is 1" O.D. tubing. For economy, where "Heat Loss Chart" indicates the requirements for multiple tracers, a single tracer with heat transfer cement should be considered.

c. When using heat transfer cement, tracers of 3/8" and 1/2" are recommended. If more tracer area is required, multiple tracers of 3/8" and 1/2" are used.

d. Maximum tracer lengths are based on tracer size and steam pressure as follows:

1. Steam pressure 15 thru 25 psig:

    200' for 3/8" and 1/2" tracers, 300' for 3/4" and 1" tracers.

2. Steam pressure 50 thru 200 psig:

    200' for 3/8" and 1/2" tracers, and 400' for 3/4" and 1" tracers.

Tracer lengths for tracing with heat transfer cement shall be based on recommendation of manufacturer.

Tracer Pocket Depth

Pocket depth is the distance the tracer rises in the direction of flow from a low point to a high point. The *total* pocket depth is the sum of all risers of the tracer.

Maximum tracer total pocket depth is equal to 40% of tracing steam gage pressure expressed in feet.

Example: Tracing steam 150 psig; 150 x .40 = 60' total pocket depth.

Figure 5-9, heat loss chart, is used to determine the size and number of steam tracers. Note that this chart is based on still air and heat loss must be increased by the factors shown for the area's normal winter wind velocity.

Example #1, water line:

A 4" water line is to be maintained at 35° minimum fluid temperature. The low ambient design temperature is −10°.

$\Delta t_1 = 35°$ minus $(-10°) = 45$.

Available steam 50 psig at 250°.

Table 5-4
Flow Diagram Legend for Winterizing

ST	Steam traced and minimum insulation
STS	Steam traced with spacers and minimum insulation
STT	Steam traced with heat transfer cement and insulated
SJ	Steam jacketed pipe and insulated
ET	Electric traced and minimum insulation
ETT	Electric traced with heat transfer cement and insulated
WD	Winter drained and no insulation
WC	Winter circulated and no insulation
WF	Winter flushed and no insulation

Equipment: List the symbol with the title.

ST	1" insulation, steam traced
ET	1" insulation, electric traced
PP	1" insulation, process protection

Instruments: List the symbol next to the instrument number circle.

ST	Steam traced and minimum insulation
ET	Electric traced and minimum insulation
WS	Winter seal

Minimum fluid temperature = 35°.

$\Delta t_2 = 250°$ minus $35° = 215°$.

Entering a 4" line and 45 for Δt_1 and 215 for Δt_2, one 3/8" O.D. tracer is indicated.

Example #2, Process Line:

An 8" line is to be maintained at 125° minimum fluid temperature. The low ambient design temperature is 25°.
$\Delta t_1 = 125°$ minus $25° = 100°$.

Available steam 150 psig at 365°.

Minimum fluid temperature = 125°.

$\Delta t_2 = 365°$ minus $125° = 240°$.

Entering an 8" line and 100° for Δt_1 and 240° for Δt_2 one 3/8" O.D. tracer is indicated.

Winterizing Flow Diagram Symbols

Winterizing is accomplished by several means, in addition to steam tracing. Whatever the means, winterizing must be shown on the flow diagram to

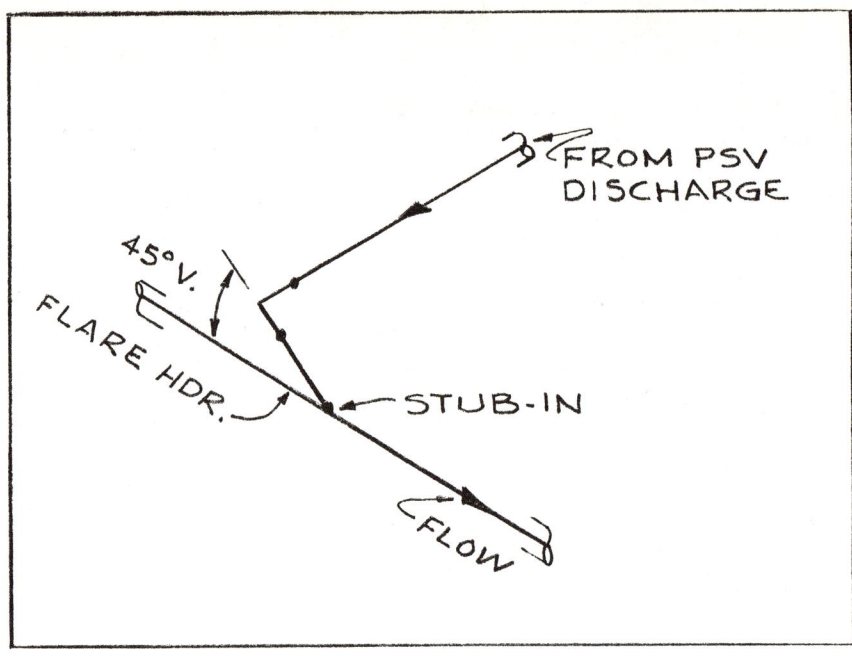

Figure 5-10. Detail of directing flow in flare header.

communicate it to the customer, the field, the operation people and the piping designer. The usual place to communicate this is immediately after the line number and size. For instance, 427A-6″ ST means that line 427 is material specification A, 6″ line size, and is steam traced. Winterizing symbols may vary between companies, but Table 5-4 depicts terminology that is fairly common.

Flare Systems

Flare is the term used for the system which disposes of a plant's waste gases. Flare stacks are the tall structures that route the gas to the atmosphere, usually with a "flare tip" which ignites the gas. Not all flare stacks are ignited. In some parts of the world, such as Lake Maracaibo in Venezuela, the government will not allow a flame and the gas is vented directly to the atmosphere.

Flare headers are the lines in units that receive relief valve discharges, vents, etc., and route this gas to a flare knock-out drum which separates any liquid that may have condensed in the line. Waste gas from the knock-out drum is then routed to the flare stack. Knock-out drums are located at the flare stack to keep the line from the drum to the stack as short as possible.

It is quite normal to have two flare headers and two knock-out drums feeding one flare stack or possibly separate stacks. One is for the low pressure flare, for relief valves set at 175 psig or below, and one is for the high pressure flare, for relief valves with set pressures above 175 psig. This pressure break point will vary from plant to plant and is determined by a process pressure drop calculation and allowable back pressure in the low pressure flare header. When two flare headers are required, the utility flow diagram will show the pressure break point and which lines will vent into each header.

Flare headers must be self-draining from origin to the flare knock-out drum. There should be no rises in the header as these will form pockets where condensed liquids will accumulate and eventually block the vapor flow.

Lines from relief valves to the flare headers must also be routed to be self-draining into the top

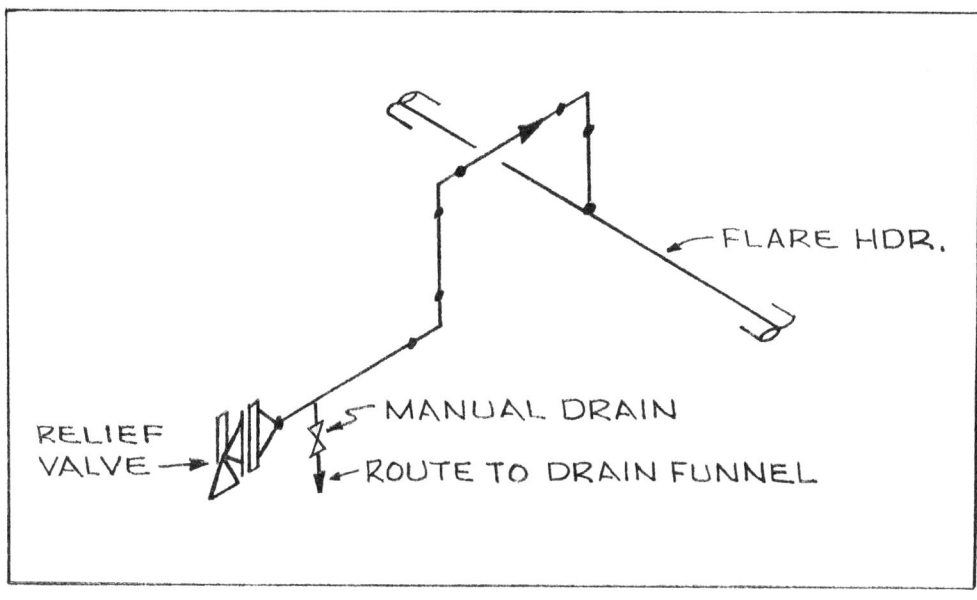

Figure 5-11. Drain valve installation for flare line.

of the header. This means relief valves will be located higher than the flare header and the unit flare header is usually one of the highest lines in the pipe rack to keep it flowing down to the knock-out drum. High pressure relief valve discharges should have the flow directed into the header. This would apply for relief valve set pressures of over 1000 psig. Figure 5-10 shows the piping detail for directing flow into a flare header.

There are cases where it is impossible or impractical to keep relief valve outlets self-draining to the flare header. But in no case can a liquid pocket be allowed to cause uncalculated back pressure on the relief valve. When pressure must be relieved, the relief valve must have a smooth flowing outlet line to rid the piping of overpressure. When the relief valve outlet must be lower than the flare header, piping shall conform to Figure 5-11, drain valve installation for flare line. The manual drain valve must be accessible from grade or platform. This installation would require regular draining by the unit operators. Note that the flare line goes into the top of the flare header. This must always go in the top to prevent any condensate which may be in the header from backing up in the lateral coming from the relief valve.

Dead legs in cold climates must be steam traced to prevent freezing. Steam tracing may evaporate all condensate in the leg and this is one method of ridding dead legs of accumulated liquids. Even when steam tracing is used the manual drain valve must be provided.

If the dead leg is in a spot where it prevents access (by ladder or platform) to the drain valve, provide two valves. One shall be located at the dead leg, as shown in Figure 5-11, which shall always be left open. The other valve shall be located at the drain funnel and this will be the operating valve. The line to the operating valve will be steam traced in cold climates. The valve at the dead leg is

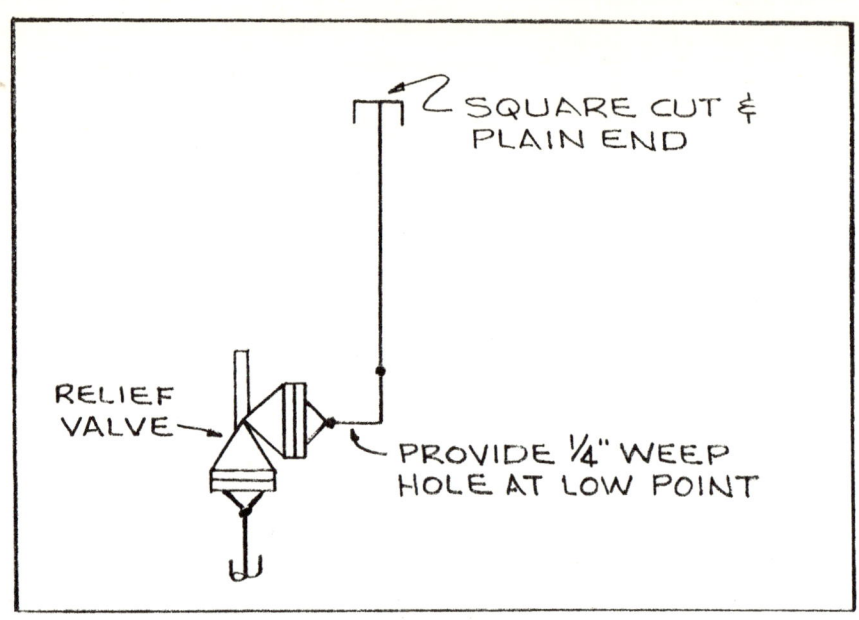

Figure 5-12. PSV discharging to atmosphere.

provided as an emergency shut-off in case of a break in this small line.

Relief valves are usually located to relieve vapors. Hydrocarbon vapors are sometimes vented to the atmosphere if located very high on top of tall towers. Figure 5-12 shows how to route this to the atmosphere. The tail pipe is to terminate a minimum of 7'-6" above any platform within a 40' radius. Relief valves must be accessible from a platform. Always mount relief valves with the spring vertical except for small liquid thermal relief valves, which may be installed with the spring in a horizontal position.

Never route liquid reliefs to a flare header. These are to be routed to a nearby drain funnel or, if none is available, run the PSV discharge line to within 6" of grade. In cryogenic piping it is normal to route liquids to a flare header; however, these light hydrocarbons usually immediately flash into vapor as they become warmer.

Tail Pipe Cuts

There is a very old myth among piping designers that calls for a 45° bevel on relief valve tail pipes discharging to the atmosphere. This myth has been handed down from generation to generation and in almost every plant today the beveled tail pipe is seen. The idea proposed by the myth is that beveling will direct the outlet velocity away from a platform or building. The square cut end shown in Figure 5-12 will direct the discharge upward, away from platforms, etc., and will do it at the minimum cost. The 45° bevel is more expensive to cut and results in extra waste pipe. The real falacy of this myth is seen when observing the actual installation where bevels have been cut which, in effect, direct the velocity toward operating platforms! This is caused by improper orientation of the bevel. The square cut end design eliminates all orientation problems.

Chapter 5
Review Test

The student should be able to answer the following questions. If three or more are missed, reread the chapter.

1. Underground _____ lines should be avoided.

2. Underground utility systems are classified in two divisions. Name them. _____
_____ and _____

3. The minimum slope for gravity sewers is _____.

4. Define storm water sewers. _____
_____

5. What is in a combined sewer? _____
_____

6. Define sanitary sewer. _____

7. What is an invert? _____

8. What is a catch basin? _____

9. Cast iron soil pipe is available in sizes _____ to _____.

10. A 90° elbow in cast iron is called _____.

11. Cast iron joints are usually sealed with _____.

12. All sewer lines are to be located below the _____.

13. To keep liquids from freezing in cold climates is one use of _____.

14. A flare system disposes of a plant's _____ _____.

15. When discharging to the atmosphere, relief valve tail pipes are to have their ends _____ cut.

6 Pipe Fabrication

Fabrication is the assembling and attaching of component pieces to make a completed item. Fabrication of piping, the joining together of weldable pipe and fittings, is done by field personnel at the job site and by "shop fabricators," a shop located at a metropolitan area with access to qualified personnel and all materials on a large scale. Shop fabrication is usually less costly than field fabrication due to modern assembly line techniques and access to the latest fabrication equipment. This savings will average 10-15% over the usual field fabrication. Because shops have instant communication with steel mills and large suppliers, they can locate special materials or fittings for their customers' needs. And as they buy huge quantities of piping materials yearly, they can get the best possible prices. Shop fabrication thus saves money on both labor and materials.

Most pipe shops employ all union personnel. The national union agreement calls for all pipe 2" and smaller to be fabricated in the field. By arrangement with the local union, smaller pipe may be shop fabricated. They usually agree to this if they cannot supply enough people to field fabricate the work.

Normally, pipe 3" and larger is shop fabricated. Long straight runs are not supplied by shops. This pipe is sent directly to the field. Underground pipe is field fabricated.

Shops fabricate pipe for all companies. Each company's drawings and specifications are sent to the shop in many different kinds of forms. Half of their drawings are supplied as fully dimensioned plans and elevations. The other half are line isometrics, referred to as "spools" by contractors. Shops do not call isometrics spools. Shops prepare a drawing of a piece that is shippable and they call this piece a spool. A line isometric may produce a dozen shop spools.

Welding

Shops have welders and engineers fully qualified to make almost any type of weld necessary. Most large contractors, such as Fluor, have complete welding specifications and procedures which shops use. However, for customers who do not furnish welding specifications, shops have their own specifications.

Welding symbols have been standardized by AWS, the American Welding Society. Figure 6-1, welding symbols, shows these standards.

Shop Details

The shop's drafting room prepares drawings of shop spools, showing every piece and every detail dimension necessary for fabrication. With each shop piece drawing a list of material is shown with cut lengths for each piece of pipe to the closest 1/16".

When plans and elevations are supplied, a "take-off" man is given the job of preparing a rough sketch of the shop spool. The sketch is checked by a "take-off checker." Then it is passed to the detail draftsman who produces a finished drawing, the shop spool. Take-off men sometimes make a line isometric to pass on to draftsmen.

When isometrics are furnished to the shop, the take-off man is usually by-passed and the line isometric goes to the draftsman. He makes the decision of where to locate shop piece break points and completes the detail drawing. The drawing is then checked by a "break-down checker." Some companies have spools checked twice to ensure accuracy.

Shop spools are drawn orthographically and as isometrics. Some shops do single line spools while some do double line. Texas Pipe Bending Co. in Houston has done all kinds and now does single line spools in isometric except for single plane spools, which are drawn orthographically.

Figure 6-2, orthographic pipe spool, is drawn orthographically because the pipe is in one plane.

Figure 6-3, isometric pipe spool, shows how shop spools are drawn to show more than one plane.

Figure 6-4, isometric spool with miters, shows detail miter dimensioning. Note that the pipe's total length is shown in the material list.

Figure 6-5, shop spool with bends, is drawn orthographically because it is in one plane. All bend data is shown in the boxes at the drawing's top. The material list gives the total length of pipe needed to complete the spool.

Pipe Bends

Pipe bends are used to make turns without using fittings. The pipe is usually filled with sand, heated and bent to a radius and angle as specified. The bend's radius should not be less than 5 pipe diameters. For a 12" line this would be 5'.

Fabrication shops have developed charts and tables to aid detail draftsmen. These charts are useful to piping designers as well. Tables 6-1 and 6-2, 30° bend data, supply dimensional data for 30° bends. Tables 6-3 and 6-4, 45° bend data, supply the same information for 45° bends. Tables 6-5 and 6-6, 60° bend data, supply dimensions for 60° bends. Table 6-7, 90° bend data, shows 90° bend dimensions. To read this table, for a 3'-6" radius, see 3'-0" at the top and go down to 6" (at the left) which would supply the arc dimension of 5'6".

Miter Welds

Miter welds are often specified in low pressure services as elbow substitutes. In very large lines fittings are unavailable and miter weld elbows are used. Where pressure drop must be held to a minimum, the four-weld miter is used. Two-weld 90° miters are used for maximum economy, but they cause the greatest pressure drop. The three-weld miter is a compromise. Table 6-8, miter welding dimensions, gives full details on miters. For angles of 45° or less the one-weld miter is common.

Small Fittings

The pipe fabricator is concerned with dimensions that affect the length of pipe he must supply. Table 6-9, screwed and socketweld fittings, gives dimensions of interest to pipe fabricators. Normal thread engagement is also shown.

The Triangle

Piping designers run pipe vertically, horizontally and at angles. The most common angles formed are 30° and 45°. By construction, piping draftsmen make 90° triangles and apply their math background to solve triangles formed by these angles. Pipe shops have developed triangle tables to aid in quick solutions. Tables 6-10 and 6-11, 30° offsets, supply solutions for 30° triangles. Tables 6-12 and 6-13, 45° offsets, show 45° triangle solutions.

The Cutback

A *cutback* is the dimension from the header centerline to the nozzle's nearest point. Cutback dimensions are needed to determine the exact length of the nozzle pipe. Table 6-14, 90° cutback for standard weight pipe and Table 6-15, 90° cutback for extra heavy pipe, give the cutback dimensions when the nozzle's ID rests on the header's OD.

Table 6-16, cutback at elbows, supplies dimensions for cutbacks occurring at 90° elbows. The formula shown can be applied for sizes not listed.

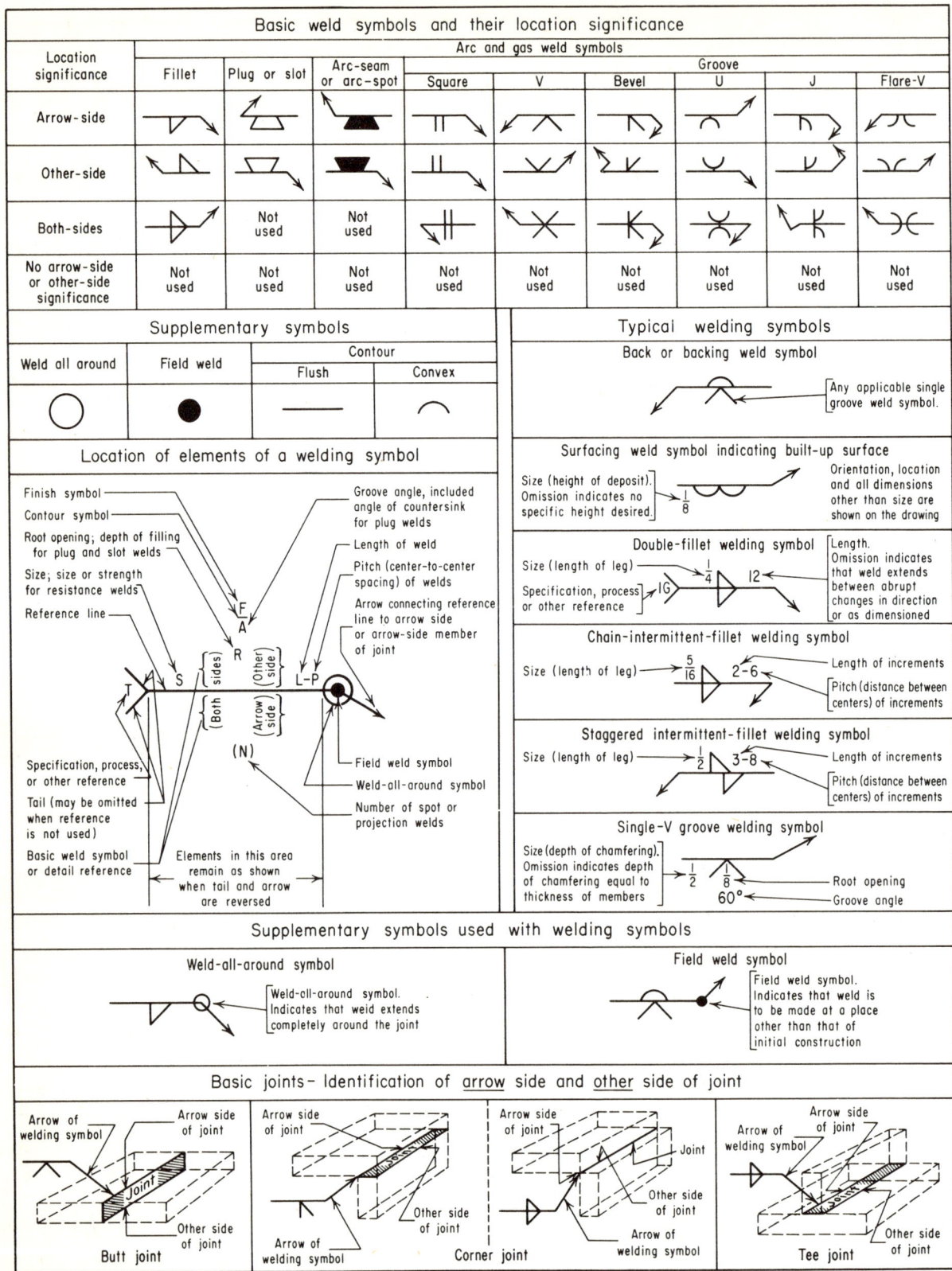

Figure 6-1. Welding Symbols. Courtesy of Texas Pipe Bending Co., Inc.

Pipe Fabrication

Figure 6-1., continued. STANDARD WELDING SYMBOLS

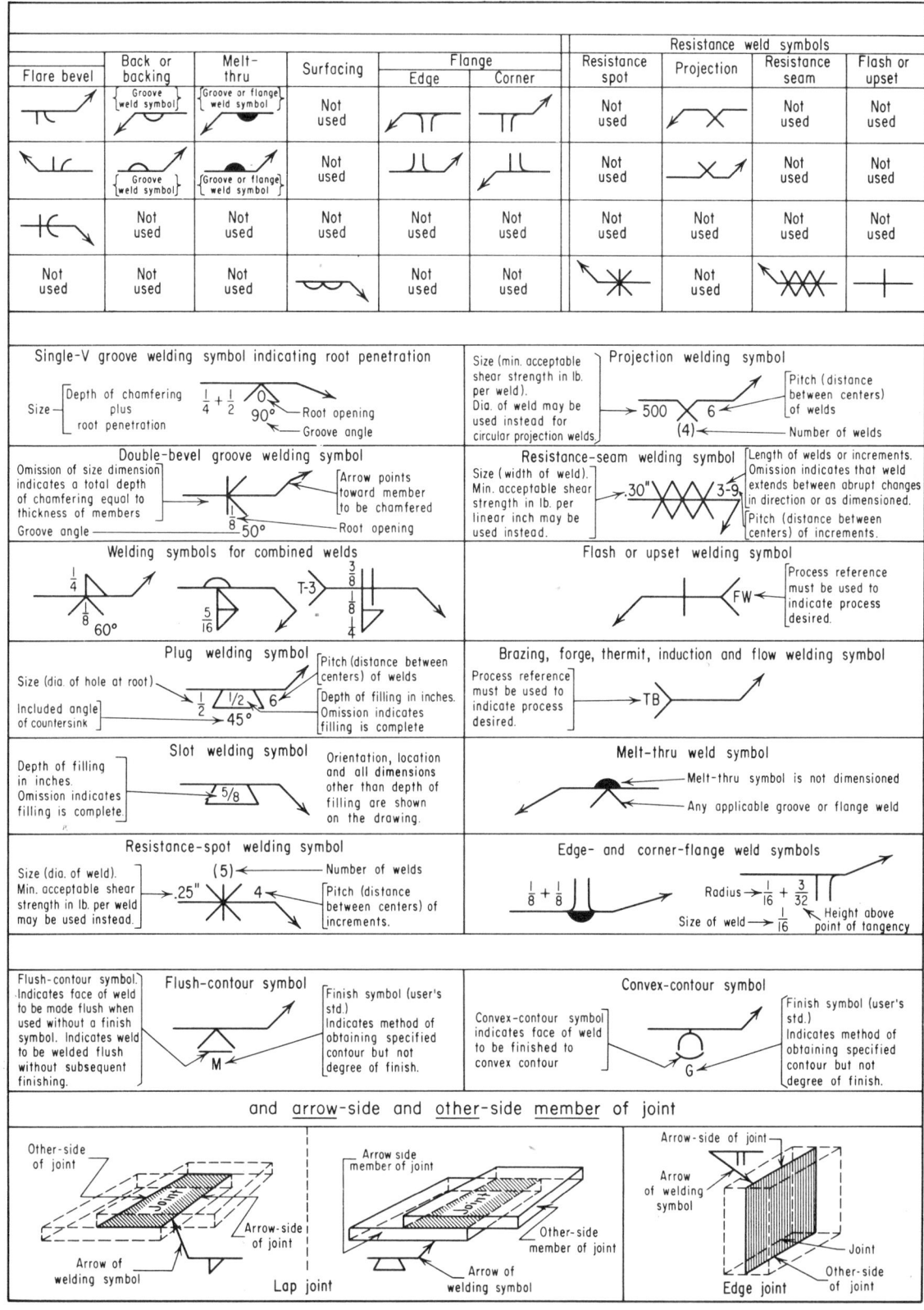

96 Process Piping Design

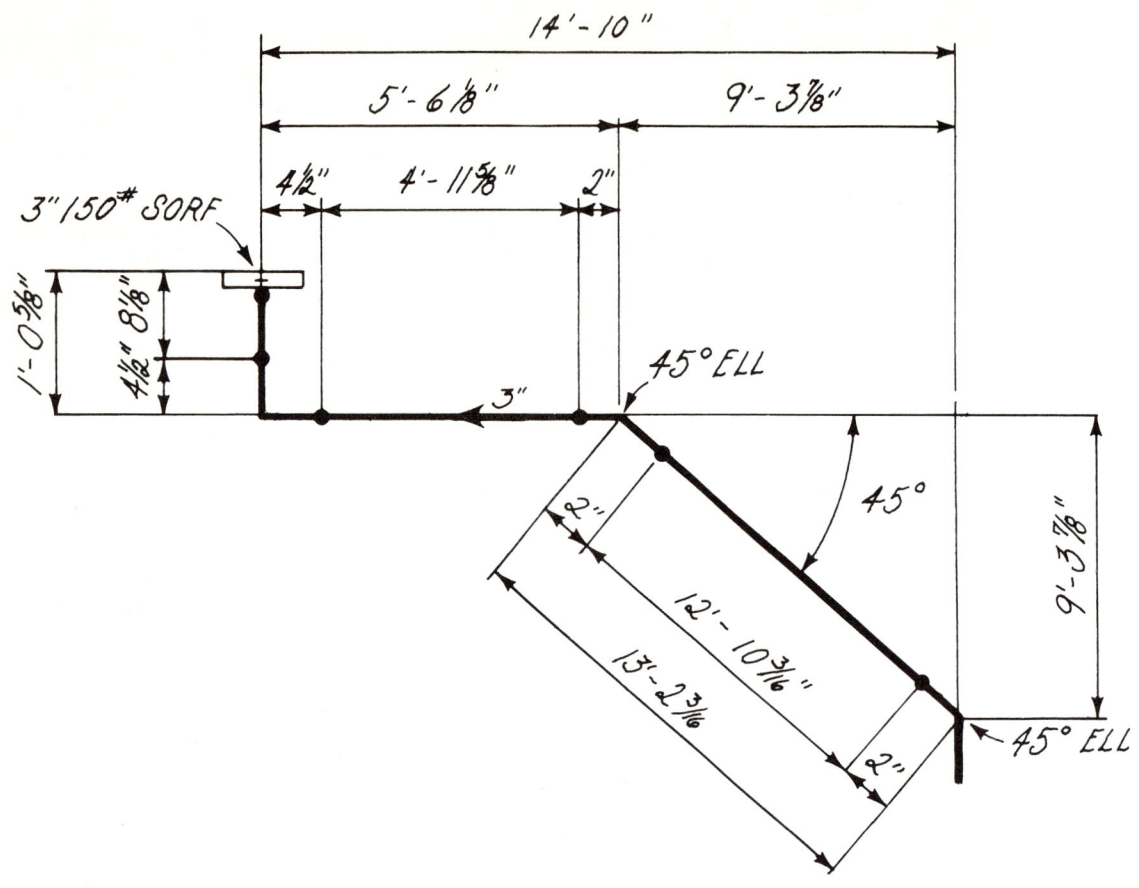

Material List

Quantity	Size	Description
1	3"	Std. Wt. LR 90° Ell
1	3"	Std. Wt. 45° Ell
1	3"	150 lb. SO RF Flg.

All pipe to be ASTM A-106 Gr. B Smls. Sch 40
1	3"	0' – 7-3/4"	IPE 1BE
1	3"	4' – 11-5/8"	2BE
1	3"	12' – 10-3/16	2BE

Total length 18' – 7"

Figure 6-2. Orthographic pipe spool. Courtesy of Texas Pipe Bending Co., Inc.

Pipe Fabrication

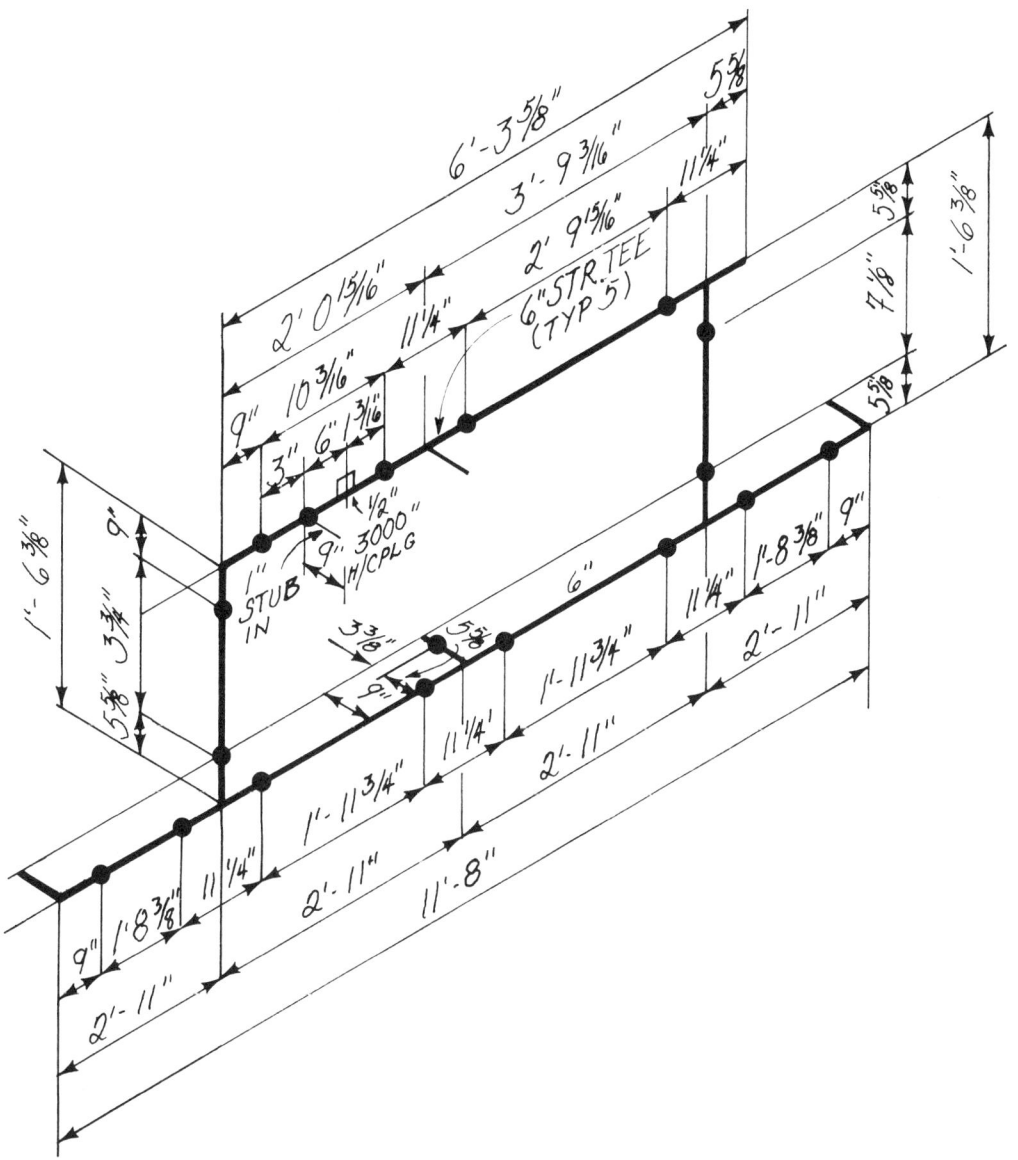

Material List

Quantity	Size	Description
3	6″	Std. Wt. LR 90° Ell
5	6″	Std. Wt. Str. Tee
1	1/2″	3000 lb. Thd'd Half Cplg.

All pipe to be ASTM A-106 Gr B. Smls Sch. 40

1	6″	0′ – 3-3/8″	2BE
1	6″	0′ – 3-3/4″	2BE
1	6″	0′ – 7-1/8″	2BE
1	6″	0′ – 10-3/16″	2BE
2	6″	1′ – 8-3/8″	2BE
2	6″	1′ – 11-3/4″	2BE
1	6″	2′ – 9-15/16″	2BE

Total length 12′ – 11″

Figure 6-3. Isometric pipe spool. Courtesy of Texas Pipe Bending Co., Inc.

98   Process Piping Design

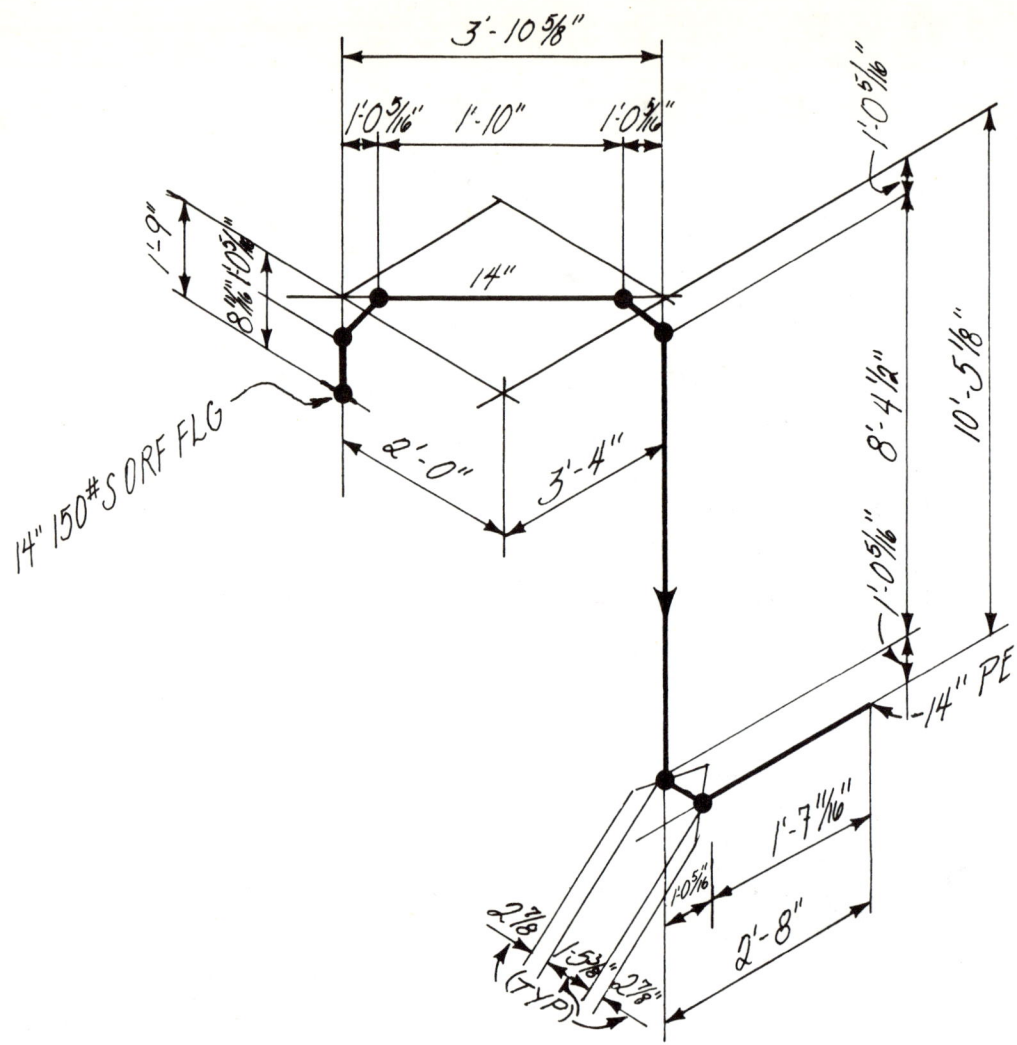

	Material List	
Quantity	Size	Description
1	14″	150 lb. SO RF Flg.

All pipe to be ASTM A-106 Gr. B. Smls. 0.375″ Wall
| 1 | 14″ | 16′ – 11-1/8″ | 2PE |

Figure 6-4. Isometric spool with mitres. Courtesy of Texas Pipe Bending Co., Inc.

Pipe Fabrication

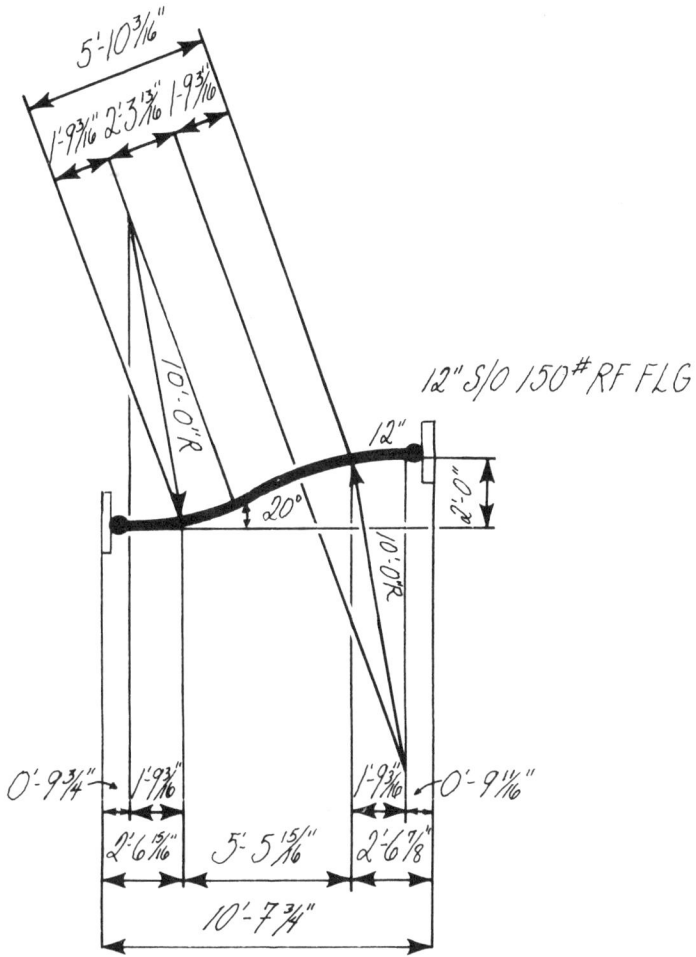

Quantity	Size	Description
2	12"	150 lb. SO RF Flg.

All pipe to be ASTM A-106 Gr. B. Smls 0.375" Wall
| 1 | 12" | 10' – 11-1/4" 2 PE |

Figure 6-5. Shop spool with bends. Courtesy of Texas Pipe Bending Co., Inc.

Process Piping Design

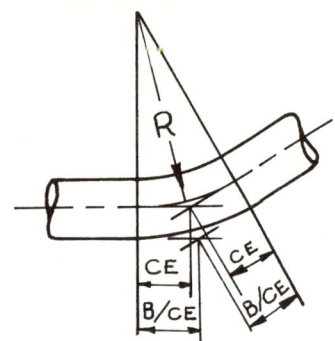

Table 6-1
Center-to-end (CE), Back Center-to-end (B/CE) and Arc Length for 30° Bends of Varying Radii and Pipe Sizes

Rad.	Feet — Inches Arc	(CE)	\(B/CE\) Nominal Pipe Sizes (inches) 2	3	4	5	6	8	10	12
6	3-1/8	1-5/8	1-15/16							
7	3-11/16	1-7/8	2-3/16							
8	4-3/16	2-1/8	2-7/16							
9	4-11/16	2-7/16	2-3/4	2-15/16						
10	5-1/4	2-11/16	3	3-3/16						
11	5-3/4	2-15/16	3-1/4	3-7/16						
1 — 0	6-5/16	3-3/16	3-1/2	3-11/16	3-13/16					
1 — 1	6-13/16	3-1/2	3-13/16	4	4-1/8					
1 — 2	7-5/16	3-3/4	4-1/16	4-1/4	4-3/8					
1 — 3	7-7/8	4	4-5/16	4-1/2	4-5/8	4-3/4				
1 — 4	8-3/8	4-5/16	4-5/8	4-13/16	4-15/16	5-1/16				
1 — 5	8-7/8	4-9/16	4-7/8	5-1/16	5-3/16	5-5/16				
1 — 6	9-7/16	4-13/16	5-1/8	5-5/16	5-7/16	5-9/16	5-11/16			
1 — 7	9-15/16	5-1/16	5-3/8	5-9/16	5-11/16	5-13/16	5-15/16			
1 — 8	10-1/2	5-3/8	5-11/16	5-7/8	6	6-1/8	6-1/4			
1 — 9	11	5-5/8	5-15/16	6-1/8	6-1/4	6-3/8	6-1/2			
1 — 10	11-1/2	5-7/8	6-3/16	6-3/8	6-1/2	6-5/8	6-3/4			
1 — 11	1 — 0-1/16	6-3/16	6-1/2	6-11/16	6-13/16	6-15/16	7-1/16			
2 — 0	1 — 0-9/16	6-7/16	6-3/4	6-15/16	7-1/16	7-3/16	7-5/16	7-9/16		
2 — 1	1 — 1-1/16	6-11/16	7	7-3/16	7-5/16	7-7/16	7-9/16	7-13/16		
2 — 2	1 — 1-5/8	6-15/16	7-1/4	7-7/16	7-9/16	7-11/16	7-13/16	8-1/16		
2 — 3	1 — 2-1/8	7-1/4	7-9/16	7-3/4	7-7/8	8	8-1/8	8-3/8		
2 — 4	1 — 2-5/8	7-1/2	7-13/16	8	8-1/8	8-1/4	8-3/8	8-5/8		
2 — 5	1 — 3-3/16	7-3/4	8-1/16	8-1/4	8-3/8	8-1/2	8-5/8	8-7/8		
2 — 6	1 — 3-11/16	8-1/16	8-3/8	8-9/16	8-11/16	8-13/16	8-15/16	9-3/16	9-1/2	
2 — 7	1 — 4-1/4	8-5/16	8-5/8	8-13/16	8-15/16	9-1/16	9-3/16	9-7/16	9-3/4	
2 — 8	1 — 4-3/4	8-9/16	8-7/8	9-1/16	9-3/16	9-5/16	9-7/16	9-11/16	10	
2 — 9	1 — 5-1/4	8-13/16	9-1/8	9-5/16	9-7/16	9-9/16	9-11/16	9-15/16	10-1/4	
2 — 10	1 — 5-13/16	9-1/8	9-7/16	9-5/8	9-3/4	9-7/8	10	10-1/4	10-9/16	
2 — 11	1 — 6-5/16	9-3/8	9-11/16	9-7/8	10	10-1/8	10-1/4	10-1/2	10-13/16	
3 — 0	1 — 6-7/8	9-5/8	9-15/16	10-1/8	10-1/4	10-3/8	10-1/2	10-3/4	11-1/16	11-5/16
3 — 1	1 — 7-3/8	9-15/16	10-1/4	10-7/16	10-9/16	10-11/16	10-13/16	11-1/16	11-3/8	11-5/8
3 — 2	1 — 7-7/8	10-3/16	10-1/2	10-11/16	10-13/16	10-15/16	11-1/16	11-5/16	11-5/8	11-7/8
3 — 3	1 — 8-7/16	10-7/16	10-3/4	10-15/16	11-1/16	11-3/16	11-5/16	11-9/16	11-7/8	1 — 0-1/8

Source: Texas Pipe Bending Co., Inc., Houston, Texas.

Pipe Fabrication

Table 6-2
Center-to-end (CE), Back Center-to-end (B/CE) and Arc Length for 30° Bends of Varying Radii and Pipe Sizes

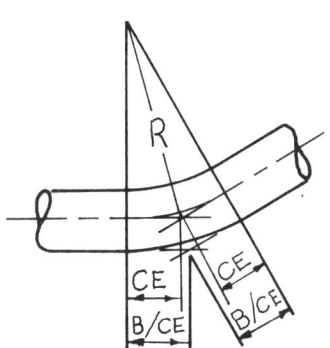

Feet – Inches			(B/CE) Nominal Pipe Sizes							
Rad.	Arc	(CE)	2	3	4	5	6	8	10	12
3 – 4	1 – 8-15/16	10-11/16	11	11-3/16	11-5/16	11-7/16	11-9/16	11-13/16	1 – 0-1/8	1 – 0-3/8
3 – 5	1 – 9-1/2	11	11-5/16	11-1/2	11-5/8	11-3/4	11-7/8	1 – 0-1/8	1 – 0-7/16	1 – 0-11/16
3 – 6	1 – 10	11-1/4	11-9/16	11-3/4	11-7/8	1 – 0	1 – 0-1/8	1 – 0-3/8	1 – 0-11/16	1 – 0-15/16
3 – 7	1 – 10-1/2	11-1/2	11-13/16	1 – 0	1 – 0-1/8	1 – 0-1/4	1 – 0-3/8	1 – 0-5/8	1 – 0-15/16	1 – 1-3/16
3 – 8	1 – 11-1/16	11-13/16	1 – 0-1/8	1 – 0-5/16	1 – 0-7/16	1 – 0-9/16	1 – 0-11/16	1 – 0-15/16	1 – 1-1/4	1 – 1-1/2
3 – 9	1 – 11-9/16	1 – 0-1/16	1 – 0-3/8	1 – 0-9/16	1 – 0-11/16	1 – 0-13/16	1 – 0-15/16	1 – 1-3/16	1 – 1-1/2	1 – 1-3/4
3 – 10	2 – 0-1/8	1 – 0-5/16	1 – 0-5/8	1 – 0-13/16	1 – 0-15/16	1 – 1-1/16	1 – 1-3/16	1 – 1-7/16	1 – 1-3/4	1 – 2
3 – 11	2 – 0-5/8	1 – 0-5/8	1 – 0-15/16	1 – 1-1/8	1 – 1-1/4	1 – 1-3/8	1 – 1-1/2	1 – 1-3/4	1 – 2-1/16	1 – 2-5/16
4 – 0	2 – 1-1/8	1 – 0-7/8	1 – 1-3/16	1 – 1-3/8	1 – 1-1/2	1 – 1-5/8	1 – 1-3/4	1 – 2	1 – 2-5/16	1 – 2-9/16
4 – 1	2 – 1-11/16	1 – 1-1/8	1 – 1-7/16	1 – 1-5/8	1 – 1-3/4	1 – 1-7/8	1 – 2	1 – 2-1/4	1 – 2-9/16	1 – 2-13/16
4 – 2	2 – 2-3/16	1 – 1-3/8	1 – 1-11/16	1 – 1-7/8	1 – 2	1 – 2-1/8	1 – 2-1/4	2 – 2-1/2	1 – 2-13/16	1 – 3-1/16
4 – 3	2 – 2-11/16	1 – 1-11/16	1 – 2	1 – 2-3/16	1 – 2-5/16	1 – 2-7/16	1 – 2-9/16	1 – 2-13/16	1 – 3-1/8	1 – 3-3/8
4 – 4	2 – 3-1/4	1 – 1-15/16	1 – 2-1/4	1 – 2-7/16	1 – 2-9/16	1 – 2-11/16	1 – 2-13/16	1 – 3-1/16	1 – 3-3/8	1 – 3-5/8
4 – 5	2 – 3-3/4	1 – 2-3/16	1 – 2-1/2	1 – 2-11/16	1 – 2-13/16	1 – 2-15/16	1 – 3-1/16	1 – 3-5/16	1 – 3-5/8	1 – 3-7/8
4 – 6	2 – 4-1/4	1 – 2-1/2	1 – 2-13/16	1 – 3	1 – 3-1/8	1 – 3-1/4	1 – 3-3/8	1 – 3-5/8	1 – 3-15/16	1 – 4-3/16
4 – 7	2 – 4-13/16	1 – 2-3/4	1 – 3-1/16	1 – 3-1/4	1 – 3-3/8	1 – 3-1/2	1 – 3-5/8	1 – 3-7/8	1 – 4-3/16	1 – 4-7/16
4 – 8	2 – 5-5/16	1 – 3	1 – 3-5/16	1 – 3-1/2	1 – 3-5/8	1 – 3-3/4	1 – 3-7/8	1 – 4-1/8	1 – 4-7/16	1 – 4-11/16
4 – 9	2 – 5-7/8	1 – 3-1/4	1 – 3-9/16	1 – 3-3/4	1 – 3-7/8	1 – 4	1 – 4-1/8	1 – 4-3/8	1 – 4-11/16	1 – 4-15/16
4 – 10	2 – 6-3/8	1 – 3-9/16	1 – 3-7/8	1 – 4-1/16	1 – 4-3/16	1 – 4-1/4	1 – 4-3/8	1 – 4-5/8	1 – 4-15/16	1 – 5-3/16
4 – 11	2 – 6-7/8	1 – 3-13/16	1 – 4-1/8	1 – 4-5/16	1 – 4-7/16	1 – 4-9/16	1 – 4-11/16	1 – 4-15/16	1 – 5-1/4	1 – 5-1/2
5 – 0	2 – 7-7/16	1 – 4-1/16	1 – 4-3/8	1 – 4-9/16	1 – 4-11/16	1 – 4-13/16	1 – 4-15/16	1 – 5-3/16	1 – 5-1/2	1 – 5-3/4
5 – 1	2 – 7-13/16	1 – 4-3/8	1 – 4-11/16	1 – 4-7/8	1 – 5	1 – 5-1/8	1 – 5-1/4	1 – 5-1/2	1 – 5-13/16	1 – 6-1/16
5 – 2	2 – 8-7/16	1 – 4-5/8	1 – 4-15/16	1 – 5-1/8	1 – 5-1/4	1 – 5-3/8	1 – 5-1/2	1 – 5-3/4	1 – 6-1/16	1 – 6-5/16
5 – 3	2 – 9	1 – 4-7/8	1 – 5-3/16	1 – 5-3/8	1 – 5-1/2	1 – 5-5/8	1 – 5-3/4	1 – 6	1 – 6-5/16	1 – 6-9/16
5 – 4	2 – 9-1/2	1 – 5-1/8	1 – 5-7/16	1 – 5-5/8	1 – 5-3/4	1 – 5-7/8	1 – 6	1 – 6-1/4	1 – 6-9/16	1 – 6-13/16
5 – 5	2 – 10-1/16	1 – 5-7/16	1 – 5-3/4	1 – 5-15/16	1 – 6-1/16	1 – 6-3/16	1 – 6-5/16	1 – 6-9/16	1 – 6-7/8	1 – 7-1/8
5 – 6	2 – 10-9/16	1 – 5-11/16	1 – 6	1 – 6-3/16	1 – 6-5/16	1 – 6-7/16	1 – 6-9/16	1 – 6-13/16	1 – 7-1/8	1 – 7-3/8
5 – 7	2 – 11-1/16	1 – 5-15/16	1 – 6-1/4	1 – 6-7/16	1 – 6-9/16	1 – 6-11/16	1 – 6-13/16	1 – 7-1/16	1 – 7-3/8	1 – 7-5/8
5 – 8	2 – 11-5/8	1 – 6-1/4	1 – 6-9/16	1 – 6-3/4	1 – 6-7/8	1 – 7	1 – 7-1/8	1 – 7-3/8	1 – 7-11/16	1 – 7-15/16
5 – 9	3 – 0-1/8	1 – 6-1/2	1 – 6-13/16	1 – 7	1 – 7-1/8	1 – 7-1/4	1 – 7-3/8	1 – 7-5/8	1 – 7-15/16	1 – 8-3/16
5 – 10	3 – 0-5/8	1 – 6-3/4	1 – 7-1/16	1 – 7-1/4	1 – 7-3/8	1 – 7-1/2	1 – 7-5/8	1 – 7-7/8	1 – 8-3/16	1 – 8-7/16
5 – 11	3 – 1-3/16	1 – 7	1 – 7-5/16	1 – 7-1/2	1 – 7-5/8	1 – 7-3/4	1 – 7-7/8	1 – 8-1/8	1 – 8-7/16	1 – 8-11/16
6 – 0	3 – 1-11/16	1 – 7-5/16	1 – 7-5/8	1 – 7-13/16	1 – 7-15/16	1 – 8-1/16	1 – 8-3/16	1 – 8-7/16	1 – 8-3/4	1 – 9

Source: Texas Pipe Bending Co., Inc., Houston, Texas.

Table 6-3
Center-to-End (CE), Back Center-to-End (B/CE) and Arch Length for 45° Bends of Varying Radii and Pipe Sizes

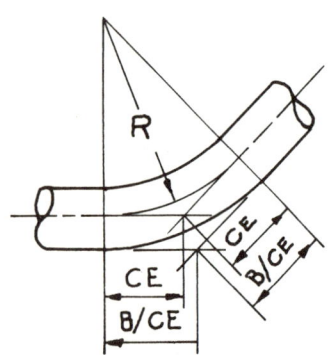

	Feet — Inches		(B/CE) Nominal Pipe Sizes							
Rad.	Arc	CE	2	3	4	5	6	8	10	12
6	4-11/16	2-1/2	3							
7	5-1/2	2-7/8	3-3/8							
8	6-5/16	3-5/16	3-13/16							
9	7-1/16	3-3/4	4-1/4	4-1/2						
10	7-7/8	4-1/8	4-5/8	4-7/8						
11	8-5/8	4-9/16	5-1/16	5-5/16						
1 – 0	9-7/16	5	5-1/2	5-3/4	5-15/16					
1 – 1	10-3/16	5-3/8	5-7/8	6-1/8	6-5/16					
1 – 2	11	5-13/16	6-5/16	6-9/16	6-3/4					
1 – 3	11-3/4	6-3/16	6-11/16	6-15/16	7-1/8	7-5/16				
1 – 4	1 – 0-9/16	6-5/8	7-1/8	7-3/8	7-9/16	7-3/4				
1 – 5	1 – 1-3/8	7-1/16	7-9/16	7-13/16	8	8-3/16				
1 – 6	1 – 2-1/8	7-7/16	7-15/16	8-3/16	8-3/8	8-9/16	8-13/16			
1 – 7	1 – 2-15/16	7-7/8	8-3/8	8-5/8	8-13/16	9	9-1/4			
1 – 8	1 – 3-11/16	8-5/16	8-13/16	9-1/16	9-1/4	9-7/16	9-11/16			
1 – 9	1 – 4-1/2	8-11/16	9-3/16	9-7/16	9-5/8	9-13/16	10-1/16			
1 – 10	1 – 5-1/4	9-1/8	9-5/8	9-7/8	10-1/16	10-1/4	10-1/2			
1 – 11	1 – 6-1/16	9-1/2	10	10-1/4	10-7/16	10-11/16	10-15/16			
2 – 0	1 – 6-7/8	9-15/16	10-7/16	10-11/16	10-7/8	11-1/16	11-5/16	11-3/4		
2 – 1	1 – 7-5/8	10-3/8	10-7/8	11-1/8	11-5/16	11-1/2	11-3/4	1 – 0-3/16		
2 – 2	1 – 8-7/16	10-3/4	11-1/4	11-1/2	11-11/16	11-7/8	1 – 0-1/8	1 – 0-9/16		
2 – 3	1 – 9-3/16	11-3/16	11-11/16	11-15/16	1 – 0-1/8	1 – 0-5/16	1 – 0-9/16	1 – 1		
2 – 4	1 – 10	11-5/8	1 – 0-1/8	1 – 0-3/8	1 – 0-9/16	1 – 0-3/4	1 – 1	1 – 1-7/16		
2 – 5	1 – 10-3/4	1 – 0	1 – 0-1/2	1 – 0-3/4	1 – 0-15/16	1 – 1-1/8	1 – 1-3/8	1 – 1-13/16		
2 – 6	1 – 11-9/16	1 – 0-7/16	1 – 0-15/16	1 – 1-3/16	1 – 1-3/8	1 – 1-9/16	1 – 1-13/16	1 – 2-1/4	1 – 2-11/16	
2 – 7	2 – 0-3/8	1 – 0-13/16	1 – 1-5/16	1 – 1-9/16	1 – 1-3/4	1 – 1-15/16	1 – 2-3/16	1 – 2-5/8	1 – 3-1/16	
2 – 8	2 – 1-1/8	1 – 1-1/4	1 – 1-3/4	1 – 2	1 – 2-3/16	1 – 2-3/8	1 – 2-5/8	1 – 3-1/16	1 – 3-1/2	
2 – 9	2 – 1-15/16	1 – 1-11/16	1 – 2-3/16	1 – 2-7/16	1 – 2-5/8	1 – 2-13/16	1 – 3-1/16	1 – 3-1/2	1 – 3-15/16	
2 – 10	2 – 2-11/16	1 – 2-1/16	1 – 2-9/16	1 – 2-13/16	1 – 3	1 – 3-3/16	1 – 3-7/16	1 – 3-7/8	1 – 4-5/16	
2 – 11	2 – 3-1/2	1 – 2-1/2	1 – 3	1 – 3-1/4	1 – 3-7/16	1 – 3-5/8	1 – 3-7/8	1 – 4-5/16	1 – 4-3/4	
3 – 0	2 – 4-1/4	1 – 2-15/16	1 – 3-7/16	1 – 3-11/16	1 – 3-7/8	1 – 4-1/16	1 – 4-5/16	1 – 4-3/4	1 – 5-3/16	1 – 5-9/16
3 – 1	2 – 5-1/16	1 – 3-5/16	1 – 3-13/16	1 – 4-1/16	1 – 4-1/4	1 – 4-7/16	1 – 4-11/16	1 – 5-1/8	1 – 5-9/16	1 – 5-15/16
3 – 2	2 – 5-7/8	1 – 3-3/4	1 – 4-1/4	1 – 4-1/2	1 – 4-11/16	1 – 4-7/8	1 – 5-1/8	1 – 5-9/16	1 – 6	1 – 6-3/8
3 – 3	2 – 6-5/8	1 – 4-1/8	1 – 4-5/8	1 – 4-7/8	1 – 5-1/16	1 – 5-1/4	1 – 5-1/2	1 – 5-15/16	1 – 6-3/8	1 – 6-3/4

Source: Texas Pipe Bending Co., Inc., Houston, Texas.

Pipe Fabrication

Table 6-4.
Center-to-end (CE), Back Center-to-end (B/CE) and Arc Length for 45° Bends of Varying Radii and Pipe Sizes

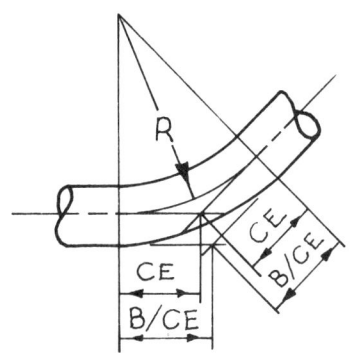

Feet — Inches			(B/CE) Nominal Pipe Sizes							
Rad.	Arc	(CE)	2	3	4	5	6	8	10	12
3 — 4	2 — 7-7/16	1 — 4-9/16	1 — 5-1/16	1 — 5-5/16	1 — 5-1/2	1 — 5-11/16	1 — 5-15/16	1 — 6-3/8	1 — 6-13/16	1 — 7-3/16
3 — 5	2 — 8-3/16	1 — 5	1 — 5-1/2	1 — 5-3/4	1 — 5-15/16	1 — 6-1/8	1 — 6-3/8	1 — 6-13/16	1 — 7-1/4	1 — 7-5/8
3 — 6	2 — 9	1 — 5-3/8	1 — 5-7/8	1 — 6-1/8	1 — 6-5/16	1 — 6-1/2	1 — 6-3/4	1 — 7-3/16	1 — 7-5/8	1 — 8
3 — 7	2 — 9-3/4	1 — 5-13/16	1 — 6-5/16	1 — 6-9/16	1 — 6-3/4	1 — 6-15/16	1 — 7-3/16	1 — 7-5/8	1 — 8-1/16	1 — 8-7/16
3 — 8	2 — 10-9/16	1 — 6-1/4	1 — 6-3/4	1 — 7	1 — 7-3/16	1 — 7-3/8	1 — 7-5/8	1 — 8-1/16	1 — 8-1/2	1 — 8-7/8
3 — 9	2 — 11-5/16	1 — 6-5/8	1 — 7-1/8	1 — 7-3/8	1 — 7-9/16	1 — 7-3/4	1 — 8	1 — 8-7/16	1 — 8-7/8	1 — 9-1/4
3 — 10	3 — 0-1/8	1 — 7-1/16	1 — 7-9/16	1 — 7-13/16	1 — 8	1 — 8-3/16	1 — 8-7/16	1 — 8-7/8	1 — 9-5/16	1 — 9-11/16
3 — 11	3 — 0-15/16	1 — 7-7/16	1 — 7-15/16	1 — 8-3/16	1 — 8-3/8	1 — 8-9/16	1 — 8-13/16	1 — 9-1/4	1 — 9-11/16	1 — 10-1/16
4 — 0	3 — 1-11/16	1 — 7-7/8	1 — 8-3/8	1 — 8-5/8	1 — 8-13/16	1 — 9	1 — 9-1/4	1 — 9-11/16	1 — 10-1/8	1 — 10-1/2
4 — 1	3 — 2-1/2	1 — 8-5/16	1 — 8-13/16	1 — 9-1/16	1 — 9-1/4	1 — 9-7/16	1 — 9-11/16	1 — 10-1/8	1 — 10-9/16	1 — 10-15/16
4 — 2	3 — 3-1/4	1 — 8-11/16	1 — 9-3/16	1 — 9-7/16	1 — 9-5/8	1 — 9-13/16	1 — 10-1/16	1 — 10-1/2	1 — 1-015/16	1 — 11-5/16
4 — 3	3 — 4-1/16	1 — 9-1/8	1 — 9-5/8	1 — 9-7/8	1 — 10-1/16	1 — 10-1/4	1 — 10-1/2	1 — 10-15/16	1 — 11-3/8	1 — 11-3/4
4 — 4	3 — 4-13/16	1 — 9-9/16	1 — 10-1/16	1 — 10-5/16	1 — 10-1/2	1 — 10-11/16	1 — 10-15/16	1 — 11-3/8	1 — 11-13/16	2 — 0-3/16
4 — 5	3 — 5-5/8	1 — 9-15/16	1 — 10-7/16	1 — 10-11/16	1 — 10-7/8	1 — 11-1/16	1 — 11-5/16	1 — 11-3/4	2 — 0-3/16	2 — 0-9/16
4 — 6	3 — 6-7/16	1 — 10-3/8	1 — 10-7/8	1 — 11-1/8	1 — 11-5/16	1 — 11-1/2	1 — 11-3/4	2 — 0-3/16	2 — 0-5/8	2 — 1
4 — 7	3 — 7-3/16	1 — 10-13/16	1 — 11-5/16	1 — 11-9/16	1 — 11-3/4	1 — 11-15/16	2 — 0-3/16	2 — 0-5/8	2 — 1-1/16	2 — 1-7/16
4 — 8	3 — 8	1 — 11-3/16	1 — 11-11/16	1 — 11-15/16	2 — 0-1/8	2 — 0-5/16	2 — 0-9/16	2 — 1	2 — 1-7/16	2 — 1-13/16
4 — 9	3 — 8-3/4	1 — 11-5/8	2 — 0-1/8	2 — 0-3/8	2 — 0-9/16	2 — 0-3/4	2 — 1	2 — 1-7/16	2 — 1-7/8	2 — 2-1/4
4 — 10	3 — 9-9/16	2 — 0	2 — 0-1/2	2 — 0-3/4	2 — 0-15/16	2 — 1-1/8	2 — 1-3/8	2 — 1-13/16	2 — 2-1/4	2 — 2-5/8
4 — 11	3 — 10-5/16	2 — 0-7/16	2 — 0-15/16	2 — 1-3/16	2 — 1-3/8	2 — 1-9/16	2 — 1-13/16	2 — 2-1/4	2 — 2-11/16	2 — 3-1/16
5 — 0	3 — 11-1/8	2 — 0-7/8	2 — 1-3/8	2 — 1-5/8	2 — 1-13/16	2 — 2	2 — 2-1/4	2 — 2-11/16	2 — 3-1/8	2 — 3-1/2
5 — 1	3 — 11-15/16	2 — 1-1/4	2 — 1-3/4	2 — 2	2 — 2-3/16	2 — 3-3/8	2 — 2-5/8	2 — 3-1/16	2 — 3-1/2	2 — 3-7/8
5 — 2	4 — 0-11/16	2 — 1-11/16	2 — 2-3/16	2 — 2-7/16	2 — 2-5/8	2 — 2-13/16	2 — 3-1/16	2 — 3-1/2	2 — 3-15/16	2 — 4-5/16
5 — 3	4 — 1-1/2	2 — 2-1/8	2 — 2-5/8	2 — 2-7/8	2 — 3-1/16	2 — 3-1/4	2 — 3-1/2	2 — 3-15/16	2 — 4-3/8	2 — 4-3/4
5 — 4	4 — 2-1/4	2 — 2-1/2	2 — 3	2 — 3-1/4	2 — 3-7/16	2 — 3-5/8	2 — 3-7/8	2 — 4-5/16	2 — 4-3/4	2 — 5-1/8
5 — 5	4 — 3-1/16	2 — 2-15/16	2 — 3-7/16	2 — 3-11/16	2 — 3-7/8	2 — 4-1/16	2 — 4-5/16	2 — 4-3/4	2 — 5-3/16	2 — 5-9/16
5 — 6	4 — 3-13/16	2 — 3-5/16	2 — 3-13/16	2 — 4-1/16	2 — 4-1/4	2 — 4-7/16	2 — 4-11/16	2 — 5-1/8	2 — 5-9/16	2 — 6-15/16
5 — 7	4 — 4-5/8	2 — 3-3/4	2 — 4-1/4	2 — 4-1/2	2 — 4-11/16	2 — 4-7/8	2 — 5-1/8	2 — 5-9/16	2 — 6	2 — 6-3/8
5 — 8	4 — 5-7/16	2 — 4-3/16	2 — 4-11/16	2 — 4-15/16	2 — 5-1/8	2 — 5-5/16	2 — 5-9/16	2 — 6	2 — 6-7/16	2 — 6-13/16
5 — 9	4 — 6-3/16	2 — 4-9/16	2 — 5-1/16	2 — 5-5/16	2 — 5-1/2	2 — 5-11/16	2 — 5-15/16	2 — 6-3/8	2 — 6-13/16	2 — 7-3/16
5 — 10	4 — 7	2 — 5	2 — 5-1/2	2 — 5-3/4	2 — 5-15/16	2 — 6-1/8	2 — 6-3/8	2 — 6-13/16	2 — 7-1/4	2 — 7-5/8
5 — 11	4 — 7-3/4	2 — 5-3/8	2 — 5-7/8	2 — 6-1/8	2 — 6-5/16	2 — 6-1/2	2 — 6-3/4	2 — 7-3/16	2 — 7-5/8	2 — 8
6 — 0	4 — 8-9/16	2 — 5-13/16	2 — 6-5/16	2 — 6-9/16	2 — 6-3/4	2 — 6-15/16	2 — 7-3/16	2 — 7-5/8	2 — 8-1/16	2 — 8-7/16

Source: Texas Pipe Bending Co., Inc., Houston, Texas.

Table 6-5
Center-to-end (CE), Back Center-to-end (B/CE) and Arc Length for 60° Bends of Varying Radii and Pipe Sizes

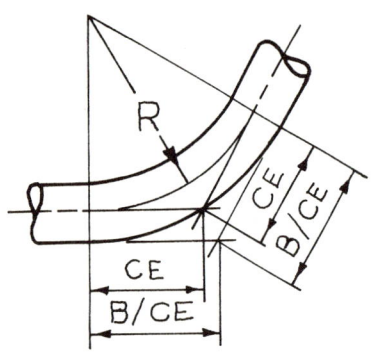

	Feet – Inches		(B/CE) Nominal Pipe Sizes							
Rad.	Arc	(CE)	2	3	4	5	6	8	10	12
6	6-5/16	3-7/16	4-1/8							
7	7-5/16	4-1/16	4-3/4							
8	8-3/8	4-5/8	5-5/16							
9	9-7/16	5-3/16	5-7/8	6-3/16						
10	10-1/2	5-3/4	6-7/16	6-3/4						
11	11-1/2	6-3/8	7-1/16	7-3/8						
1 – 0	1 – 0-9/16	6-15/16	7-5/8	7-15/16	8-1/4					
1 – 1	1 – 1-5/8	7-1/2	8-3/16	8-1/2	8-13/16					
1 – 2	1 – 2-5/8	8-1/16	8-3/4	9-1/16	9-3/8					
1 – 3	1 – 3-11/16	8-11/16	9-3/8	9-11/16	10	10-5/16				
1 – 4	1 – 4-3/4	9-1/4	9-15/16	10-1/4	10-9/16	10-7/8				
1 – 5	1 – 5-13/16	9-13/16	10-1/2	10-13/16	11-1/8	11-7/16				
1 – 6	1 – 6-7/8	10-3/8	11-1/16	11-3/8	11-11/16	1 – 0	1 – 0-5/16			
1 – 7	1 – 7-7/8	11	11-11/16	1 – 0	1 – 0-5/16	1 – 0-5/8	1 – 0-15/16			
1 – 8	1 – 8-15/16	11-9/16	1 – 0-1/4	1 – 0-9/16	1 – 0-7/8	1 – 1-3/16	1 – 1-1/2			
1 – 9	1 – 10	1 – 0-1/8	1 – 0-13/16	1 – 1-1/8	1 – 1-7/16	1 – 1-3/4	1 – 2-1/16			
1 – 10	1 – 11-1/16	1 – 0-11/16	1 – 1-3/8	1 – 1-11/16	1 – 2	1 – 2-5/16	1 – 2-5/8			
1 – 11	2 – 0-1/8	1 – 1-1/4	1 – 1-15/16	1 – 2-1/4	1 – 2-9/16	1 – 2-7/8	1 – 3-3/16			
2 – 0	2 – 1-1/8	1 – 1-7/8	1 – 2-9/16	1 – 2-7/8	1 – 3-3/16	1 – 3-1/2	1 – 3-13/16	1 – 4-3/8		
2 – 1	2 – 2-3/16	1 – 2-7/16	1 – 3-1/16	1 – 3-7/16	1 – 3-3/4	1 – 4-1/16	1 – 4-3/8	1 – 4-15/16		
2 – 2	2 – 3-1/4	1 – 3	1 – 3-11/16	1 – 4	1 – 4-5/16	1 – 4-5/8	1 – 4-15/16	1 – 5-1/2		
2 – 3	2 – 4-1/4	1 – 3-9/16	1 – 4-1/4	1 – 4-9/16	1 – 4-7/8	1 – 5-3/16	1 – 5-1/2	1 – 6-1/16		
2 – 4	2 – 5-5/16	1 – 4-3/16	1 – 4-7/8	1 – 5-3/16	1 – 5-1/2	1 – 5-13/16	1 – 6-1/8	1 – 6-11/16		
2 – 5	2 – 6-3/8	1 – 4-3/4	1 – 5-7/16	1 – 5-3/4	1 – 6-1/16	1 – 6-3/8	1 – 6-11/16	1 – 7-1/4		
2 – 6	2 – 7-7/16	1 – 5-5/16	1 – 6	1 – 6-5/16	1 – 6-5/8	1 – 6-15/16	1 – 7-1/4	1 – 7-13/16	1 – 8-7/16	
2 – 7	2 – 8-7/16	1 – 5-7/8	1 – 6-9/16	1 – 6-7/8	1 – 7-3/16	1 – 7-1/2	1 – 7-13/16	1 – 8-3/8	1 – 9	
2 – 8	2 – 9-1/2	1 – 6-1/2	1 – 7-3/16	1 – 7-1/2	1 – 7-13/16	1 – 8-1/8	1 – 8-7/16	1 – 9	1 – 9-5/8	
2 – 9	2 – 10-9/16	1 – 7-1/16	1 – 7-3/4	1 – 8-1/16	1 – 8-3/8	1 – 8-11/16	1 – 9	1 – 9-9/16	1 – 10-3/16	
2 – 10	2 – 11-5/8	1 – 7-5/8	1 – 8-5/16	1 – 8-5/8	1 – 8-15/16	1 – 9-1/4	1 – 9-9/16	1 – 10-1/8	1 – 10-3/4	
2 – 11	3 – 0-5/8	1 – 8-3/16	1 – 8-7/8	1 – 9-3/16	1 – 9-1/2	1 – 9-13/16	1 – 10-1/8	1 – 10-11/16	1 – 11-5/16	
3 – 0	3 – 1-11/16	1 – 8-13/16	1 – 9-1/2	1 – 9-13/16	1 – 10-1/8	1 – 10-7/16	1 – 10-3/4	1 – 11-5/16	1 – 11-15/16	2 – 0-1/2
3 – 1	3 – 2-3/4	1 – 9-3/8	1 – 10-1/16	1 – 10-3/8	1 – 10-11/16	1 – 11	1 – 11-5/16	1 – 11-7/8	2 – 0-1/2	2 – 1-1/16
3 – 2	3 – 3-13/16	1 – 9-15/16	1 – 10-5/8	1 – 10-15/16	1 – 11-1/4	1 – 11-9/16	1 – 11-7/8	2 – 0-7/16	2 – 1-1/16	2 – 1-5/8
3 – 3	3 – 4-13/16	1 – 10-1/2	1 – 11-3/16	1 – 11-1/2	1 – 11-13/16	2 – 0-1/8	2 – 0-7/16	2 – 1	2 – 1-5/8	2 – 2-3/16

Source: Texas Pipe Bending Co., Inc., Houston, Texas.

Pipe Fabrication

Table 6-6
Center-to-end (CE), Back Center-to-end (B/CE), and Arc Length for 60° Bends of Varying Radii and Pipe Size

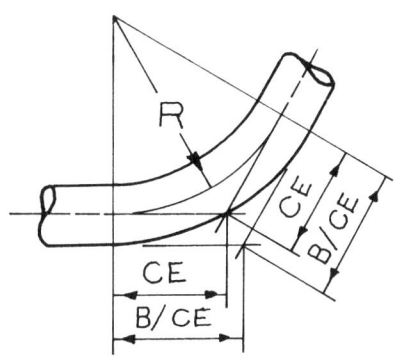

	Feet – Inches		(B/CE) Nominal Pipe Sizes							
Rad.	Arc	(CE)	2	3	4	5	6	8	10	12
3 – 4	3 – 5-7/8	1 – 11-1/8	1 – 11-13/16	2 – 0-1/8	2 – 0-7/16	2 – 0-3/4	2 – 1-1/16	2 – 1-5/8	1 – 2-1/4	2 – 2-13/16
3 – 5	3 – 6-15/16	1 – 11-11/16	2 – 0-3/8	2 – 0-11/16	2 – 1	2 – 1-5/16	2 – 1-5/8	2 – 2-3/16	2 – 2-13/16	2 – 3-3/8
3 – 6	3 – 8	2 – 0-1/4	2 – 0-15/16	2 – 1-1/4	2 – 1-9/16	2 – 1-7/8	2 – 2-3/16	2 – 2-3/4	2 – 3-3/8	2 – 3-15/16
3 – 7	3 – 9	2 – 0-13/16	2 – 1-1/2	2 – 1-13/16	2 – 2-1/8	2 – 2-7/16	2 – 2-3/4	2 – 3-5/16	2 – 3-15/16	2 – 4-1/2
3 – 8	3 – 10-1/16	2 – 1-3/8	2 – 2-1/16	2 – 2-3/8	2 – 2-11/16	2 – 3	2 – 3-5/16	2 – 3-7/8	2 – 4-1/2	2 – 5-1/16
3 – 9	3 – 11-1/8	2 – 2	2 – 2-11/16	2 – 3	2 – 3-5/16	2 – 3-5/8	2 – 3-15/16	2 – 4-1/2	2 – 5-1/8	2 – 5-11/16
3 – 10	4 – 0-3/16	2 – 2-9/16	2 – 3-1/4	2 – 3-9/16	2 – 3-7/8	2 – 4-3/16	2 – 4-1/2	2 – 5-1/16	2 – 5-11/16	2 – 6-1/4
3 – 11	4 – 1-3/16	2 – 3-1/8	2 – 3-13/16	2 – 4-1/8	2 – 4-7/16	2 – 4-3/4	2 – 5-1/16	2 – 5-5/8	2 – 6-1/4	2 – 6-13/16
4 – 0	4 – 2-1/4	2 – 3-11/16	2 – 4-3/8	2 – 4-11/16	2 – 5	2 – 5-5/16	2 – 5-5/8	2 – 6-3/16	2 – 6-13/16	2 – 7-3/8
4 – 1	4 – 3-5/16	2 – 4-5/16	2 – 5	2 – 5-5/16	2 – 5-5/8	2 – 5-15/16	2 – 6-1/4	2 – 6-13/16	2 – 7-7/16	2 – 8
4 – 2	4 – 4-3/8	2 – 4-7/8	2 – 5-9/16	2 – 5-7/8	2 – 6-3/16	2 – 6-1/2	2 – 6-13/16	2 – 7-3/8	2 – 8	2 – 8-9/16
4 – 3	4 – 5-7/16	2 – 5-7/16	2 – 6-1/8	2 – 6-7/16	2 – 6-3/4	2 – 7-1/16	2 – 7-3/8	2 – 7-15/16	2 – 8-9/16	2 – 9-1/8
4 – 4	4 – 6-7/16	2 – 6	2 – 6-11/16	2 – 7	2 – 7-5/16	2 – 7-5/8	2 – 7-15/16	2 – 8-1/2	2 – 9-1/8	2 – 9-11/16
4 – 5	4 – 7-1/2	2 – 6-5/8	2 – 7-5/16	2 – 7-5/8	2 – 7-15/16	2 – 8-1/4	2 – 8-9/16	2 – 9-1/8	2 – 9-3/4	2 – 10-5/16
4 – 6	4 – 8-9/16	2 – 7-3/16	2 – 7-7/8	2 – 8-3/16	2 – 8-1/2	2 – 8-13/16	2 – 9-1/8	2 – 9-11/16	2 – 10-5/16	2 – 10-7/8
4 – 7	4 – 9-5/8	2 – 7-3/4	2 – 8-7/16	2 – 8-3/4	2 – 9-1/16	2 – 9-3/8	2 – 9-11/16	2 – 10-1/4	2 – 10-7/8	2 – 11-7/16
4 – 8	4 – 10-5/8	2 – 8-5/16	2 – 9	2 – 9-5/16	2 – 9-5/8	2 – 9-15/16	2 – 10-1/4	2 – 10-13/16	2 – 11-7/16	3 – 0
4 – 9	4 – 11-11/16	2 – 8-15/16	2 – 9-5/8	2 – 9-15/16	2 – 10-1/4	2 – 10-9/16	2 – 10-7/8	2 – 11-7/16	2 – 0-1/16	3 – 0-5/8
4 – 10	5 – 0-3/4	2 – 9-1/2	2 – 10-3/16	2 – 10-1/2	2 – 10-13/16	2 – 11-1/8	2 – 11-7/16	3 – 0	3 – 0-5/8	3 – 1-3/16
4 – 11	5 – 1-13/16	2 – 10-1/16	2 – 10-3/4	2 – 11-1/16	2 – 11-3/8	2 – 11-11/16	3 – 0	3 – 0-9/16	3 – 1-3/16	3 – 1-3/4
5 – 0	5 – 2-13/16	2 – 10-5/8	2 – 11-5/16	2 – 11-5/8	2 – 11-15/16	3 – 0-1/4	3 – 0-9/16	3 – 1-1/8	3 – 1-3/4	3 – 2-5/16
5 – 1	5 – 3-7/8	2 – 11-1/4	2 – 11-15/16	3 – 0-1/4	3 – 0-9/16	3 – 0-7/8	3 – 1-3/16	3 – 1-3/4	3 – 2-3/8	3 – 2-15/16
5 – 2	5 – 4-15/16	2 – 11-13/16	3 – 0-1/2	3 – 0-13/16	3 – 1-1/8	3 – 1-7/16	3 – 1-3/4	3 – 2-5/16	3 – 2-15/16	3 – 3-1/2
5 – 3	5 – 6	3 – 0-3/8	3 – 1-1/16	3 – 1-3/8	3 – 1-11/16	3 – 2	3 – 2-5/16	3 – 2-7/8	3 – 3-1/2	3 – 4-1/16
5 – 4	5 – 7	3 – 0-15/16	3 – 1-5/8	3 – 1-15/16	3 – 2-1/4	3 – 2-9/16	3 – 2-7/8	3 – 3-7/16	3 – 4-1/16	3 – 4-5/8
5 – 5	5 – 8-1/16	3 – 1-1/2	3 – 2-3/16	3 – 2-1/2	2 – 2-13/16	3 – 3-1/8	3 – 3-7/16	3 – 4	3 – 4-5/8	3 – 5-3/16
5 – 6	5 – 9-1/8	3 – 2-1/8	3 – 2-13/16	3 – 3-1/8	3 – 3-7/16	3 – 3-3/4	3 – 4-1/16	3 – 4-5/8	3 – 5-1/4	3 – 5-13/16
5 – 7	5 – 10-3/16	3 – 2-11/16	3 – 3-3/8	3 – 3-11/16	3 – 4	3 – 4-5/16	3 – 4-5/8	3 – 5-3/16	3 – 5-13/16	3 – 6-3/8
5 – 8	5 – 11-3/16	3 – 3-1/4	3 – 3-15/16	3 – 4-1/4	3 – 4-9/16	3 – 4-7/8	3 – 5-3/16	3 – 5-3/4	3 – 6-3/8	3 – 6-15/16
5 – 9	6 – 0-1/4	3 – 3-13/16	3 – 4-1/2	3 – 4-13/16	3 – 5-1/8	3 – 5-7/16	3 – 5-3/4	3 – 6-5/16	3 – 6-15/16	3 – 7-1/2
5 – 10	6 – 1-5/16	3 – 4-7/16	3 – 5-1/8	3 – 5-7/16	3 – 5-3/4	3 – 6-1/16	3 – 6-3/8	3 – 6-15/16	3 – 7-9/16	3 – 8-1/8
5 – 11	6 – 2-3/8	3 – 5	3 – 5-11/16	3 – 6	3 – 6-5/16	3 – 6-5/8	3 – 6-15/16	3 – 7-1/2	3 – 8-1/8	3 – 8-11/16
6 – 0	6 – 3-3/8	3 – 5-9/16	3 – 6-1/4	3 – 6-9/16	3 – 6-7/8	3 – 7-3/16	3 – 7-1/2	3 – 8-1/16	3 – 8-11/16	3 – 9-1/4

Source: Texas Pipe Bending Co., Inc., Houston, Texas.

Table 6-7
Arc for 90° Bends

Radius (feet)

Radius (inches)	0 – 0	1 – 0	2 – 0	3 – 0	4 – 0	5 – 0	6 – 0	7 – 0	8 – 0	Rad. (inches)	Arc (inches)
0		1 – 6-7/8	3 – 1-11/16	4 – 8-9/16	6 – 3-3/8	7 – 10-1/4	9 – 5-1/8	10 – 11-15/16	12 – 6-13/16	1-1/4	1-15/16
1	1-9/16	1 – 8-7/16	3 – 3-1/4	4 – 10-1/8	6 – 5	7 – 11-13/16	9 – 6-11/16	11 – 1-1/2	12 – 8-3/8	1-1/2	2-3/8
2	3-1/8	1 – 10	3 – 4-13/16	4 – 11-11/16	6 – 6-9/16	8 – 1-3/8	9 – 8-1/4	11 – 3-1/16	12 – 9-15/16	1-3/4	2-3/4
3	4-11/16	1 – 11-9/16	3 – 6-7/16	5 – 1-1/4	6 – 8-1/8	8 – 2-15/16	9 – 9-13/16	11 – 4-11/16	12 – 11-1/2	2-1/4	3-9/16
4	6-5/16	2 – 1-1/8	3 – 8	5 – 2-13/16	6 – 9-11/16	8 – 4-1/2	9 – 11-3/8	11 – 6-1/4	13 – 1-1/16	2-1/2	3-15/16
5	7-7/8	2 – 2-11/16	3 – 9-9/16	5 – 4-3/8	6 – 11-1/4	8 – 6-1/8	10 – 0-15/16	11 – 7-13/16	13 – 2-5/8	2-3/4	4-5/16
6	9-7/16	2 – 4-1/4	3 – 11-1/8	5 – 6	7 – 0-13/16	8 – 7-11/16	10 – 2-1/2	11 – 9-3/8	13 – 4-1/4	3-1/4	5-1/8
7	11	2 – 5-7/8	4 – 0-11/16	5 – 7-9/16	7 – 2-3/8	8 – 9-1/4	10 – 4-1/16	11 – 10-15/16	13 – 5-13/16	3-1/2	5-1/2
8	1 – 0-9/16	2 – 7-7/16	4 – 2-1/4	5 – 9-1/8	7 – 3-15/16	8 – 10-13/16	10 – 5-11/16	12 – 0-1/2	13 – 7-3/8	3-3/4	5-7/8
9	1 – 2-1/8	2 – 9	4 – 3-13/16	5 – 10-11/16	7 – 5-9/16	9 – 0-3/8	10 – 7-1/4	12 – 2-1/16	13 – 8-15/16	4-1/4	6-11/16
10	1 – 3-11/16	2 – 10-9/16	4 – 5-7/16	6 – 0-1/4	7 – 7-1/8	9 – 1-15/16	10 – 8-13/16	12 – 3-5/8	13 – 10-1/2	4-1/2	7-1/16
11	1 – 5-1/4	3 – 0-1/8	4 – 7	6 – 1-13/16	7 – 8-11/16	9 – 3-1/2	10 – 10-3/8	12 – 5-1/4	14 – 0-1/16	4-3/4	7-7/16

Radius (inches)	9 – 0	10 – 0	11 – 0	12 – 0	13 – 0	14 – 0	15 – 0	16 – 0	17 – 0	Rad. (inches)	Arc (inches)
0	14 – 1-5/8	15 – 8-1/2	17 – 3-3/8	18 – 10-3/16	20 – 5-1/16	21 – 11-7/8	23 – 6-3/4	25 – 1-5/8	26 – 8-7/16	5-1/4	8-1/4
1	14 – 3-3/16	15 – 10-1/16	17 – 4-15/16	18 – 11-3/4	20 – 6-5/8	22 – 1-7/16	23 – 8-5/16	25 – 3-3/16	26 – 10	5-1/2	8-5/8
2	14 – 4-13/16	15 – 11-5/8	17 – 6-1/2	19 – 1-5/16	20 – 8-3/16	22 – 3-1/16	23 – 9-7/8	25 – 4-3/4	26 – 11-9/16	5-3/4	9-1/16
3	14 – 6-3/8	16 – 1-3/16	17 – 8-1/16	19 – 2-15/16	20 – 9-3/4	22 – 4-5/8	23 – 11-7/16	25 – 6-5/16	27 – 1-1/8	6-1/2	10-3/16
4	14 – 7-15/16	16 – 2-3/4	17 – 9-5/8	19 – 4-1/2	20 – 11-5/16	22 – 6-3/16	24 – 1	25 – 7-7/8	27 – 2-3/4	7-1/2	11-3/4
5	14 – 9-1/2	16 – 4-3/8	17 – 11-3/16	19 – 6-1/16	21 – 0-7/8	22 – 7-3/4	24 – 2-5/8	25 – 9-7/16	27 – 4-5/16	8-1/2	13-3/8
6	14 – 11-1/16	16 – 5-15/16	18 – 0-3/4	19 – 7-5/8	21 – 2-1/2	22 – 9-5/16	24 – 4-3/16	25 – 11	27 – 5-7/8	9-1/2	14-15/16
7	15 – 0-5/8	16 – 7-1/2	18 – 2-5/16	19 – 9-3/16	21 – 4-1/16	22 – 10-7/8	24 – 5-3/4	26 – 0-9/16	27 – 7-7/16	10-1/2	16-1/2
8	15 – 2-3/16	16 – 9-1/16	18 – 3-15/16	19 – 10-3/4	21 – 5-5/8	23 – 0-7/16	24 – 7-5/16	26 – 2-3/16	27 – 9	11-1/2	18-1/16
9	15 – 3-13/16	16 – 10-5/8	18 – 5-1/2	20 – 0-5/16	21 – 7-3/16	23 – 2-1/16	24 – 8-7/8	26 – 3-3/4	27 – 10-9/16	12-1/2	19-5/8
10	15 – 5-3/8	17 – 0-3/16	18 – 7-1/16	20 – 1-7/8	21 – 8-3/4	23 – 3-5/8	24 – 10-7/16	26 – 5-5/16	28 – 0-1/8	13-1/2	21-3/16
11	15 – 6-15/16	17 – 1-3/4	18 – 8-5/8	20 – 3-1/2	21 – 10-5/16	23 – 5-3/16	25 – 0	26 – 6-7/8	28 – 1-3/4	14-1/2	22-3/4
										15-1/2	24-3/8

Source: Texas Pipe Bending Co., Inc., Houston, Texas.

Pipe Fabrication

Table 6-8A
Miter Welding

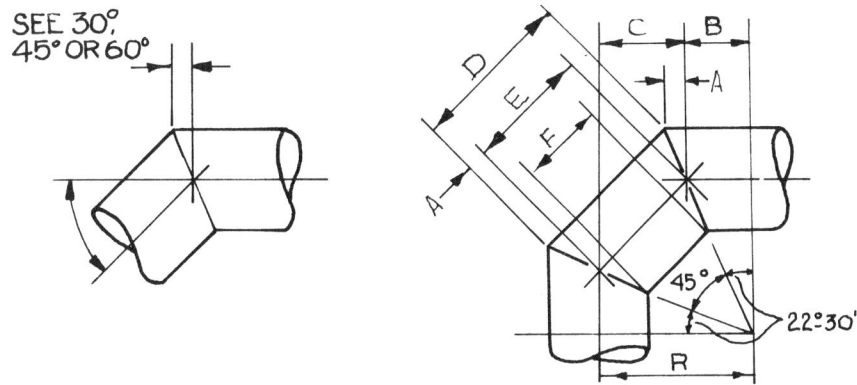

Size	30°	45°	60°	R	A	B	C	D	E	F
3	1/2	3/4	1	4-1/2	3/4	1-7/8	2-5/8	5-1/4	3-3/4	2-1/4
4	5/8	15/16	1-5/16	6	15/16	2-1/2	3-1/2	6-7/8	5	3-1/8
6	7/8	1-3/8	1-15/16	9	1-3/8	3-3/4	5-1/4	10-3/16	7-7/16	4-11/16
8	1-1/8	1-13/16	2-1/2	1 – 0	1-13/16	5	7	1 – 1-9/16	9-15/16	6-5/16
10	1-7/16	2-1/4	3-1/8	1 – 3	2-1/4	6-3/16	8-13/16	1 – 4-15/16	1 – 0-7/16	7-15/16
12	1-11/16	2-5/8	3-11/16	1 – 6	2-5/8	7-7/16	10-9/16	1 – 8-3/16	1 – 2-15/16	9-11/16
14	1-7/8	2-7/8	4-1/16	1 – 9	2-7/8	8-11/16	1 – 0-5/16	1 – 11-1/8	1 – 5-3/8	11-5/8
16	2-1/8	3-5/16	4-5/8	2 – 0	3-5/16	9-15/16	1 – 2-1/16	2 – 2-1/2	1 – 7-7/8	1 – 1-1/4
18	2-7/16	3-3/4	5-3/16	2 – 3	3-3/4	11-3/16	1 – 3-13/16	2 – 5-7/8	1 – 10-3/8	1 – 2-7/8
20	2-11/16	4-1/8	5-3/4	2 – 6	4-1/8	1 – 0-7/16	1 – 5-9/16	2 – 9-1/8	2 – 0-7/8	1 – 4-5/8
22	2-15/16	4-9/16	6-3/8	2 – 9	4-9/16	1 – 1-11/16	1 – 7-5/16	3 – 0-7/16	2 – 3-5/16	1 – 6-3/16
24	3-3/16	5	6-15/16	3 – 0	5	1 – 2-15/16	1 – 9-1/16	3 – 3-13/16	2 – 5-13/16	1 – 7-13/16
26	3-1/2	5-3/8	7-1/2	3 – 3	5-3/8	1 – 4-1/8	1 – 10-7/8	3 – 7-1/16	2 – 8-5/16	1 – 9-9/16
28	3-3/4	5-13/16	8-1/16	3 – 6	5-13/16	1 – 5-3/8	2 – 0-5/8	3 – 10-7/16	2 – 10-13/16	1 – 11-3/16
30	4	6-3/16	8-5/8	3 – 9	6-3/16	1 – 6-5/8	2 – 2-3/8	4 – 1-5/8	3 – 1-1/4	2 – 0-7/8
32	4-5/16	6-5/8	9-1/4	4 – 0	6-5/8	1 – 7-7/8	2 – 4-1/8	4 – 5	3 – 3-3/4	2 – 2-1/2
34	4-9/16	7-1/16	9-13/16	4 – 3	7-1/16	1 – 9-1/8	2 – 5-7/8	4 – 8-3/8	3 – 6-1/4	2 – 4-1/8
36	4-13/16	7-7/16	10-3/8	4 – 6	7-7/16	1 – 10-3/8	2 – 7-5/8	4 – 11-5/8	3 – 8-3/4	2 – 5-7/8
38	5-1/16	7-7/8	11	4 – 9	7-7/8	1 – 11-5/8	2 – 9-3/8	5 – 3	3 – 11-1/4	2 – 7-1/2
40	5-3/8	8-5/16	11-9/16	5 – 0	8-5/16	2 – 0-7/8	2 – 11-1/8	5 – 6-5/16	4 – 1-11/16	2 – 9-1/16
42	5-5/8	8-11/16	1 – 0-1/8	5 – 3	8-11/16	2 – 2-1/8	3 – 0-7/8	5 – 9-9/16	4 – 4-3/16	2 – 10-13/16
48	6-7/16	9-15/16	1 – 1-7/8	6 – 0	9-15/16	2 – 5-13/16	3 – 6-3/16	6 – 7-1/2	4 – 11-5/8	3 – 3-3/4
54	7-1/4	11-3/16	1 – 3-9/16	6 – 9	11-3/16	2 – 9-9/16	3 – 11-7/8	7 – 5-1/2	5 – 7-1/8	3 – 8-3/4
60	8-1/16	1 – 0-7/16	1 – 5-5/16	7 – 6	1 – 0-7/16	3 – 1-1/4	4 – 4-3/4	8 – 3-7/16	6 – 2-9/16	4 – 1-11/16
72	9-5/8	1 – 2-15/16	1 – 8-13/16	9 – 0	1 – 2-15/16	3 – 8-3/4	5 – 3-1/4	9 – 11-3/8	7 – 5-1/2	4 – 11-5/8

Source: Texas Pipe Bending Co., Houston, Texas.

Table 6-8B
Miter Welding Dimensions

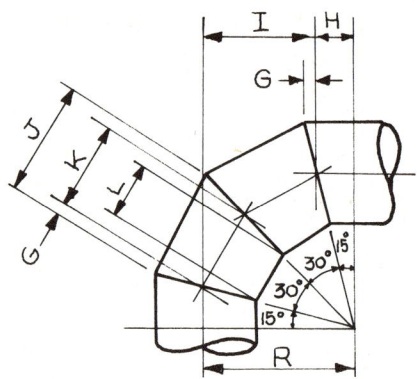 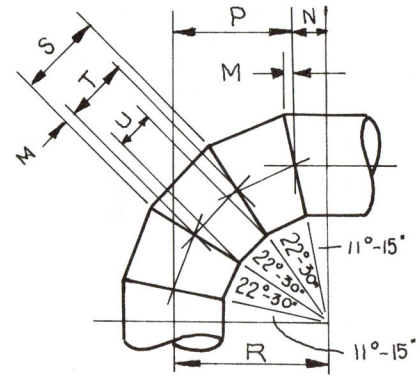

G	H	I	J	K	L	M	N	P	S	T	U
1/2	1-3/16	3-5/16	3-7/16	2-7/16	1-7/16	3/8	7/8	3-5/8	2-9/16	1-13/16	1-1/16
5/8	1-5/8	4-3/8	4-7/16	3-3/16	1-15/16	7/16	1-3/16	4-13/16	3-1/4	2-3/8	1-1/2
7/8	2-7/16	6-9/16	6-9/16	4-13/16	3-1/16	11/16	1-13/16	7-3/16	4-15/16	3-9/16	2-3/16
1-1/8	3-3/16	8-13/16	8-11/16	6-7/16	4-3/16	7/8	2-3/8	9-5/8	6-1/2	4-3/4	3
1-7/16	4	11	10-15/16	8-1/16	5-3/16	1-1/16	3	1 – 0	8-1/16	5-15/16	3-13/16
1-11/16	4-13/16	1 – 1-3/16	1 – 1	9-5/8	6-1/4	1-1/4	3-9/16	1 – 2-7/16	9-11/16	7-3/16	4-11/16
1-7/8	5-5/8	1 – 3-3/8	1 – 3	11-1/4	7-1/2	1-3/8	4-3/16	1 – 4-13/16	11-1/8	8-3/8	5-5/8
2-1/8	6-7/16	1 – 5-9/16	1 – 5-1/8	1 – 0-7/8	8-5/8	1-9/16	4-3/4	1 – 7-1/4	1 – 0-11/16	9-9/16	6-7/16
2-7/16	7-1/4	1 – 7-3/4	1 – 7-5/16	1 – 2-7/16	9-9/16	1-13/16	5-3/8	1 – 9-5/8	1 – 2-3/8	10-3/4	7-1/4
2-11/16	8-1/16	1 – 9-15/16	1 – 9-7/16	1 – 4-1/16	10-11/16	2	5-15/16	2 – 0-1/16	1 – 3-15/16	11-15/16	7-15/16
2-15/16	8-13/16	2 – 0-3/16	1 – 11-9/16	1 – 5-11/16	11-13/16	2-3/16	6-9/16	2 – 2-7/16	1 – 5-1/2	1 – 1-1/8	8-3/4
3-3/16	9-5/8	2 – 2-3/8	2 – 1-11/16	1 – 7-5/16	1 – 0-15/16	2-3/8	7-3/16	2 – 4-13/16	1 – 7-1/16	1 – 2-5/16	9-9/16
3-1/2	10-7/16	2 – 4-9/16	2 – 3-7/8	1 – 8-7/8	1 – 1-7/8	2-9/16	7-3/4	2 – 7-1/4	1 – 8-5/8	1 – 3-1/2	10-3/8
3-3/4	11-1/4	2 – 6-3/4	2 – 6	1 – 10-1/2	1 – 3	2-13/16	8-3/8	2 – 9-5/8	1 – 10-5/16	1 – 4-11/16	11-1/16
4	1 – 0-1/16	2 – 8-15/16	2 – 8-1/8	2 – 0-1/8	1 – 4-1/8	3	8-15/16	3 – 0-1/16	1 – 11-7/8	1 – 5-7/8	11-7/8
4-5/16	1 – 0-7/8	2 – 11-1/8	2 – 10-5/16	2 – 1-11/16	1 – 5-1/16	3-3/16	9-9/16	3 – 2-7/16	2 – 1-1/2	1 – 7-1/8	1 – 0-3/4
4-9/16	1 – 1-11/16	3 – 1-5/16	3 – 0-7/16	2 – 3-5/16	1 – 6-3/16	3-3/8	10-1/8	3 – 4-7/8	2 – 3-1/16	1 – 8-5/16	1 – 1-9/16
4-13/16	1 – 2-7/16	3 – 3-9/16	3 – 2-9/16	2 – 4-15/16	1 – 7-5/16	3-9/16	10-3/4	3 – 7-1/4	2 – 4-5/8	1 – 9-1/2	1 – 2-3/8
5-1/16	1 – 3-1/4	3 – 5-3/4	3 – 4-11/16	2 – 6-9/16	1 – 8-7/16	3-3/4	11-5/16	3 – 9-11/16	2 – 6-3/16	1 – 10-11/16	1 – 3-3/16
5-3/8	1 – 4-1/16	3 – 7-15/16	3 – 6-7/8	2 – 8-1/8	1 – 9-3/8	4	11-15/16	4 – 0-1/16	2 – 7-7/8	1 – 11-7/8	1 – 3-7/8
5-5/8	1 – 4-7/8	3 – 10-1/8	3 – 9	2 – 9-3/4	1 – 10-1/2	4-3/16	1 – 0-1/2	4 – 2-1/2	2 – 9-7/16	2 – 1-1/16	1 – 4-11/16
6-7/16	1 – 7-5/16	4 – 4-11/16	4 – 3-7/16	3 – 2-9/16	2 – 1-11/16	4-3/4	1 – 2-5/16	4 – 9-11/16	3 – 2-1/8	2 – 4-5/8	1 – 7-1/8
7-1/4	1 – 9-11/16	4 – 11-5/16	4 – 9-7/8	3 – 7-3/8	2 – 4-7/8	5-3/8	1 – 4-1/8	5 – 4-7/8	3 – 7	2 – 8-1/4	1 – 9-1/2
8-1/16	2 – 0-1/8	5 – 5-7/8	5 – 4-3/8	4 – 0-1/4	2 – 8-1/8	5-15/16	1 – 5-7/8	6 – 0-1/8	3 – 11-5/8	2 – 11-3/4	1 – 11-7/8
9-5/8	2 – 5	6 – 7	6 – 5-1/4	4 – 10	3 – 2-3/4	7-3/16	1 – 9-1/2	7 – 2-1/2	4 – 9-3/8	3 – 7	2 – 4-5/8

Source: Texas Pipe Bending Co., Inc., Houston, Texas

Pipe Fabrication

Table 6-9A
Socket Weld Fittings

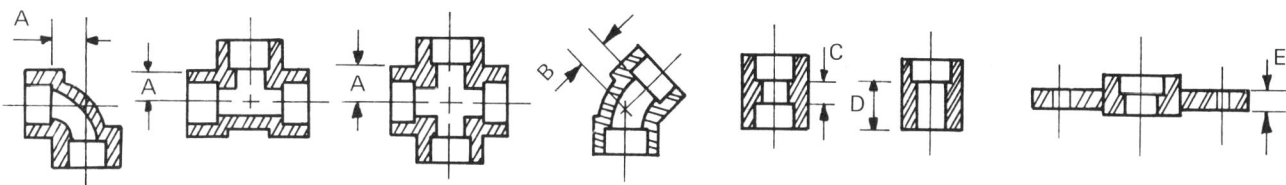

90° Ell   Tee   Cross   45° Ell   Coupling   Half Coupling   Flange

Nominal Pipe Size	2000 lb. & 3000 lb.		4000 lb.		6000 lb.		All Wts.		150 lb.	300 lb.	600 lb.	Nominal Pipe Size
	A	B	A	B	A	B	C	D	E	E	E	
1/8	7/16	5/16					1/4	5/8				1/8
1/4	7/16	5/16					1/4	5/8	1/4	1/2	3/4	1/4
3/8	9/16	5/16			9/16	3/8	1/4	11/16	1/4	1/2	3/4	3/8
1/2	5/8	7/16	3/4	1/2	5/8	3/8	3/8	7/8	1/4	1/2	3/4	1/2
3/4	3/4	1/2	7/8	9/16	3/4	7/16	3/8	15/16	3/16	9/16	13/16	3/4
1	7/8	9/16	1-1/16	11/16	7/8	1/2	1/2	1-1/8	3/16	9/16	13/16	1
1-1/4	1-1/16	11/16	1-1/4	13/16	1-1/16	5/8	1/2	1-3/16	1/4	1/2	13/16	1-1/4
1-1/2	1-1/4	13/16	1-1/2	1	1-1/4	5/8	1/2	1-1/4	1/4	9/16	7/8	1-1/2
2	1-1/2	1	1-5/8	1-1/8	1-1/2	7/8	3/4	1-5/8	5/16	5/8	1	2

Table 6-9B
Threaded Fittings

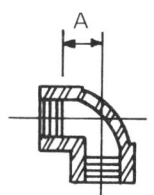 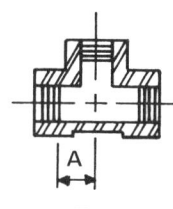 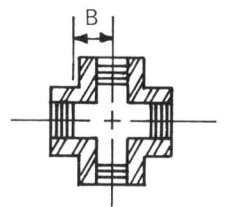 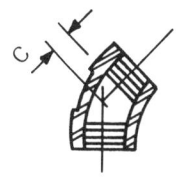 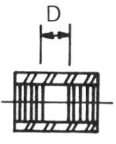

90° Ell   Tee   Cross   45° Ell   Coupling

Nominal Pipe Size	2000 lb.			3000 lb.			6000 lb.			3000 lb. 6000 lb.	Normal Thread Engagement	Nominal Pipe Size
	A	B	C	A	B	C	A	B	C	D		
1/8	9/16	3/4	7/16	9/16	3/4	7/16	3/4	3/4	1/2	3/4	1/4	1/8
1/4	7/16	5/8	5/16	5/8	5/8	3/8	3/4	3/4	1/2	5/8	3/8	1/4
3/8	5/8	5/8	3/8	3/4	3/4	1/2	15/16	15/16	5/8	3/4	3/8	3/8
1/2	5/8	5/8	3/8	13/16	13/16	1/2	1	1	5/8	7/8	1/2	1/2
3/4	3/4	3/4	7/16	15/16	15/16	9/16	1-3/16	1-3/16	3/4	7/8	9/16	3/4
1	13/16	13/16	7/16	1-1/16	1-1/16	5/8	1-5/16	1-5/16	11/16	1	11/16	1
1-1/4	1-1/16	1-1/16	5/8	1-5/16	1-5/16	11/16	1-11/16	1-11/16	1	1-1/4	11/16	1-1/4
1-1/2	1-5/16	1-5/16	11/16	1-11/16	1-11/16	1	1-13/16	1-13/16	1-1/16	1-3/4	11/16	1-1/2
2	1-5/8	1-5/8	15/16	1-3/4	1-3/4	1	2-1/2	2-1/2	1-5/16	1-7/8	3/4	2

Source: Texas Pipe Bending Co., Inc., Houston, Texas

Table 6-10
Table of 30° Offsets

0 ft. — 0-1/4 in. to 0 ft. — 11-3/4 in.

O	H	A	O	H	A	O	H	A
			0 — 4	0 — 8	0 — 6-15/16	0 — 8	1 — 4	1 — 1-15/16
0 — 0-1/4	0 — 0-1/2	0 — 0-7/16	0 — 4-1/4	0 — 8-1/2	0 — 7-3/8	0 — 8-1/4	1 — 4-1/2	1 — 2-5/16
0 — 0-1/2	0 — 1	0 — 0-7/8	0 — 4-1/2	0 — 9	0 — 7-13/16	0 — 8-1/2	1 — 5	1 — 2-3/4
0 — 0-3/4	0 — 1-1/2	0 — 1-5/16	0 — 4-3/4	0 — 9-1/2	0 — 8-1/4	0 — 8-3/4	1 — 5-1/2	1 — 3-1/8
0 — 1	0 — 2	0 — 1-3/4	0 — 5	0 — 10	0 — 8-11/16	0 — 9	1 — 6	1 — 3-9/16
0 — 1-1/4	0 — 2-1/2	0 — 2-3/16	0 — 5-1/4	0 — 10-1/2	0 — 9-1/16	0 — 9-1/4	1 — 6-1/2	1 — 4
0 — 1-1/2	0 — 3	0 — 2-5/8	0 — 5-1/2	0 — 11	0 — 9-1/2	0 — 9-1/2	1 — 7	1 — 4-7/16
0 — 1-3/4	0 — 3-1/2	0 — 3	0 — 5-3/4	0 — 11-1/2	0 — 9-15/16	0 — 9-3/4	1 — 7-1/2	1 — 4-7/8
0 — 2	0 — 4	0 — 3-7/16	0 — 6	1 — 0	0 — 10-3/8	0 — 10	1 — 8	1 — 5-5/16
0 — 2-1/4	0 — 4-1/2	0 — 3-7/8	0 — 6-1/4	1 — 0-1/2	0 — 10-13/16	0 — 10-1/4	1 — 8-1/2	1 — 5-3/4
0 — 2-1/2	0 — 5	0 — 4-5/16	0 — 6-1/2	1 — 1	0 — 11-1/4	0 — 10-1/2	1 — 9	1 — 6-3/16
0 — 2-3/4	0 — 5-1/2	0 — 4-3/4	0 — 6-3/4	1 — 1-1/2	0 — 11-11/16	0 — 10-3/4	1 — 9-1/2	1 — 6-5/8
0 — 3	0 — 6	0 — 5-3/16	0 — 7	1 — 2	1 — 0-1/8	0 — 11	1 — 10	1 — 7-1/16
0 — 3-1/4	0 — 6-1/2	0 — 5-5/8	0 — 7-1/4	1 — 2-1/2	1 — 0-9/16	0 — 11-1/4	1 — 10-1/2	1 — 7-1/2
0 — 3-1/2	0 — 7	0 — 6-1/16	0 — 7-1/2	1 — 3	1 — 1	0 — 11-1/2	1 — 11	1 — 7-15/16
0 — 3-3/4	0 — 7-1/2	0 — 6-1/2	0 — 7-3/4	1 — 3-1/2	1 — 1-7/16	0 — 11-3/4	1 — 11-1/2	1 — 8-3/8

1 ft. — 0 in. to 1 ft. — 11-3/4 in.

O	H	A	O	H	A	O	H	A
1 — 0	2 — 0	1 — 8-13/16	1 — 4	2 — 8	2 — 3-11/16	1 — 8	3 — 4	2 — 10-5/8
1 — 0-1/4	2 — 0-1/2	1 — 9-3/16	1 — 4-1/4	2 — 8-1/2	2 — 4-1/8	1 — 8-1/4	3 — 4-1/2	2 — 11-1/16
1 — 0-1/2	2 — 1	1 — 9-5/8	1 — 4-1/2	2 — 9	2 — 4-9/16	1 — 8-1/2	3 — 5	2 — 11-1/2
1 — 0-3/4	2 — 1-1/2	1 — 10-1/16	1 — 4-3/4	2 — 9-1/2	2 — 5	1 — 8-3/4	3 — 5-1/2	2 — 11-15/16
1 — 1	2 — 2	1 — 10-1/2	1 — 5	2 — 10	2 — 5-7/16	1 — 9	3 — 6	3 — 0-3/8
1 — 1-1/4	x2 — 2-1/2	1 — 10-15/16	1 — 5-1/4	2 — 10-1/2	2 — 5-7/8	1 — 9-1/4	3 — 6-1/2	3 — 0-13/16
1 — 1-1/2	2 — 3	1 — 11-3/8	1 — 5-1/2	2 — 11	2 — 6-5/16	1 — 9-1/2	3 — 7	3 — 1-1/4
1 — 1-3/4	2 — 3-1/2	1 — 11-13/16	1 — 5-3/4	2 — 11-1/2	2 — 6-3/4	1 — 9-3/4	3 — 7-1/2	3 — 1-11/16
1 — 2	2 — 4	2 — 0-1/4	1 — 6	3 — 0	3 — 7-3/16	1 — 10	3 — 8	3 — 2-1/8
1 — 2-1/4	2 — 4-1/2	2 — 0-11/16	1 — 6-1/4	3 — 0-1/2	2 — 7-5/8	1 — 10-1/4	3 — 8-1/2	3 — 2-9/16
1 — 2-1/2	2 — 5	2 — 1-1/8	1 — 6-1/2	3 — 1	2 — 8-1/16	1 — 10-1/2	3 — 9	3 — 3
1 — 2-3/4	2 — 5-1/2	2 — 1-9/16	1 — 6-3/4	3 — 1-1/2	2 — 8-1/2	1 — 10-3/4	3 — 9-1/2	3 — 3-3/8
1 — 3	2 — 6	2 — 2	1 — 7	3 — 2	2 — 8-15/16	1 — 11	3 — 10	3 — 3-13/16
1 — 3-1/4	2 — 6-1/2	2 — 2-7/16	1 — 7-1/4	3 — 2-1/2	2 — 9-5/16	1 — 11-1/4	3 — 10-1/2	3 — 4-1/4
1 — 3-1/2	2 — 7	2 — 2-7/8	1 — 7-1/2	3 — 3	2 — 9-3/4	1 — 11-1/2	3 — 11	3 — 4-11/16
1 — 3-3/4	2 — 7-1/2	2 — 3-1/4	1 — 7-3/4	3 — 3-1/2	2 — 10-3/16	1 — 11-3/4	3 — 11-1/2	3 — 5-1/8

Source: Texas Pipe Bending Co., Inc., Houston, Texas.

Pipe Fabrication

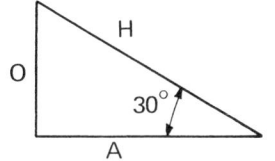

Table 6-11
Table of 30° Offsets

2' – 0" to 2' – 11-3/4"

O	H	A	O	H	A	O	H	A
2 – 0	4 – 0	3 – 5-9/16	2 – 4	4 – 8	4 – 0-1/2	2 – 8	5 – 4	4 – 7-7/16
2 – 0-1/4	4 – 0-1/2	3 – 6	2 – 4-1/4	4 – 8-1/2	4 – 0-15/16	2 – 8-1/4	5 – 4-1/2	4 – 7-7/8
2 – 0-1/2	4 – 1	3 – 6-7/16	2 – 4-1/2	4 – 9	4 – 1-3/8	2 – 8-1/2	5 – 5	4 – 8-5/16
2 – 0-3/4	4 – 1-1/2	3 – 6-7/8	2 – 4-3/4	4 – 9-1/2	4 – 1-13/16	2 – 8-3/4	5 – 5-1/2	4 – 8-3/4
2 – 1	4 – 2	3 – 7-5/16	2 – 5	4 – 10	4 – 2-1/4	2 – 9	5 – 6	4 – 9-3/16
2 – 1-1/4	4 – 2-1/2	3 – 7-3/4	2 – 5-1/4	4 – 10-1/2	4 – 2-11/16	2 – 9-1/4	5 – 6-1/2	4 – 9-9/16
2 – 1-1/2	4 – 3	3 – 8-3/16	2 – 5-1/2	4 – 11	4 – 3-1/8	2 – 9-1/2	5 – 7	4 – 10
2 – 1-3/4	4 – 3-1/2	3 – 8-5/8	2 – 5-3/4	4 – 11-1/2	4 – 3-1/2	2 – 9-3/4	5 – 7-1/2	4 – 10-7/16
2 – 2	4 – 4	3 – 9-1/16	2 – 6	5 – 0	4 – 3-15/16	2 – 10	5 – 8	4 – 10-7/8
2 – 2-1/4	4 – 4-1/2	3 – 9-1/2	2 – 6-1/4	5 – 0-1/2	4 – 4-3/8	2 – 10-1/4	5 – 8-1/2	4 – 11-5/16
2 – 2-1/2	4 – 5	3 – 9-7/8	2 – 6-1/2	5 – 1	4 – 4-13/16	2 – 10-1/2	5 – 9	4 – 11-3/4
2 – 2-3/4	4 – 5-1/2	3 – 10-5/16	2 – 6-3/4	5 – 1-1/2	4 – 5-1/4	2 – 10-3/4	5 – 9-1/2	5 – 0-3/16
2 – 3	4 – 6	3 – 10-3/4	2 – 7	5 – 2	4 – 5-11/16	2 – 11	5 – 10	5 – 0-5/8
2 – 3-1/4	4 – 6-1/2	3 – 11-3/16	2 – 7-1/4	5 – 2-1/2	4 – 6-1/8	2 – 11-1/4	5 – 10-1/2	5 – 1-1/16
2 – 3-1/2	4 – 7	3 – 11-5/8	2 – 7-1/2	5 – 3	4 – 6-9/16	2 – 11-1/2	5 – 11	5 – 1-1/2
2 – 3-3/4	4 – 7-1/2	4 – 0-1/16	2 – 7-3/4	5 – 3-1/2	4 – 7	2 – 11-3/4	5 – 11-1/2	5 – 1-13/16

3' – 0" to 3' – 11-3/4"

O	H	A	O	H	A	O	H	A
3 – 0	6 – 0	5 – 2-3/8	3 – 4	6 – 8	5 – 9-5/16	3 – 8	7 – 4	6 – 4-3/16
3 – 0-1/4	6 – 0-1/2	5 – 2-13/16	3 – 4-1/4	6 – 8-1/2	5 – 9-11/16	3 – 8-1/4	7 – 4-1/2	6 – 4-5/8
3 – 0-1/2	6 – 1	5 – 3-1/4	3 – 4-1/2	6 – 9	5 – 10-1/8	3 – 8-1/2	7 – 5	6 – 5-1/16
3 – 0-3/4	6 – 1-1/2	5 – 3-5/8	3 – 4-3/4	6 – 9-1/2	5 – 10-9/16	3 – 8-3/4	7 – 5-1/2	6 – 6-1/2
3 – 1	6 – 2	5 – 4-1/16	3 – 5	6 – 10	5 – 11	3 – 9	7 – 6	6 – 5-15/16
3 – 1-1/4	6 – 2-1/2	5 – 4-1/2	3 – 5-1/4	6 – 10-1/2	5 – 11-7/16	3 – 9-1/4	7 – 6-1/2	6 – 6-3/8
3 – 1-1/2	6 – 3	5 – 4-15/16	3 – 5-1/2	6 – 11	5 – 11-7/8	3 – 9-1/2	7 – 7	6 – 6-13/16
3 – 1-3/4	6 – 3-1/2	5 – 5-3/8	3 – 5-3/4	6 – 11-1/2	6 – 0-5/16	3 – 9-3/4	7 – 7-1/2	6 – 7-1/4
3 – 2	6 – 4	5 – 5-13/16	3 – 6	7 – 0	6 – 0-3/4	3 – 10	7 – 8	6 – 7-11/16
3 – 2-1/4	6 – 4-1/2	5 – 6-1/4	3 – 6-1/4	7 – 0-1/2	6 – 1-3/16	3 – 10-1/4	7 – 8-1/2	6 – 8-1/8
3 – 2-1/2	6 – 5	5 – 6-11/16	3 – 6-1/2	7 – 1	6 – 1-5/8	3 – 10-1/2	7 – 9	6 – 8-9/16
3 – 2-3/4	6 – 5-1/2	5 – 7-1/8	3 – 6-3/4	7 – 1-1/2	6 – 2-1/16	3 – 10-3/4	7 – 9-1/2	6 – 9
3 – 3	6 – 6	5 – 7-9/16	3 – 7	7 – 2	6 – 2-1/2	3 – 11	7 – 10	6 – 9-7/16
3 – 3-1/4	6 – 6-1/2	5 – 8	3 – 7-1/4	7 – 2-1/2	6 – 2-15/16	3 – 11-1/4	7 – 10-1/2	6 – 9-13/16
3 – 3-1/2	6 – 7	5 – 8-7/16	3 – 7-1/2	7 – 3	6 – 3-3/8	3 – 11-1/2	7 – 11	6 – 10-1/4
3 – 3-3/4	6 – 7-1/2	5 – 8-7/8	3 – 7-3/4	7 – 3-1/2	6 – 3-3/4	3 – 11-3/4	7 – 11-1/2	6 – 10-11/16

Source: Texas Pipe Bending Co., Inc., Houston, Texas.

Table 6-12A
45° Offsets

	0	1	2	3	4	5	6	7	8	9	10	11	
0 1/16	0 1/16	1-7/16 1-1/2	2-13/16 2-7/8	4-1/4 4-5/16	5-11/16 5-3/4	7-1/16 7-3/16	8-1/2 8-9/16	9-7/8 10	11-5/16 11-3/8	12-3/4 12-13/16	14-1/8 14-1/4	15-9/16 15-5/8	0 1/16
1/8 3/16	3/16 1/4	1-9/16 1-11/16	3 3-1/16	4-7/16 4-1/2	5-13/16 5-7/8	7-1/4 7-5/16	8-11/16 8-3/4	10-1/16 10-3/16	11-1/2 11-9/16	12-7/8 13	14-5/16 14-7/16	15-3/4 15-13/16	1/8 3/16
1/4 5/16	3/8 7/16	1-3/4 1-7/8	3-3/16 3-1/4	4-5/8 4-11/16	6 6-0/8	7-7/16 7-1/2	8-13/16 8-15/16	10-1/4 10-3/8	11-11/16 11-3/4	13-1/16 13-3/16	14-1/2 14-9/16	15-15/16 16	1/4 5/16
3/8 7/16	1/2 5/8	1-15/16 2-1/16	3-3/8 3-7/16	4-3/4 4-7/8	6-3/16 6-1/4	7-5/8 7-11/16	9 9-1/8	10-7/16 10-1/2	11-7/8 11-15/16	13-1/4 13-3/8	14-11/16 14-3/4	16-1/16 16-3/16	3/8 7/16
1/2 9/16	11/16 13/16	2-1/8 2-3/16	3-9/16 3-5/8	4-15/16 5-1/16	6-3/8 6-7/16	7-3/4 7-7/8	9-3/16 9-1/4	10-5/8 10-11/16	12 12-1/8	13-7/16 13-9/16	14-7/8 14-15/16	16-1/4 16-3/8	1/2 9/16
5/8 11/16	7/8 1	2-5/16 2-3/8	3-11/16 3-13/16	5-1/8 5-3/16	6-9/16 6-5/8	7-15/16 8-1/16	9-3/8 9-7/16	10-13/16 10-7/8	12-3/16 12-5/16	13-5/8 13-11/16	15 15-1/8	16-7/16 16-1/2	5/8 11/16
3/4 13/16	1-1/16 1-1/8	2-1/2 2-9/16	3-7/8 4	5-5/16 5-3/8	6-11/16 6-13/16	8-1/8 8-1/4	9-9/16 9-5/8	11 11-1/16	12-3/8 12-7/16	13-13/16 13-7/8	15-3/16 15-5/16	16-5/8 16-11/16	3/4 13/16
7/8 15/16	1-1/4 1-5/16	2-5/8 2-3/4	4-1/16 4-1/8	5-1/2 5-9/16	6-7/8 7	8-5/16 8-3/8	9-11/16 9-13/16	11-1/8 11-1/4	12-9/16 12-5/8	13-15/16 14-1/16	15-3/8 15-7/16	16-13/16 16-7/8	7/8 15/16

Source: Texas Pipe Bending Co., Inc., Houston, Texas.

Table 6-12B
45° TRIANGLES – BASE TO HYPOTENUSE

	12	13	14	15	16	17	18	19	20	21	22	23	
0 1/16	17 17-1/16	18-3/8 18-1/2	19-13/16 19-7/8	21-3/16 21-5/16	22-5/8 22-11/16	2 – 0-1/16 2 – 0-1/8	2 – 1-7/16 2 – 1-9/16	2 – 2-7/8 2 – 2-15/16	2 – 4-5/16 2 – 4-3/8	2 – 5-11/16 2 – 5-13/16	2 – 7-1/8 2 – 7-3/16	2 – 8-1/2 2 – 8-5/8	0 1/16
1/8 3/16	17-1/8 17-1/4	18-9/16 18-5/8	20 20-1/16	21-3/8 21-1/2	22-13/16 22-7/8	2 – 0-1/4 2 – 0-5/16	2 – 1-5/8 2 – 1-3/4	2 – 3-1/16 2 – 3-1/8	2 – 4-1/2 2 – 4-9/16	2 – 5-7/8 2 – 5-15/16	2 – 7-5/16 2 – 7-3/8	2 – 8-11/16 2 – 8-13/16	1/8 3/16
1/4 5/16	17-5/16 17-7/8	18-3/4 18-13/16	20-1/8 20-1/4	21-9/16 21-5/8	23 23-1/16	2 – 0-3/8 2 – 0-1/2	2 – 1-13/16 2 – 1-7/8	2 – 3-1/4 2 – 3-5/16	2 – 4-5/8 2 – 4-3/4	2 – 6-1/16 2 – 6-1/8	2 – 7-7/16 2 – 7-9/16	2 – 8-7/8 2 – 9	1/4 5/16
3/8 7/16	17-1/2 17-9/16	18-7/8 19	20-5/16 20-7/16	21-3/4 21-13/16	23-3/16 23-1/4	2 – 0-9/16 2 – 0-11/16	2 – 2 2 – 2-1/16	2 – 3-3/8 2 – 3-1/2	2 – 4-13/16 2 – 4-7/8	2 – 6-1/4 2 – 6-5/16	2 – 7-5/8 2 – 7-3/4	2 – 9-1/16 2 – 9-1/8	3/8 7/16
1/2 9/16	17-11/16 17-3/4	19-1/8 19-3/16	20-1/2 20-5/8	21-15/16 22	23-5/16 23-7/16	2 – 0-3/4 2 – 0-13/16	2 – 2-3/16 2 – 2-1/4	2 – 3-9/16 2 – 3-11/16	2 – 5 2 – 5-1/16	2 – 6-3/8 2 – 6-7/16	2 – 7-13/16 2 – 7-15/16	2 – 9-1/4 2 – 9-5/16	1/2 9/16
5/8 11/16	17-7/8 17-15/16	19-1/4 19-3/8	20-11/16 20-3/4	22-1/8 22-3/16	23-1/2 23-5/8	2 – 0-15/16 2 – 1	2 – 2-5/16 2 – 2-7/16	2 – 3-3/4 2 – 3-13/16	2 – 5-3/16 2 – 5-1/4	2 – 6-9/16 2 – 6-11/16	2 – 8 2 – 8-1/16	2 – 9-7/16 2 – 9-1/2	5/8 11/16
3/4 13/16	18-1/16 18-1/8	19-7/16 19-9/16	20-7/8 20-15/16	22-1/4 22-3/8	23-11/16 22-3/4	2 – 1-1/16 2 – 1-3/16	2 – 2-1/2 2 – 2-5/8	2 – 3-15/16 2 – 4	2 – 5-3/8 2 – 5-7/16	2 – 6-3/4 2 – 6-7/8	2 – 8-3/16 2 – 8-1/4	2 – 9-9/16 2 – 9-11/16	3/4 13/16
7/8 15/16	18-3/16 18-5/16	19-5/8 19-11/16	21-1/16 21-1/8	22-7/16 22-9/16	23-7/8 23-15/16	2 – 1-1/4 2 – 1-3/8	2 – 2-11/16 2 – 2-13/16	2 – 4-1/8 2 – 4-3/16	2 – 5-1/2 2 – 5-5/8	2 – 6-15/16 2 – 7	2 – 8-3/8 2 – 8-7/16	2 – 9-3/4 2 – 9-7/8	7/8 15/16

Source: Texas Pipe Bending Co., Inc., Houston, Texas.

Table 6-13
45° Offsets

	2 – 4	2 – 1	2 – 2	2 – 3	2 – 4	2 – 5	2 – 6	2 – 7	2 – 8	2 – 9	2 – 10	2 – 11	
0	2 – 9-15/16	2 – 11-3/8	3 – 0-3/4	3 – 2-3/16	3 – 3-5/8	3 – 5	3 – 6-7/16	3 – 7-13/16	3 – 9-1/4	3 – 10-11/16	4 – 0-1/16	4 – 1-1/2	0
1/16	2 – 10	2 – 11-7/16	3 – 0-7/8	3 – 2-1/4	3 – 3-11/16	3 – 5-1/8	3 – 6-1/2	3 – 7-15/16	3 – 9-5/16	3 – 10-3/4	4 – 0-3/16	4 – 1-9/16	1/16
1/8	2 – 10-1/8	2 – 11-9/16	3 – 0-15/16	3 – 2-3/8	3 – 3-3/4	3 – 5-3/16	3 – 6-5/8	3 – 8	3 – 9-7/16	3 – 10-7/8	4 – 0-1/4	4 – 4-11/16	1/8
3/16	2 – 10-3/16	2 – 11-5/8	3 – 1-1/16	3 – 2-7/16	3 – 3-7/8	3 – 5-1/4	3 – 6-11/16	3 – 8-1/8	3 – 9-1/2	3 – 10-15/16	4 – 0-3/8	4 – 1-3/4	3/16
1/4	2 – 10-5/16	2 – 11-11/16	3 – 1-1/8	3 – 2-9/16	3 – 3-15/16	3 – 5-3/8	3 – 6-3/4	3 – 8-3/16	3 – 9-5/8	3 – 11	4 – 0-7/16	4 – 1-7/8	1/4
3/16	2 – 10-3/8	2 – 11-13/16	3 – 1-3/16	3 – 2-5/8	3 – 4-1/16	3 – 5-7/16	3 – 6-7/8	3 – 8-5/16	3 – 9-11/16	3 – 11-1/8	4 – 0-1/2	4 – 1-15/16	3/16
3/8	2 – 10-1/2	2 – 11-7/8	3 – 1-5/16	3 – 2-11/16	3 – 4-1/8	3 – 5-9/16	3 – 6-15/16	3 – 8-3/8	3 – 9-3/16	3 – 11-3/16	4 – 0-5/8	4 – 2	3/8
7/16	2 – 10-9/16	3 – 0	3 – 1-3/8	3 – 2-13/16	3 – 4-3/16	3 – 5-5/8	3 – 7-1/16	3 – 8-7/16	3 – 9-7/8	3 – 11-5/16	4 – 0-11/16	4 – 2-1/16	7/16
1/2	2 – 10-5/8	3 – 0-1/16	3 – 1-1/2	3 – 2-7/8	3 – 4-5/16	3 – 5-3/4	3 – 7-1/8	3 – 8-9/16	3 – 9-15/16	3 – 11-3/8	4 – 0-13/16	4 – 2-3/16	1/2
9/16	2 – 10-3/4	3 – 0-1/8	3 – 1-9/16	3 – 3	3 – 4-3/8	3 – 5-13/16	3 – 7-1/4	3 – 8-5/8	3 – 10-1/16	3 – 11-7/16	4 – 0-7/8	4 – 2-5/16	9/16
3/8	2 – 10-13/16	3 – 0-1/4	3 – 1-11/16	3 – 3-1/16	3 – 4-1/2	3 – 5-7/8	3 – 7-5/16	3 – 8-3/4	3 – 10-1/8	3 – 11-9/16	4 – 0-15/16	4 – 2-3/8	3/8
11/16	2 – 10-15/16	3 – 0-5/16	3 – 1-3/4	3 – 3-1/8	3 – 4-9/16	3 – 6	3 – 7-3/8	3 – 8-13/16	3 – 10-1/4	3 – 11-5/8	4 – 1-1/16	4 – 2-1/2	11/16
3/4	2 – 11	3 – 0-7/16	3 – 1-13/16	3 – 3-1/4	3 – 4-11/16	3 – 6-1/16	3 – 7-1/2	3 – 8-15/16	3 – 10-5/16	3 – 11-11/16	4 – 1-1/8	4 – 2-9/16	3/4
13/16	2 – 11-1/16	3 – 0-1/2	3 – 1-15/16	3 – 3-5/16	3 – 4-3/4	3 – 6-3/16	3 – 7-9/16	3 – 9	3 – 10-3/8	3 – 11-13/16	4 – 1-1/4	4 – 2-5/8	13/16
7/8	2 – 11-3/16	3 – 0-9/16	3 – 2	3 – 3-7/16	3 – 4-13/16	3 – 6-1/4	3 – 7-11/16	3 – 9-1/16	3 – 10-1/2	3 – 11-13/16	4 – 1-3/16	4 – 2-3/4	7/8
15/16	2 – 11-1/4	3 – 0-11/16	3 – 2-1/8	3 – 3-1/2	3 – 4-15/16	3 – 6-5/16	3 – 7-3/4	3 – 9-3/16	3 – 10-9/16	4 – 0	4 – 1-7/16	4 – 2-13/16	15/16

Source: Texas Pipe Bending Co., Inc., Houston, Texas.

Table 6-13 cont
45° Triangles — Base to Hypotenuse

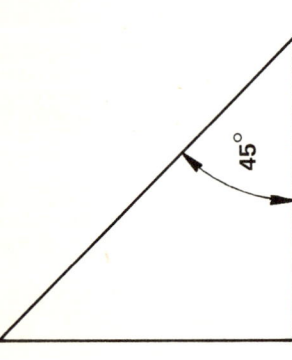

	3 — 0	3 — 1	3 — 2	3 — 3	3 — 4	3 — 5	3 — 6	3 — 7	3 — 8	3 — 9	3 — 10	3 — 11	
0 1/16	4 — 2-13/16 4 — 3	4 — 4-3/16 4 — 4-7/16	4 — 5-3/4 4 — 5-13/16	4 — 7-1/8 4 — 7-1/4	4 — 8-9/16 4 — 8-5/8	4 — 10 4 — 10-1/16	4 — 11-3/8 4 — 11-1/2	5 — 0-13/16 5 — 0-7/8	5 — 2-1/4 5 — 2-5/16	5 — 3-5/8 5 — 3-3/4	5 — 5-1/16 5 — 5-1/8	5 — 6-7/16 5 — 6-9/16	0 1/16
1/8 3/16	4 — 3-1/8 4 — 3-3/16	4 — 4-1/2 4 — 4-9/16	4 — 5-15/16 4 — 6	4 — 7-5/16 4 — 7-7/16	4 — 8-3/4 4 — 8-13/16	4 — 10-3/16 4 — 10-1/4	4 — 11-9/16 4 — 11-11/16	5 — 1 5 — 1-1/16	5 — 2-3/8 5 — 2-1/2	5 — 3-13/16 5 — 3-7/8	5 — 5-3/16 5 — 5-5/16	5 — 6-5/8 5 — 6-3/4	1/8 3/16
1/4 5/16	4 — 3-1/4 4 — 3-3/8	4 — 4-11/16 4 — 4-3/4	4 — 6-1/8 4 — 6-3/8	4 — 7-1/2 4 — 7-5/8	4 — 8-13/16 4 — 9	4 — 10-5/16 4 — 10-7/16	4 — 11-3/4 4 — 11-13/16	5 — 1-3/16 5 — 1-1/4	5 — 2-9/16 5 — 2-11/16	5 — 4 5 — 4-1/16	5 — 5-7/16 5 — 5-1/2	5 — 6-13/16 5 — 6-15/16	1/4 5/16
3/8 7/16	4 — 3-7/16 4 — 3-1/2	4 — 4-7/8 4 — 4-13/16	4 — 6-1/4 4 — 6-3/8	4 — 7-11/16 4 — 7-3/4	4 — 9-1/8 4 — 9-3/16	4 — 10-1/2 4 — 10-5/8	4 — 11-13/16 5 — 0	5 — 1-5/16 5 — 1-7/16	5 — 2-3/4 5 — 2-7/8	5 — 4-3/16 5 — 4-1/4	5 — 5-9/16 5 — 5-11/16	5 — 7 5 — 7-1/16	3/8 7/16
1/2 9/16	4 — 3-5/8 4 — 3-11/16	4 — 5-1/16 4 — 5-1/8	4 — 6-7/16 4 — 6-9/16	4 — 7-7/8 4 — 7-15/16	4 — 9-1/4 4 — 9-3/8	4 — 10-11/16 4 — 10-3/4	5 — 0-1/8 5 — 0-3/16	5 — 1-1/2 5 — 1-5/8	5 — 2-15/16 5 — 3	5 — 4-3/8 5 — 4-7/16	5 — 5-3/4 5 — 5-7/8	5 — 7-3/16 5 — 7-1/4	1/2 9/16
5/8 11/16	4 — 3-13/16 4 — 3-7/8	4 — 5-3/16 4 — 5-5/16	4 — 6-3/8 4 — 6-11/16	4 — 8-1/16 4 — 8-1/8	4 — 9-7/16 4 — 9-9/16	4 — 10-7/8 4 — 10-15/16	5 — 0-1/4 5 — 0-3/8	5 — 1-11/16 5 — 1-13/16	5 — 3-1/8 5 — 3-3/16	5 — 4-1/2 5 — 4-5/8	5 — 5-15/16 5 — 6	5 — 7-3/8 5 — 7-7/16	5/8 11/16
3/4 13/16	4 — 4 4 — 4-1/16	4 — 5-3/8 4 — 5-1/2	4 — 6-13/16 4 — 6-7/8	4 — 8-3/16 4 — 8-3/16	4 — 9-5/8 4 — 9-11/16	4 — 11-1/16 4 — 11-1/8	5 — 0-7/16 5 — 0-9/16	5 — 1-7/8 5 — 1-15/16	5 — 3-5/16 5 — 3-3/8	5 — 4-11/16 5 — 4-13/16	5 — 6-1/8 5 — 6-3/16	5 — 7-1/2 5 — 7-5/16	3/4 13/16
7/8 15/16	4 — 4-1/8 4 — 4-1/4	4 — 5-9/16 4 — 5-5/8	4 — 7 4 — 7-1/16	4 — 8-3/8 4 — 8-1/2	4 — 9-13/16 4 — 9-7/8	4 — 11-1/4 4 — 11-5/16	5 — 0-5/8 5 — 0-3/4	5 — 2-1/16 5 — 2-1/8	5 — 3-7/16 5 — 3-9/16	5 — 4-7/8 5 — 4-15/16	5 — 6-5/16 5 — 6-3/8	5 — 7-11/16 5 — 7-13/16	7/8 15/16

Source: Texas Pipe Bending Co., Inc., Houston, Texas.

Table 6-14
Cutback at 90° of Nozzle to OD of Header
Standard Weight Pipe

Nozzle → Header ↓	3/4	1	1-1/2	2	2-1/2	3	3-1/2	4	5	6	8	10	12	14	16	18	20	22	24	30
3/4	0	1/2	7/8	1-1/8	1-3/8	1-11/16	1-15/16	2-3/16	2-3/4	3-5/16	4-5/16	5-3/8	6-3/8	7	8	9	10	11	12	15
1		0	13/16	1-1/16	1-5/16	1-11/16	1-15/16	2-3/16	2-3/4	3-1/4	4-1/4	5-3/8	6-3/8	7	8	9	10	11	12	15
1-1/2			0	7/8	1-3/16	1-9/16	1-13/16	2-1/8	2-11/16	3-3/16	4-1/4	5-5/16	6-5/16	6-15/16	7-15/16	8-15/16	9-15/16	11	12	15
2				0	1	1-7/16	1-11/16	2	2-9/16	3-1/8	4-3/16	5-1/4	6-5/16	6-15/16	7-15/16	8-15/16	9-15/16	10-15/16	11-15/16	14-15/16
2-1/2					0	1-1/4	1-9/16	1-7/8	2-1/2	3-1/16	4-1/8	5-1/4	6-1/4	6-7/8	7-7/8	8-15/16	9-15/16	10-15/16	11-15/16	14-15/16
3						0	1-5/16	1-5/8	2-5/16	2-15/16	4-1/16	5-1/8	6-3/16	6-13/16	7-7/8	8-7/8	9-7/8	10-7/8	11-7/8	14-15/16
3-1/2							0	1-3/8	2-1/8	2-13/16	3-15/16	5-1/16	6-1/8	6-3/4	7-13/16	8-13/16	9-13/16	10-7/8	11-7/8	14-7/8
4								0	1-15/16	2-5/8	3-13/16	5	6-1/16	6-11/16	7-3/4	8-3/4	9-13/16	10-13/16	11-13/16	14-7/8
5									0	2-1/8	3-1/2	4-3/4	5-7/8	6-1/2	7-9/16	8-5/8	9-11/16	10-11/16	11-3/4	14-13/16
6										0	3-1/16	4-7/16	5-5/8	6-5/16	7-3/8	8-1/2	9-1/2	10-1/2	11-5/8	14-11/16
8											0	3-9/16	4-15/16	5-3/4	6-15/16	8-1/16	9-3/16	10-1/4	11-5/16	14-7/8
10												0	3-15/16	4-7/8	6-1/4	7-1/2	8-11/16	9-13/16	10-15/16	14-1/8
12													0	3-5/8	5-5/16	6-11/16	8	9-1/2	10-3/8	13-3/4
14														0	4-1/2	6-1/16	7-1/2	8-3/16	10	13-7/16
16															0	4-3/4	6-1/2	7-15/16	9-1/4	12-15/16
18																0	5-1/16	6-13/16	8-5/16	12-1/4
20																	0	5-5/16	7-3/16	11-1/2
22																		0	5-9/16	10-9/16
24																			0	9-1/2
30																				0

Source: Texas Pipe Bending Co., Inc., Houston, Texas.

Table 6-15
Cutback at 90° for ID of Nozzle to OD of Header
Extra Heavy Pipe

	1/2	3/4	1	1-1/2	2	2-1/2	3	3-1/2	4	5	6	8	10	12	14	16	18	20	24	30
1/2	0	7/16	9/16	15/16	1-1/8	1-7/16	1-3/4	2	2-1/4	2-3/4	3-5/16	4-5/16	5-3/8	6-3/8	7	8	9	10	12	15
3/4		0	9/16	7/8	1-1/8	1-3/8	1-11/16	1-15/16	2-3/16	2-3/4	3-5/16	4-5/16	5-3/8	6-3/8	7	8	9	10	12	15
1			0	13/16	1-1/16	1-3/8	1-11/16	1-15/16	2-3/16	2-3/4	3-1/4	4-5/16	5-3/8	6-3/8	7	8	9	10	12	15
1-1/2				0	15/16	1-1/4	1-9/16	1-7/8	2-1/8	2-11/16	3-1/4	4-1/4	5-5/16	6-5/16	6-15/16	7-15/16	8-15/16	10	12	15
2					0	1-1/16	1-7/16	1-3/4	2	2-5/8	3-3/16	4-3/16	5-3/16	6-5/16	6-15/16	7-15/16	8-15/16	9-15/16	11-15/16	14-15/16
2-1/2						0	1-5/16	1-5/8	1-15/16	2-1/2	3-1/8	4-1/8	5-1/4	6-1/4	6-7/8	7-7/8	8-15/16	9-15/16	11-15/16	14-15/16
3							0	1-3/8	1-11/16	2-3/8	3	4-1/16	5-3/16	6-3/16	6-7/8	7-7/8	8-7/8	9-7/8	11-15/16	14-15/16
3-1/2								0	1-1/2	2-3/16	2-7/8	3-15/16	5-1/8	6-1/8	6-13/16	7-13/16	8-13/16	9-7/8	11-7/8	14-7/8
4									0	2	2-11/16	3-7/8	5	6-1/16	6-3/4	7-3/4	8-13/16	9-13/16	11-7/8	14-13/16
5										0	2-1/4	3-9/16	4-13/16	5-7/8	6-9/16	7-5/8	8-11/16	9-11/16	11-3/4	14-13/16
6											0	3-3/16	4-9/16	5-11/16	6-3/8	7-7/16	8-1/2	9-9/16	11-5/8	14-3/4
8												0	3-13/16	5-1/8	5-7/8	7-1/16	8-1/8	9-1/4	11-3/8	14-1/2
10													0	4-1/8	5	6-5/16	7-9/16	8-3/4	10-15/16	14-3/16
12														0	3-13/16	5-7/16	6-13/16	8-1/16	10-7/16	13-13/16
14															0	4-11/16	6-1/4	7-5/8	10-1/16	13-1/2
16																0	5	6-5/8	9-3/8	13
18																	0	5-1/4	8-1/2	12-3/8
20																		0	7-5/16	11-5/8
24																			0	9-5/8

Source: Texas Pipe Bending Co., Inc., Houston, Texas.

Pipe Fabrication

Table 6-16
ID of Nozzle to OD of LR Ell on Centerline
Standard Weight Pipe

HEADER ELL	1/2	3/4	1	1-1/2	2	2-1/2	3	3-1/2	4	5	6	8	10	12	14	16	18	20	24	NOZZLE
	13/16	5/16	5/16	5/16	7/16	1/2	1/2	5/8	11/16	13/16	15/16	1-5/16	1-9/16	1-15/16	2-13/16	3-3/16	3-9/16	3-7/8	4-5/8	1/2
		1/2	1/2	7/16	9/16	5/8	5/8	11/16	13/16	15/16	1-1/16	1-7/16	1-11/16	2	2-15/16	3-5/16	3-11/16	4	4-11/16	3/4
			9/16	11/16	3/4	13/16	13/16	7/8	15/16	1-1/16	1-3/16	1-9/16	1-13/16	2-3/16	3-1/8	3-7/16	3-13/16	4-3/16	4-7/8	1
				1-5/16	1-1/4	1-5/16	1-3/16	1-1/4	1-3/8	1-7/16	1-9/16	1-7/8	2-1/8	2-1/2	3-1/8	3-13/16	4-1/8	4-1/2	5-3/16	1-1/2
					1-7/8	1-3/4	1-9/16	1-5/8	1-11/16	1-3/4	1-7/8	2-3/16	2-7/16	2-3/4	3-7/16	4-1/16	4-7/16	4-3/4	5-7/16	2
						2-5/16	2	2	2-1/16	2-1/16	2-1/8	2-7/16	2-11/16	3-1/16	3-3/4	4-5/16	4-11/16	5	5-3/4	2-1/2
							2-7/8	2-11/16	2-5/8	2-9/16	2-5/8	2-7/8	3-1/8	3-7/16	4	4-11/16	5-1/16	5-3/8	6-1/16	3
								3-1/2	3-1/2	3-1/16	3-1/16	3-1/4	3-7/16	3-3/4	4-3/8	5-1/16	5-3/8	5-11/16	6-3/8	3-1/2
									4	3-9/16	3-7/16	3-5/8	3-3/4	4-3/4	4-3/4	5-1/8	5-11/16	6	6-11/16	4
										5-3/16	4-5/8	4-9/16	4-5/8	5	5-3/4	6-1/16	6-3/8	6-11/16	7-5/16	5
											6-3/8	5-11/16	5-1/2	5-11/16	6-5/8	6-7/8	7-1/8	7-7/16	8-1/16	6
												8-13/16	7-5/8	7-1/2	8-3/8	8-1/2	8-11/16	8-15/16	9-7/16	8
													11-1/8	9-15/16	10-5/8	10-1/2	10-1/2	10-5/8	11-1/16	10
														13-3/4	13-9/16	12-7/8	12-5/8	12-9/16	12-3/4	12
															16-7/16	14-3/4	14-1/8	13-15/16	13-15/16	14
																19-1/8	17-1/8	16-7/16	16	16
																	21-13/16	19-5/8	18-5/16	18
																		2–0-9/16	21-1/16	20
																			2–6	24

FORMULA:

$Z = R + \dfrac{O.D. \text{ Of Ell}}{2}$

$S = 1/2 \text{ I.D. Nozzle}$

$Z^2 - (R + S)^2 = Y^2$

$R - Y = X$

Source: Texas Pipe Bending Co., Inc., Houston, Texas.

Chapter 6
Review Test

1. Define pipe "fabrication." _____

   _____

2. Normally, pipe _____ inch and larger is shop fabricated.

3. Define AWS. _____

4. Shop spools are dimensioned to the closest _____ inch.

5. A pipe bends radius should not be less than _____ pipe diameters.

6. Miter welds are specified in place of _____ for low pressure services and for use in _____.

7. For minimum pressure drop, a _____ weld miter is used.

8. Branches and nozzles are specified as OD to ID and ID to OD types. Prepare a cross-sectional view explaining these two types.

9. What is a "cutback?" _____

   _____

10. Define how a shop spool differs from a contractor's spool. _____

    _____

7 Vessels

Definitions

Vessels are the heart, or main piece, of a refinery or chemical plant. Things happen inside vessels. In reactors, a chemical change is taking place. In fractionating towers, a separation is occurring. The vessel orientation, or location of nozzles on the shell, is critical to every piping layout. While orienting a vessel, the competent piping designer will call upon his past experience and his good judgment.

Vessels have many names. A fractionating tower is a vessel and is sometimes called a fractionating column or just a column. Fractionating towers are then given names according to the function they perform. A Depropanizer fractionates out propane, which leaves the Depropanizer as a vapor coming out of the top part of the tower. A Debutanizer fractionates out butane. A Deisobutanizer fractionates out isobutanes, etc. In short, the name of the fractionating tower will tell you its purpose in the petrochemical complex.

Another vessel is a reflux accumulator. This is a horizontal vessel that is sometimes called an overhead accumulator because it accumulates the overhead product from the fractionators noted in the paragraph above. The overhead leaves the fractionating tower as a vapor, is cooled and condensed in an exchanger and flows as a liquid (sometimes with some vapor) to the overhead accumulator.

This vessel then holds a level of liquid used as "reflux" liquid, which is pumped back to the fractionating tower's top tray. Any liquid, above the fractionator's set need for reflux, is pumped out as product. A reflux accumulator usually has little, if any, internals.

A reactor is usually a vertical vessel containing a reactive catalyst used to rearrange the molecular structure of the stream being fed to it. The stream coming from the reactor can then be fractionated into the desired salable products. A reactor is where a chemical change occurs. A catalyst causes this reaction but does not take part in the reaction itself.

Separators may be either horizontal or vertical vessels. They are used to separate usually a vapor from a liquid. They also can be used to separate two liquids which have different specific gravities. To separate two liquids, a long vessel with very little velocity and long retention time is specified by the process engineer.

Horizontal Vessels

Horizontal Vessels Above 15'-0" Elevation

Figure 7-1 is a horizontal vessel, a reflux accumulator. To orient the nozzles on this vessel, the elevation must be known. Since reflux accumulators have pumps which take suction from them,

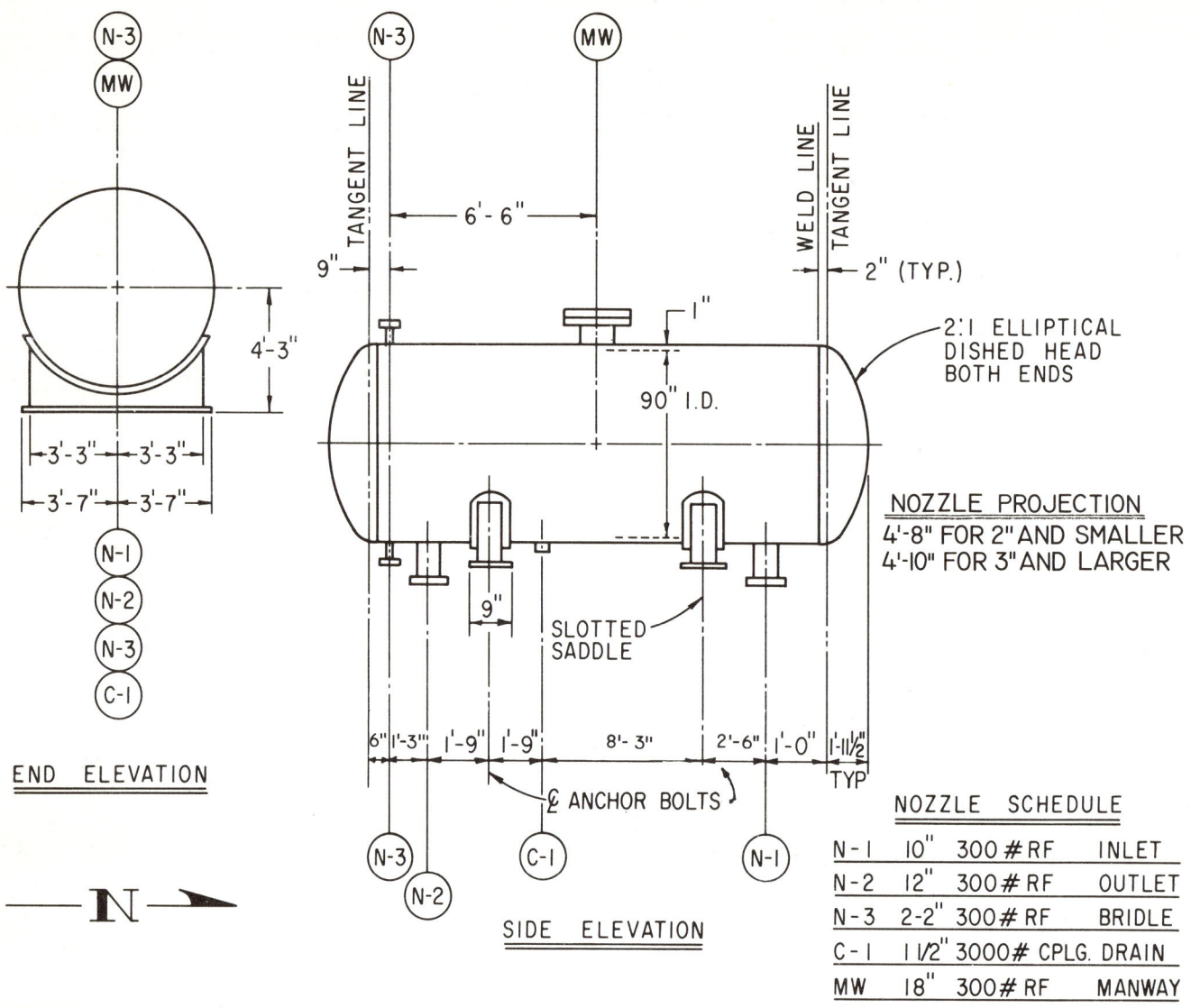

Figure 7-1. Horizontal vessel, reflux accumulator. Courtesy of Fluor Corp.

they must be elevated to supply adequate "head" (NPSH, see "Pumps" in Volume 2). This usually means the bottom of the vessel will be about 14'-0" above the high point of the finish surface. Since piping drawings always work to centerlines, and this vessel is 6'-0" ID, the centerline of this vessel will be at 3'-0" plus 14'-0" or 17'-0" above the high point of the finish surface.

Platforms

When the elevation of the centerline is over 15,'-0" above grade, ladders and platforms must be provided for access to the manhole, instruments and any valves which may be operating valves. Fifteen feet is an acceptable break point for permanent platforming. Below this, portable platforming or ladders are generally used. Some operating companies will build temporary scaffolding for access to items under 15'-0" elevation. Ladders and platforms are expensive for first installation cost as well as for continual maintenance. It is good design to keep them to a minimum. Platforms weigh about 30 pounds per square foot. Ladders without cages weigh about 12 pounds per linear foot and ladders with cages weigh about 18 pounds per linear foot.

Structural steel, fabricated but not assembled, will vary in cost from year to year. To that cost, the shipping and field erection costs must be added. Using $1.00 per pound installed as a rule of thumb, the cost of a 4'-0" x 10'-0" platform would be 40 square feet at 30 pounds per square foot or 1200 pounds. Seventeen feet of ladder without cage would add another 204 pounds to total 1404 pounds. This, then, would bring the cost to $1404. But these are not all the costs. Someone had to design the platform, make a drawing and purchase it. Vendor's prints had to be checked and returned. Letters had to be written to the vendor outlining the corrections required to his drawing. By the time all expenditures are added in, that little platform will have cost over $2000! So use platforms only where absolutely necessary and then keep them to a minimum size.

Platforms for horizontal and vertical vessels should be made from ¼" checkered plate except in areas of heavy snow where floor grating should be used. See Figure 7-2 for typical ladder and platform details for horizontal vessels.

Nozzle Orientation

Orienting the nozzles (locating them along the shell) on a horizontal vessel is basically a simple job. The pump suction must be located on the same end that the pumps are located to keep the pump suction line a minimum length so that pressure drop will be at the minimum. Every foot of pipe and every fitting causes pressure drop in any line. This is called "line loss."

Besides the pump suction (usually called liquid outlet on the vessel drawing), there are inlet, vapor outlet, level control, level gage and manhole connections. There could be other nozzles for special cases which will be shown on the mechanical flow diagram where required.

The main rule in orienting horizontal vessels is to locate the inlet and outlet nozzles a maximum distance apart. Since you may have two outlets, the liquid and vapor outlet, this means that the inlet will go at the extreme far end of the vessel (away from the rack) and the vapor and liquid outlets will be located at the extreme near end. Inlets will be located on the top part of the vessel. Liquid outlets are set on the bottom. Vapor outlets will be located on the top part of the shell.

The vessel length is determined by the process engineer to suit (*a*) an economical diameter-to-length ratio, and (*b*) a length to satisfy "disengaging requirements." What is "disengaging"? The inlet may enter the horizontal vessel as liquid and vapor. In the vapor phase, there are small droplets of liquid. This is called "entrained liquid." Gravity will settle out most of the droplets if the vapor is kept in the vessel long enough in a basically low flow condition. This time is calculated and is called "retention time" and helps determine the vessel length. The retention time is then determined by the disengaging requirements. A vapor, containing more droplets of liquid, will require more disengaging, which in turn means a longer vessel.

If the piping designer located the inlet and the vapor outlet side by side, the whole purpose of the horizontal vessel would be destroyed. The vapor coming in would just duck in to the vessel and right out the outlet, taking with it all the entrained liquid. Disengaging could not occur.

Also in the liquid portion of the inlet feed a small quantity of vapor bubbles may be found. These bubbles will rise out of the liquid into the vapor area if the vessel has adequate retention time. Again, the liquid outlet must be located at the opposite end from the inlet nozzle to enable this separation to occur.

The liquid level of these accumulators is usually maintained at the vessel centerline. It is kept there by a liquid level controller. A level gage is provided so the operator can visually check the level to ensure that the level controller is functioning properly. The level gage and level controller

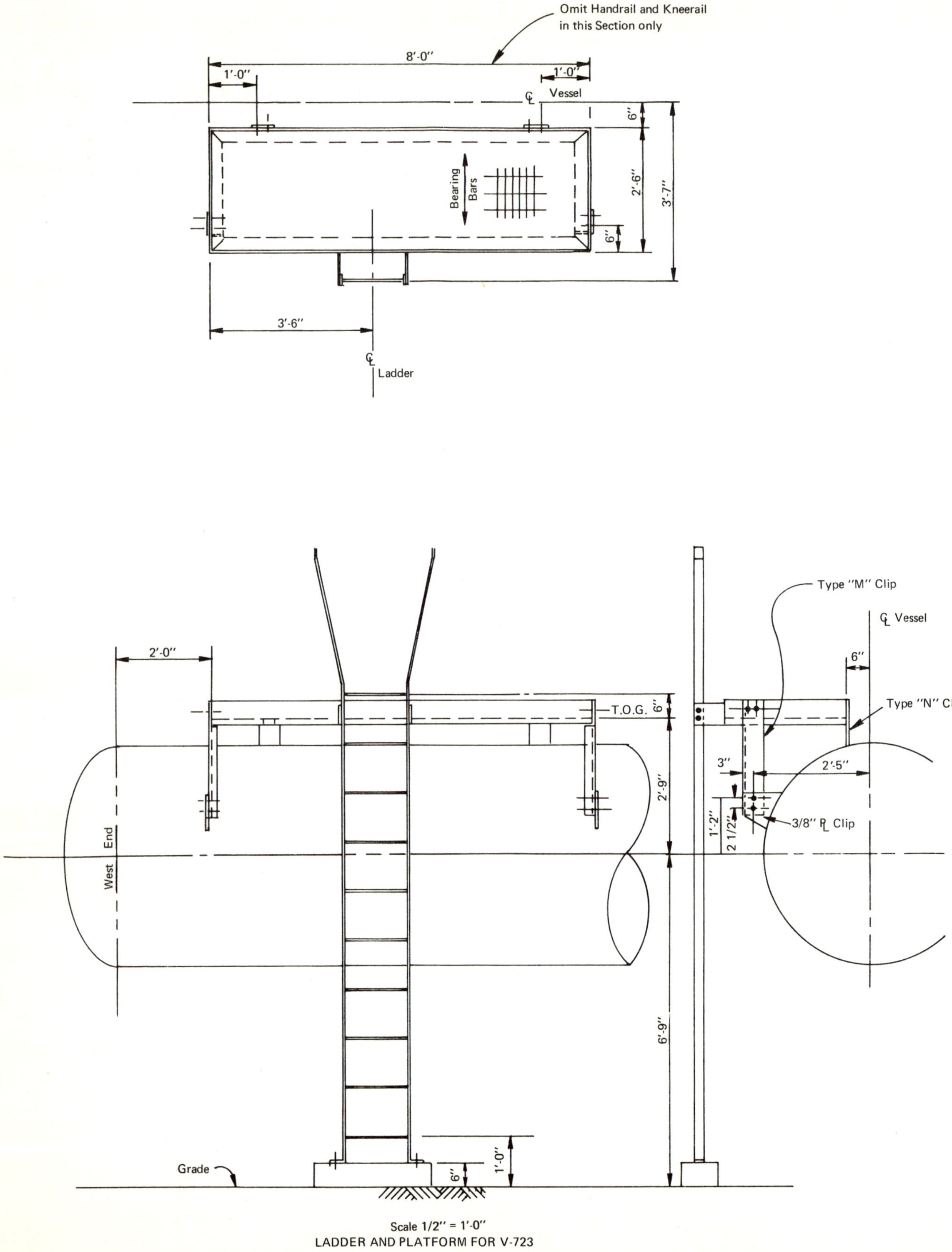

122   Process Piping Design

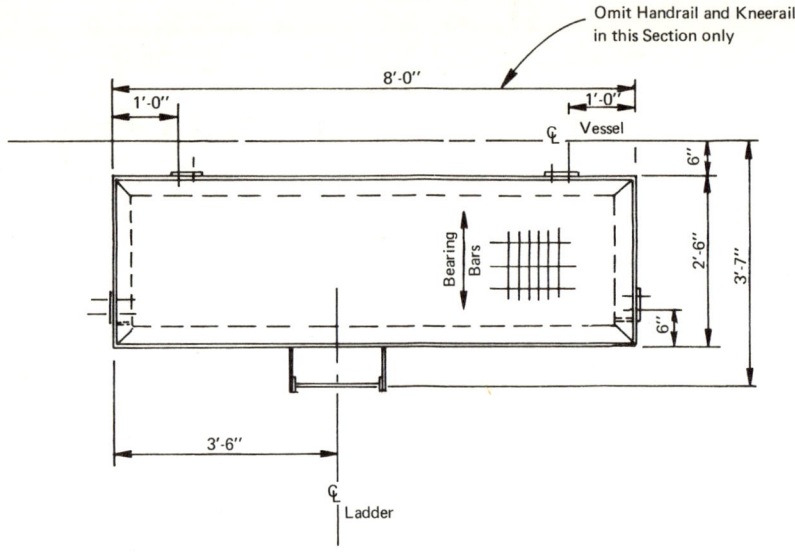

Figure 7-2. Typical ladder and platform details. Courtesy of Fluor Corp.

Vessels

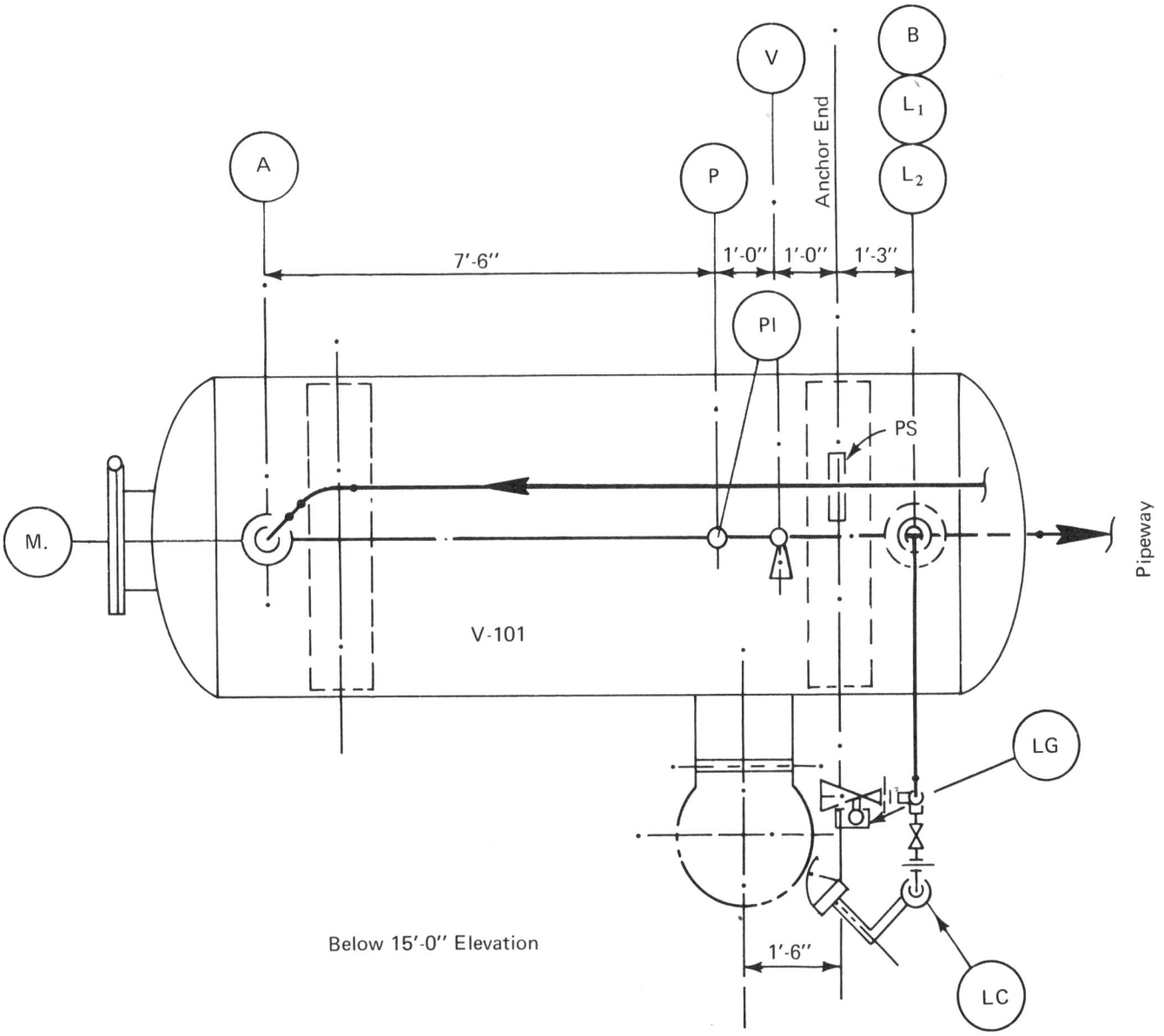

Figure 7-3. Horizontal accumulator, elevation below 15'-0". Courtesy of Fluor Corp.

should be oriented on the same end as the outlets, visible to the operator as he walks down the operating aisle. They should also be located together. The operator can adjust the level controller and change the level point in the vessel. While doing this, he needs to view the level gage to check his adjustment.

Why not locate these instruments on the inlet end? The inlet is coming in as liquid and vapor at a higher velocity than the flow through the vessel. This will cause higher turbulence at this end of the vessel. This motion of the liquid would move the float of the level controller and the level control would not function properly. Always locate level instruments in the least possible area of turbulence.

Saddles

Horizontal vessels are supported by saddles welded to the vessel shell. Except for special design cases, these saddles should be located about 15% of the vessel length in from each end, or about 70% of the vessel length apart. For a vessel which is 20'-0" tangent to tangent, the vessel saddles would be located 14'-0" apart and 3'-0" in from each tangent line. For special thin wall and/or long vessels, the vessel design department should be consulted for the best location of these saddles.

This 20'-0" long vessel is going to expand and contract due to heat. The designer wants to control this so as not to put excess strain on his pumps. The saddle near the rack will have round holes drilled in it to receive the anchor bolts. This end will be bolted down tightly and is designated as the "anchor" end, causing the vessel to grow away from the rack and the pumps.

The other saddle is provided with slotted holes for the anchor bolt and is *not* bolted down tightly, which enables the saddle to slide on the foundation as expansion or contraction occur. This is called the "sliding" end.

If the vessel is very hot or very long and a large amount of growth could occur or if the vessel is in cyclic service, meaning that the temperature rises and falls during normal operation, then "slide plates" are embedded in the concrete grout for the sliding saddle to rest on. If sliding was extensive on bare concrete, it would eventually chip away the concrete due to the friction resistance. This slide plate is installed to reduce the resistance and protect the concrete. Slide plates are usually made from sheet steel plate of ½" thickness, but for some cases special "lubrite" plate, an oil impregnated steel, is used. Teflon slide plates are now used for special cases.

In cryogenic piping, which involves subfreezing operating temperatures, supporting horizontal vessels presents a different problem. If the vessel is operating at -50°F the saddle would freeze the concrete. The slide plate would freeze to the saddle and as the vessel would then contract, the concrete would be harmed. In this case, wood blocks about 6-8" thick would be installed on both concrete piers. The anchor bolts would project through the wood blocks and the saddles would rest on these wood blocks. Wood will serve as an insulation barrier, keeping the freezing temperatures from reaching the concrete. Both saddles could have round holes for anchor bolts, as the contraction is usually small, and with bolt holes about 5/16" larger than the anchor bolts, this slop will usually take care of any contraction. If the contraction is expected to be large, a slotted saddle would be provided and would be located on the end away from the pipe rack and pumps. Another reason for locating the slotted saddle on this end is that the vessel and the inlet line both grow or contract together, reducing expansion forces. The inlet line is usually the largest line connecting to the horizontal vessel and consequently will exert the greatest forces.

Miscellaneous Connections

Other connections to horizontal accumulators will be the vent and/or drain nozzles. Both require valves which must be accessible. If the liquid outlet to the pumps does not have projection inside the vessel, it is preferable to install the drain on the lowest possible part of the pump suction line, routing the drain to the drain funnel which services the pump. This will save the cost of supplying a drain funnel to the vessel. The vent may go to the atmosphere or to the flare header. The flow diagram will define this routing.

Relief valves are sometimes located on top of these accumulators. This is costly and the competent designer will avoid this installation. It is more desirable to locate relief valves on piping going to the vessel. Adding a nozzle to the pipe is much cheaper than installing the same size nozzle on a vessel. The vessel is built to the ASME section VIII code and requires special welding. This makes vessel nozzles expensive.

The operator will need to know the temperature of the fluid in this vessel. This means a thermowell must be installed. The thermowell could be installed in the pump suction line or in the discharge line, but since any obstruction in the suction piping would cause pressure drop, it should be installed in the discharge piping. If there is an operating pump and a spare pump, locate the thermowell after the discharges are combined so the thermowell will be in the flowing fluid, regardless of which pump is operating. Never install a temperature connection in a "dead leg" or nonflowing portion of a line. The reading would be false and therefore of no value. The temperature of the pump suction and discharge is the same, and is also the temperature of the fluid in the vessel.

Inspection openings are required to be located on vessels by the ASME section VIII code. These are usually 18" manways and most horizontal vessels have at least one. The best place to locate inspection openings is in the head, on the centerline of the vessel, if platforming is not required. If platforms are required, the openings will be located on the vessel shell, either on the top or side, depending on platforming required for other vessel connections. These manways may be 18" OD pipe with WN flange and blind, with hinge or davit as required. Hinges should be used for manholes mounted on the side of the shell. Davits work better for top mounted manholes. Any manhole which is top mounted and hinged should have a "hinge stop" installed to keep the blind flange (manhole cover) from slamming down on the vessel shell. The hinge stop should be set so that the cover will open only 135°.

Horizontal Vessel Below 15'-0" Elevation

Figure 7-3 is a piping plan of a horizontal vessel located below 15'-0" elevation. No platforms are provided. The manhole is located in the head, opposite the pipeway, near the inlet. The anchor support is located on the end near the pipeway. The LC and LG are shown installed on a "bridle." A bridle is a fabricated piece of pipe which has a liquid and vapor connection to the vessel. The liquid level in the bridle is the same as in the vessel, as liquid will seek its own level. Instruments then are connected to the bridle.

Bridles

Bridles have a definite use but are costly and so should be specified only where required. In most cases it is much cheaper to install LC and LG connections on the vessel and eliminate the bridle. When bridles are mandatory, they should always be installed "self-draining." The lower connection, in the liquid section of the vessel, should be located about 6" above the bottom of the vessel. Never locate the bridle connection on the bottom of the vessel. It will accumulate anything that may enter the vessel and stop-up the bridle, and the instruments will not function. The author remembers a case where a clogged bridle was found to have old rags in it. No one could offer an explanation as to where they came from.

Miscellaneous Connections

The connections shown in Figures 7-3 are *M*-Manhole, *A*-Inlet, *P*-Pressure Gage, *V*-Vent, *B*-Liquid outlet and *L-1* and *2* are the bridle connections. The *PS* is a pipe support, built up from the vessel to support the inlet line. This vessel has all liquid coming into it so there is no vapor outlet.

Grade Mounted Vessel

Figure 7-4 shows a general arrangement for a horizontal vessel which is grade mounted. Its height is determined by the elevation necessary to clear the piping out of the bottom of the bootleg. A bootleg is installed to separate and draw off water. Since water is heavier than hydrocarbons it will drop out in the boot and can be seen in the boot's LG. It will then be drawn off through the LC valve. Some similar designs use a manually operated draw-off valve. This would be a globe body valve used for throttling. The operator would view the gage glass on a regular schedule. As more water accumulated in the boot, he would slightly open the globe valve and and draw off the water until it was below the viewing area of the LG. With the manual design, the LG is visible while the draw-off valve is operating. With the automatic design, the LG is visible while the control valve by-pass is operating.

Bootlegs are located as close to the calm end of the vessel as possible, away from the inlet end. This gives ample time for the liquid to settle to the bottom part of the vessel and drop into the boot.

Note the different possible locations for the manhole. The most desirable location is noted by *1.5* in Figure 7-4. Also shown dotted, the manway could be located below the centerline, making it easier to get in and out of the vessel. This is called a "hillside" connection and the welding is much more expensive. Use this only where dropping the manhole is specified by the customer or where doing so would eliminate the need for platforming.

Also note that the bridle bottom nozzle is specified as tangential. It will be self-draining.

Piping Arrangement for Elevated Vessels

Figure 7-5 depicts a general arrangement for a horizontal vessel which is elevated above 15'-0". The platform shown is at an elevation of 125'-3", 25'-3" above the grade elevation. The manhole is located on top of the vessel, with a hinge and hinge stop. Platforming extends all around the manhole, enabling the bolts to be removed. Also located on the platform is a control station, PIC. The inlet is

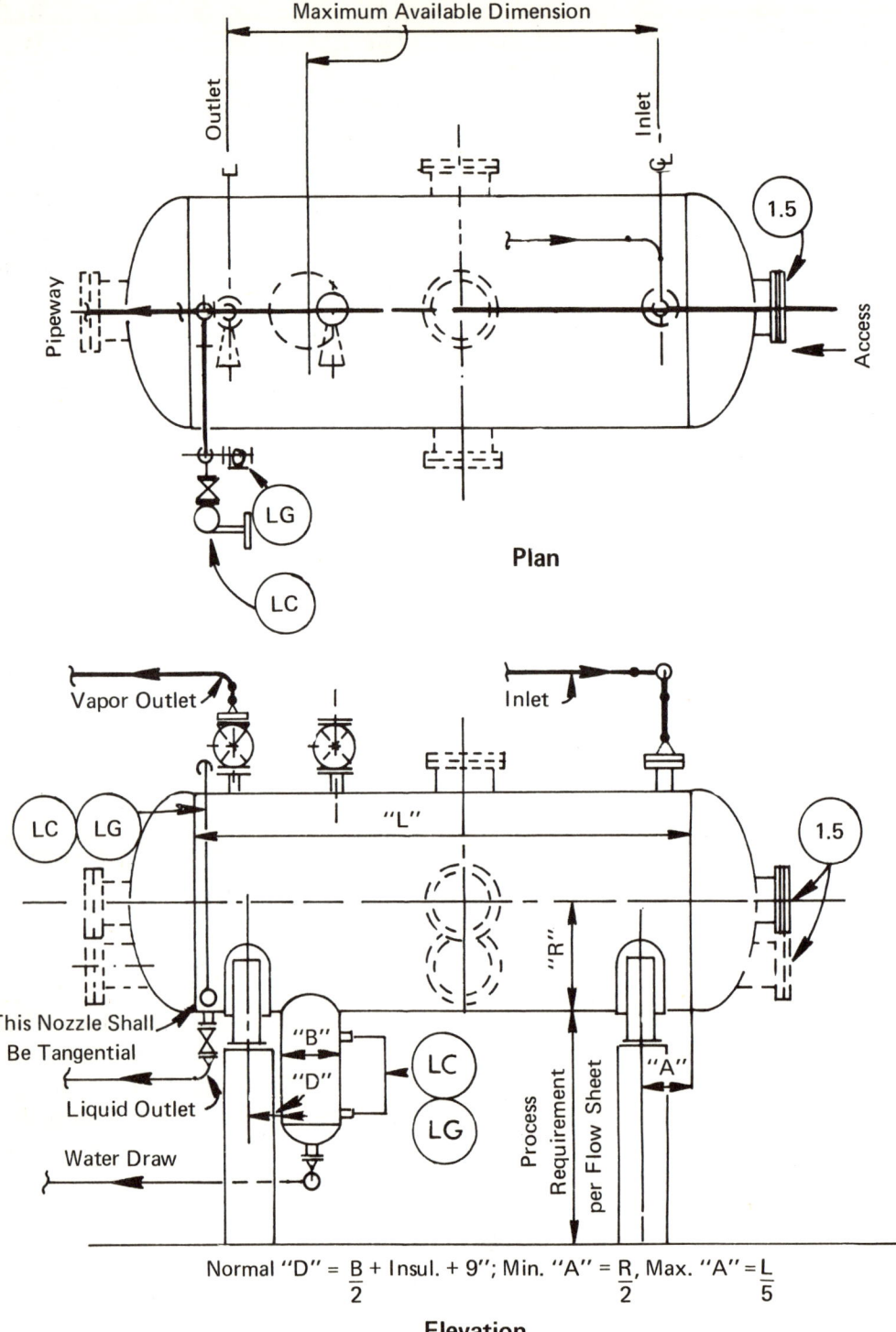

Figure 7-4. Horizontal vessel, general arrangement. Courtesy of Fluor Corp.

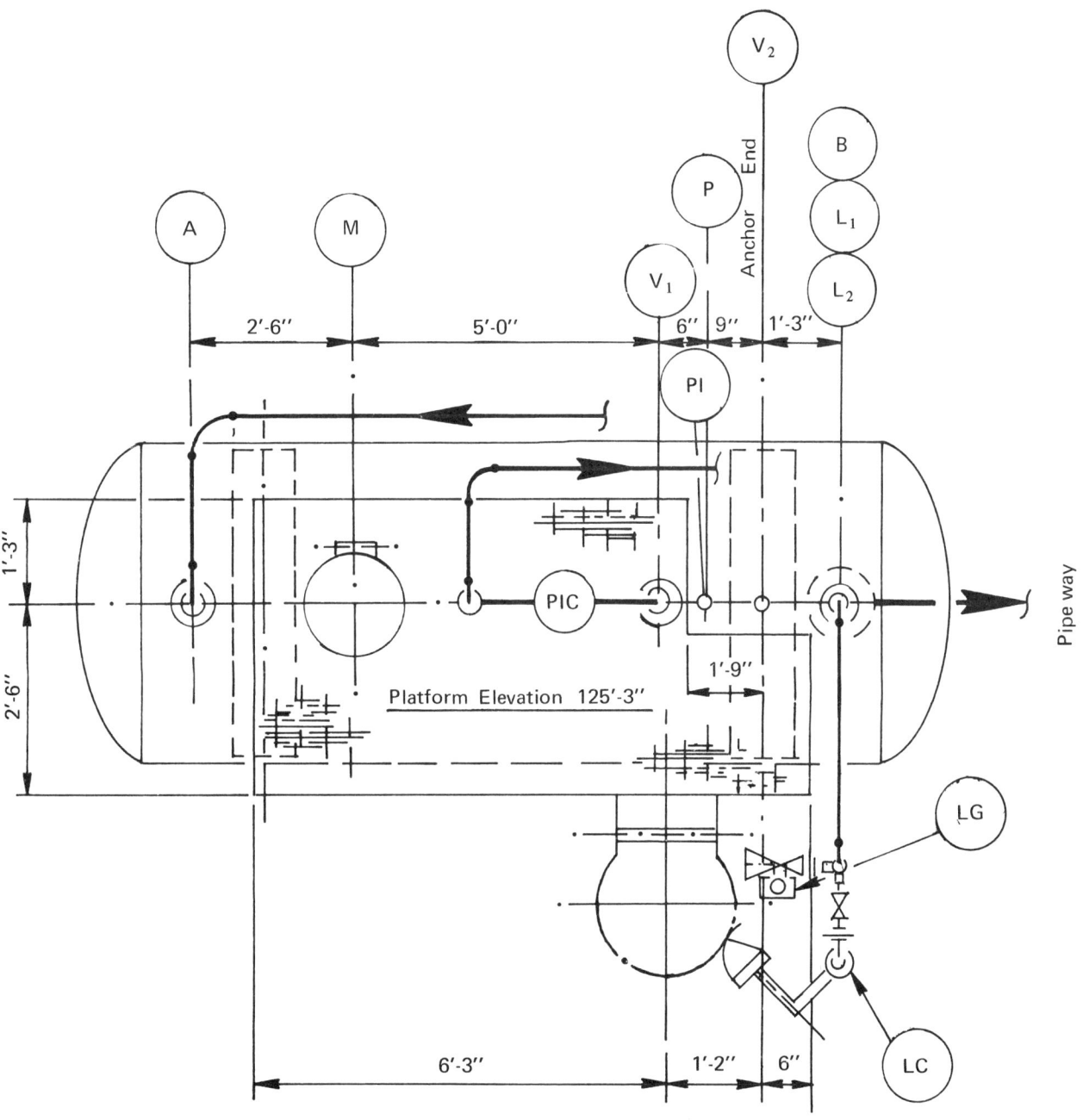

Figure 7-5. Horizontal vessel, elevation above 15'-0". Courtesy of Fluor Corp.

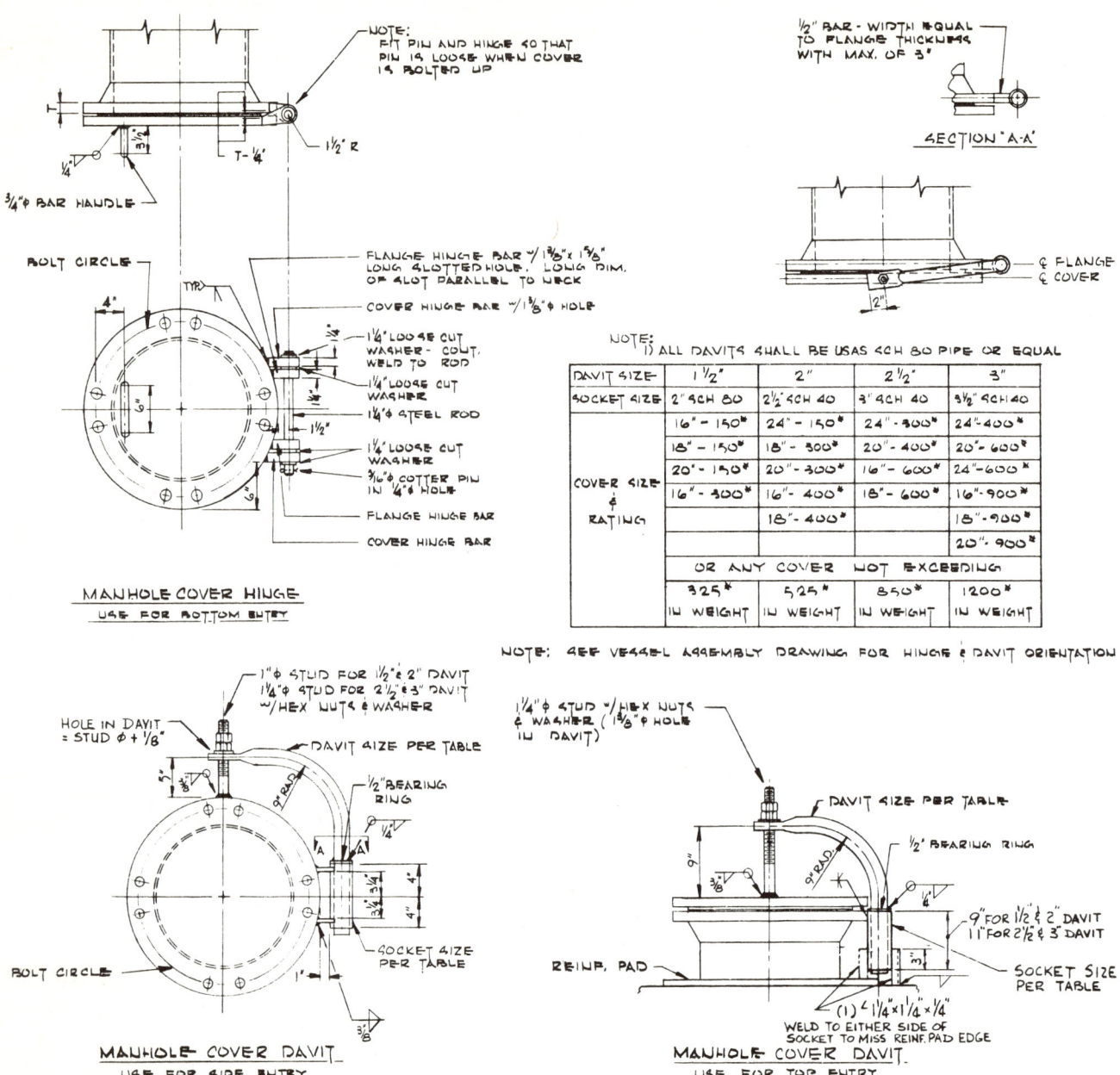

Figure 7-6. Manhole davits and hinges. Courtesy of Fluor Corp.

routed to the end away from the pipeway and the liquid outlet is on the pipeway end.

Dimensions shown are to locate nozzles and to outline extremities of the platform. The vessel group must show this data on their vessel drawing and on their platform detail drawing. Dimensions will appear on the piping layout drawing but will not necessarily be shown on the finished piping drawing. Final piping drawings will show the platform drawn to scale and piping will be dimensioned as required.

Manhole Davits and Hinges

Figure 7-6 shows Fluor's standard manhole hinge and davit details. As the cover gets larger and the flange rating increases the davits get larger. These blind flanges are heavy. An 18"-300 pound blind flange weighs 390 pounds!

The piping designer will always locate hinges and swing an arc from the hinge to make sure the blind flange will open wide. He does *not* want to locate small instrument connections behind the swinging cover. That's a good way to start a fire in the plant!

Saddle Details

Figure 7-7 is Fluor's typical saddle detail for horizontal vessels. *H* is the width of the "wear plate" which is continuously welded to the outside of the vessel to protect the vessel, distributing the load over a large area of the shell. "Ribs" are vertical plates added as reinforcement for vessels of a heavier load. As the load gets larger the ribs get thicker.

Piping Arrangement

Figure 7-8 shows how a section across a typical pipeway might look. The pump suction line, *A4*, coming from the horizontal vessel must not be pocketed but still must be routed above the aisleway, *C5*. Other points of this figure will be discussed in other related chapters.

Vertical Vessels

Fractionating Towers

Vertical vessels are many and varied. Reactors are usually vertical; separators may be either horizontal or vertical and fractionating towers must be vertical. There are countless other vertical vessels, but we will cover the ones that designers come in contact with most.

Fractionating towers are the inexperienced designer's Waterloo. At an employment interview the applicant will be asked if he can orient a tower. In other words, can the designer locate all the nozzles correctly around the circumference of the tower and in the proper elevation, including all manholes, instruments, ladders and platforming where required, skirt connections and all the other connections pertaining to the monster. To do this, the designer must have a good knowledge of the vessel trays. Trays are the horizontal plates which are located about 2′ apart and always seem to be in exactly the place where the designer wants to put a nozzle.

The most common types of trays are called "one pass" and "two pass." Three and four pass trays are also used and become even more difficult to lay out.

What does the designer need to start orienting a tower?

1. Mechanical and utility flow diagrams
2. Vessel outline drawing
3. Plot plan drawing
4. Job data such as customer specifications, platform requirements, etc.
5. Instrument data on LG, LC, PSV, etc.
6. Tray design such as number of passes, dimension of downcomer area and tray spacing
7. Line list showing pressures, temperatures and whether the fluid is liquid or vapor

When a designer orients a tower he assumes the responsibility for the tower and its related equipment. This responsibility includes the equipment coordinates, location of tray downcomers, piping equipment connections, instrument connections, manholes, ladders and platforms, and the necessary pipe supports and guides on the tower.

Manhole Orientation

Figure 7-9 shows the single pass and the two pass tray and possible locations for manholes. The downcomer is the vertical part of the tray which extends above the horizontal tray plate to form a dam, forcing liquid to have a constant level on the tray. The downcomer also drops down below the liquid level of the tray below, forming a vapor seal.

Tray Types

Trays come in several types, all with the same general purpose—to fractionate out lighter hydrocarbons from a liquid. The heavier hydrocarbons will remain in their liquid state and flow down the downcomer to the next tray. The lighter hydrocarbons will vaporize and flow up through the holes in the tray plate, making contact with the liquids on that tray. Since the downcomer drops below the level of liquid on the tray plate, vapor cannot by-pass the tray, and since vapor is light it must flow up through the bottom of the tray.

Tray types are:

1. *Bubble cap trays* were used for many years but are now obsolete. A standpipe was attached to the tray for vapor to pass

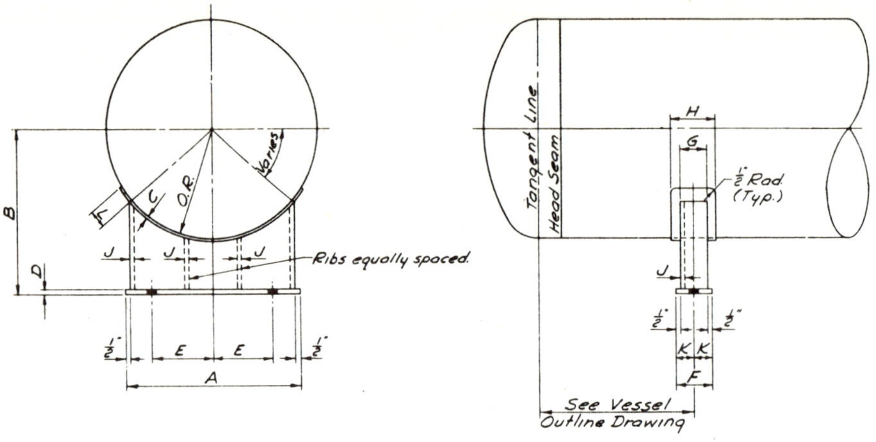

TYPICAL SADDLE DETAILS–EXCLUSIVE OF MARK 24a, 24b & 30a

VESSEL DIAM. IN INCHES	SADDLE MARK No.	MAX. VESSEL OPER. WEIGHT LBS.	A	B	C	D	E	F	G	H	J	K	L	BOLT DIAM.	No. OF RIBS	WEIGHT PER SET LBS.
24	24a	1,070	1'-10"	1'-3"	1/4"	1/2"	7"	4"	—	3"	3/8"	2"	1 1/4"	1"		70
	24b	3,770	"	"	"	"	"	"	—	"	1/2"	"	"	"		80
	24c	20,000	"	"	"	"	9"	"	3"	5"	1/4"	"	"	"		80
	24d	31,000	"	"	"	"	"	"	"	"	3/8"	"	"	"		90
	24e	42,000	"	"	"	"	"	"	"	"	1/2"	"	"	"		100
30	30a	2,840	2'-2"	1'-6"	1/4"	1/2"	9"	4"	—	3"	1/2"	2"	1 1/2"	1"		120
	30b	20,000	"	"	"	"	11"	"	3"	5"	1/4"	"	"	"		100
	30c	31,000	"	"	"	"	"	"	"	"	3/8"	"	"	"		130
	30d	42,000	"	"	"	"	"	"	"	"	1/2"	"	"	"		150
36	36a	20,500	2'-8"	2'-0"	1/4"	1/2"	1'-0"	6"	5"	7"	1/4"	3"	2"	1"		170
	36b	32,500	"	"	"	"	"	"	"	"	3/8"	"	"	"		200
	36c	42,500	"	"	"	"	"	"	"	"	1/2"	"	"	"		220
	36d	52,500	"	"	"	5/8"	"	"	"	"	5/8"	"	"	"		260
42	42a	25,000	3'-2"	2'-3"	1/4"	1/2"	1'-3"	6"	5	7"	1/4"	3"	2 1/4"	1"		200
	42b	34,000	"	"	"	"	"	"	"	"	5/16"	"	"	"		220
	42c	48,000	"	"	"	"	"	"	"	"	1/2"	"	"	"		270
	42d	97,500	"	"	"	5/8"	"	"	"	"	5/8"	"	"	"	1	330
48	48a	27,500	3'-8"	2'-6"	1/4"	1/2"	1'-6"	6"	5	7"	1/4"	3"	2 1/2"	1"		230
	48b	39,000	"	"	"	"	"	"	"	"	5/16"	"	"	"		250
	48c	62,500	"	"	"	"	"	"	"	"	1/2"	"	"	"		310
	48d	116,250	"	"	"	5/8"	"	"	"	"	5/8"	"	"	"	1	380
54	54a	25,000	4-2"	2'-9"	1/4"	1/2"	1'-9"	6	5	7	1/4"	3	2 3/4"	1"		270
	54b	47,500	"	"	"	"	"	"	"	"	3/8"	"	"	"		310
	54c	75,000	"	"	"	"	"	"	"	"	1/2"	"	"	"		360
	54d	135,000	"	"	"	5/8"	"	"	"	"	5/8"	"	"	"	1	440
60	60a	22,000	4'-8"	3'-0"	1/4"	1/2"	2'-0"	6"	5	7"	1/4"	3"	3"	1"		310
	60b	51,000	"	"	"	"	"	"	"	"	3/8"	"	"	"		370
	60c	87,500	"	"	"	"	"	"	"	"	1/2"	"	"	"		400
	60d	154,000	"	"	"	5/8"	"	"	"	"	3/8"	"	"	"	1	510
66	66a	21,400	5'-2"	3'-3"	1/4"	1/2"	2'-3"	6"	5"	7"	1/4"	3"	3 1/2"	1"		350
	66b	36,000	"	"	"	"	"	.	"	"	3/16"	"	"	"		380
	66c	55,000	"	"	"	"	"	"	"	"	3/8"	"	"	"		410
	66d	90,000	"	"	"	"	"	"	"	"	1/2"	"	"	"		470
	66e	124,000	"	"	"	5/8"	"	"	"	"	5/8"	"	"	"		570
	66f	185,000	"	"	"	"	"	"	"	"	"	"	"	"	1	580
	66g	248,000	"	"	"	"	"	"	"	"	"	"	"	"	2	590
72	72a	33,500	5'-8"	3'-6"	1/4"	1/2"	2'-6"	6"	5	7"	5/16"	3"	3 3/4"	1"		420
	72b	49,250	"	"	"	"	"	"	"	"	3/8"	"	"	"		460
	72c	99,000	"	"	"	"	"	"	"	"	1/2"	"	"	"		540
	72d	121,500	"	"	"	5/8"	"	"	"	"	5/8"	"	"	"		640
	72e	192,500	"	"	"	"	"	"	"	"	"	"	"	"	1	650
	72f	305,000	"	"	"	"	"	"	"	"	"	"	"	"	3	680

VESSEL DIAM. IN INCHES	SADDLE MARK No.	MAX. VESSEL OPER. WEIGHT LBS.	A	B	C	D	E	F	G	H	J	K	L	BOLT DIAM.	No. OF RIBS	WEIGHT PER SET LBS.
78	78a	53,000	6'-2"	3'-9"	5/16"	3/4"	2'-9"	8"	7"	9"	3/8"	4"	4"	1"		710
	78b	110,000	"	"	"	"	"	"	"	"	1/2"	"	"	"		790
	78c	141,000	"	"	"	"	"	"	"	"	5/8"	"	"	"		870
	78d	210,000	"	"	"	"	"	"	"	"	"	"	"	"	1	880
	78e	352,000	"	"	"	"	"	"	"	"	"	"	"	"	3	910
84	84a	50,000	6'-8"	4'-0"	5/16"	3/4"	3'-0"	8"	7"	9"	3/8"	4"	4 1/4"	1"		810
	84b	116,000	"	"	"	"	"	"	"	"	1/2"	"	"	"		910
	84c	150,000	"	"	"	"	"	"	"	"	5/8"	"	"	"		1020
	84d	225,000	"	"	"	"	"	"	"	"	"	"	"	"	1	1030
	84e	375,000	"	"	"	"	"	"	"	"	"	"	"	"	3	1050
90	90a	47,500	7'-2"	4'-3"	5/16"	3/4"	3'-3"	8	7	9"	3/8"	4"	4 1/2"	1"		880
	90b	115,000	"	"	"	"	"	"	"	"	1/2"	"	"	"		990
	90c	162,500	"	"	"	"	"	"	"	"	5/8"	"	"	"		1100
	90d	243,000	"	"	"	"	"	"	"	"	"	"	"	"	1	1110
	90e	325,000	"	"	"	"	"	"	"	"	"	"	"	"	2	1130
	90f	487,500	"	"	"	"		"	"	"	"	"	"	"	4	1160
96	96a	44,250	7'-8"	4'-6"	5/16"	3/4"	3'-6"	8"	7"	9"	3/8"	4"	5"	1"		940
	96b	100,000	"	"	"	"	"	"	"	"	1/2"	"	"	"		1050
	96c	175,000	"	"	"	"	"	"	"	"	5/8"	"	"	"		1170
	96d	262,500	"	"	"	"	"	"	"	"	"	"	"	"	1	1180
	96e	350,000	"	"	"	"	"	"	"	"	"	"	"	"	2	1200
	96f	532,000	"	"	"	"	"	"	"	"	"	"	"	"	4	1230
102	102a	94,500	8'-2"	4'-9"	5/16"	3/4"	3'-9"	10"	9"	11"	1/2"	5"	5 1/4"	1 1/4"		1350
	102b	199,000	"	"	"	"	"	"	"	"	5/8"	"	"	"		1500
	102c	298,000	"	"	"	"	"	"	"	"	"	"	"	"	1	1510
	102d	398,000	"	"	"	"	"	"	"	"	"	"	"	"	2	1530
	102e	476,000	"	"	"	"	"	"	"	"	3/4"	"	"	"	"	1690
	102f	714,000	"	"	"	"	"	"	"	"	"	"	"	"	4	1730
108	108a	85,000	8'-8"	5'-0"	5/16"	3/4"	4'-0"	10"	9"	11"	1/2"	5"	5 1/2"	1 1/4"		1430
	108b	187,000	"	"	"	"	"	"	"	"	5/8"	"	"	"		1590
	108c	290,000	"	"	"	"	"	"	"	"	"	"	"	"	1	1610
	108d	386,000	"	"	"	"	"	"	"	"	"	"	"	"	2	1630
	108e	466,000	"	"	"	"	"	"	"	"	3/4"	"	"	"	"	1810
	108f	699,000	"	"	"	"	"	"	"	"	"	"	"	"	4	1870
114	114a	164,000	9'-2"	5'-3"	5/16"	3/4"	4'-3"	10"	9"	11"	5/8"	5"	5 3/4"	1 1/4"		1760
	114b	300,000	"	"	"	"	"	"	"	"	"	"	"	"	1	1780
	114c	400,000	"	"	"	"	"	"	"	"	"	"	"	"	2	1800
	114d	514,000	"	"	"	"	"	"	"	"	3/4"	"	"	"	"	1980
	114e	600,000	"	"	"	7/8"	"	"	"	"	7/8"	"	"	"	"	2260
	114f	900,000	"	"	"	"	"	"	"	"	"	"	"	"	4	2330

Figure 7-7. Typical saddle details. Courtesy of Fluor Corp.

Vessels 131

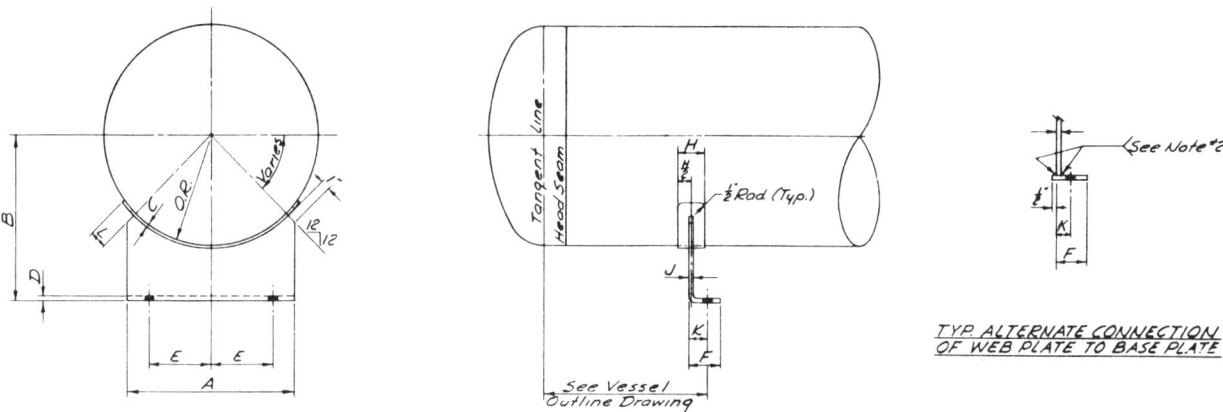

SADDLE DETAILS FOR MARK 24a, 24b & 30a ONLY

VESSEL DIAM. IN INCHES	SADDLE MARK N°	MAX. VESSEL OPER. WEIGHT LBS	A	B	C	D	E	F	G	H	J	K	L	BOLT DIAM	ANCHOR	WEIGHT PER SET LBS.	GENERAL NOTES
120	120a	150,000	9'-8"	5'-6"	5/16	3/4	4'-6"	10"	9"	11"	5/8	5"	6"	1 1/4		1860	1. Vessel diameters listed are inside diameters. The listed standard dimensions may change in the case of very thick vessel shells, or if very thick insulation is employed. Dimensions shown on vessel drawing shall take precedent.
	120b	300,000	"	"	"	"	"	"	"	"	"	"	"	"	1	1880	
	120c	410,000	"	"	"	"	"	"	"	"	"	"	"	"	2	1900	
	120d	520,000	"	"	"	"	"	"	"	"	3/4	"	"	"	"	2100	
	120e	666,000	"	"	"	7/8	"	"	"	"	7/8	"	"	"	"	2380	
	120f	999,000	"	"	"	"	"	"	"	"	"	"	"	"	4	2440	
126	126a	146,000	10'-2"	5'-9"	5/16	3/4	4'-9"	10"	9"	11"	5/8	5"	6 1/2"	1 1/4		2080	2. Saddle to vessel attachment shall be a continuous full fillet weld all around. Other welds shall be 3/8 continuous fillet all around, up to & including 72" diameter vessel & 3/8 continuous fillet all around for 78" diameter vessel & larger.
	126b	294,000	"	"	"	"	"	"	"	"	"	"	"	"	1	2100	
	126c	380,000	"	"	"	"	"	"	"	"	"	"	"	"	2	2120	
	126d	536,000	"	"	"	"	"	"	"	"	3/4	"	"	"	"	2340	
	126e	646,000	"	"	"	7/8	"	"	"	"	7/8	"	"	"	"	2640	
	126f	969,000	"	"	"	"	"	"	"	"	"	"	"	"	4	2700	
132	132a	127,500	10'-8"	6'-0"	3/8	3/4	5'-0"	10"	9"	11"	5/8	5"	6 3/4"	1 1/4		2180	3. Provide 3/8 pipe tap in wear plate if vessel is stress relieved or where plate overlaps longitudinal seams.
	132b	292,000	"	"	"	"	"	"	"	"	"	"	"	"	1	2220	
	132c	400,000	"	"	"	"	"	"	"	"	"	"	"	"	2	2240	4. Anchor bolt holes
	132d	550,000	"	"	"	"	"	"	"	"	3/4	"	"	"	"	2480	4a. Provide holes in anchor end as follows:
	132e	708,000	"	"	"	7/8	"	"	"	"	7/8	"	"	"	"	2820	1 1/4 ⌀ for 1"⌀ bolts.
	132f	1,058,000	"	"	"	"	"	"	"	"	"	"	"	"	4	2880	1 1/2 ⌀ for 1 1/4"⌀ bolts.
138	138a	124,500	11'-2"	6'-3"	3/8	3/4	5'-3"	10"	9"	11"	5/8	5"	7"	1 1/4		2340	4b. Provide slotted holes in expansion end as follows:
	138b	289,000	"	"	"	"	"	"	"	"	"	"	"	"	1	2360	1 1/4 ⌀ x 2" long for 1"⌀ bolts
	138c	386,000	"	"	"	"	"	"	"	"	"	"	"	"	2	2400	1 1/2 ⌀ x 2 1/2" long for 1 1/4"⌀ bolts
	138d	544,000	"	"	"	"	"	"	"	"	3/4	"	"	"	"	2640	4c. Slots not required for cold vessels less than 10'-0" long.
	138e	768,000	"	"	"	7/8	"	"	"	"	7/8	"	"	"	"	3000	
	138f	1,182,000	"	"	"	"	"	"	"	"	"	"	"	"	4	3060	
144	144a	280,000	11'-8"	6'-6"	3/8	3/4	5'-6"	10"	9"	11"	5/8	5"	7 1/2"	1 1/4	1	2500	
	144b	365,000	"	"	"	"	"	"	"	"	"	"	"	"	2	2540	
	144c	540,000	"	"	"	"	"	"	"	"	3/4	"	"	"	"	2800	
	144d	728,000	"	"	"	7/8	"	"	"	"	7/8	"	"	"	"	3160	
	144e	1,092,000	"	"	"	"	"	"	"	"	"	"	"	"	4	3200	
	144f	1,290,000	"	"	"	1"	"	"	"	"	1"	"	"	"	"	3400	
150	150a	275,000	12'-2"	6'-9"	3/8	3/4	5'-9"	10"	9"	11"	5/8	5"	7 1/2"	1 1/4	1	2580	
	150b	385,000	"	"	"	"	"	"	"	"	"	"	"	"	2	2620	
	150c	560,000	"	"	"	"	"	"	"	"	3/4	"	"	"	"	2900	
	150d	764,000	"	"	"	7/8	"	"	"	"	7/8	"	"	"	"	3260	
	150e	1,156,000	"	"	"	"	"	"	"	"	"	"	"	"	4	3300	
	150f	1,282,000	"	"	"	1"	"	"	"	"	1"	"	"	"	"	3700	
156	156a	266,000	12'-8"	7'-0"	3/8	3/4	6'-0"	10"	9"	11"	5/8	5"	8"	1 1/4	1	2730	
	156b	398,000	"	"	"	"	"	"	"	"	"	"	"	"	2	2760	
	156c	576,000	"	"	"	"	"	"	"	"	3/4	"	"	"	"	3120	
	156d	864,000	"	"	"	"	"	"	"	"	"	"	"	"	4	3160	
	156e	1,150,000	"	"	"	7/8	"	"	"	"	7/8	"	"	"	"	3560	
	156f	1,276,000	"	"	"	1"	"	"	"	"	1"	"	"	"	"	3980	

Figure 7.7. Typical saddle details.

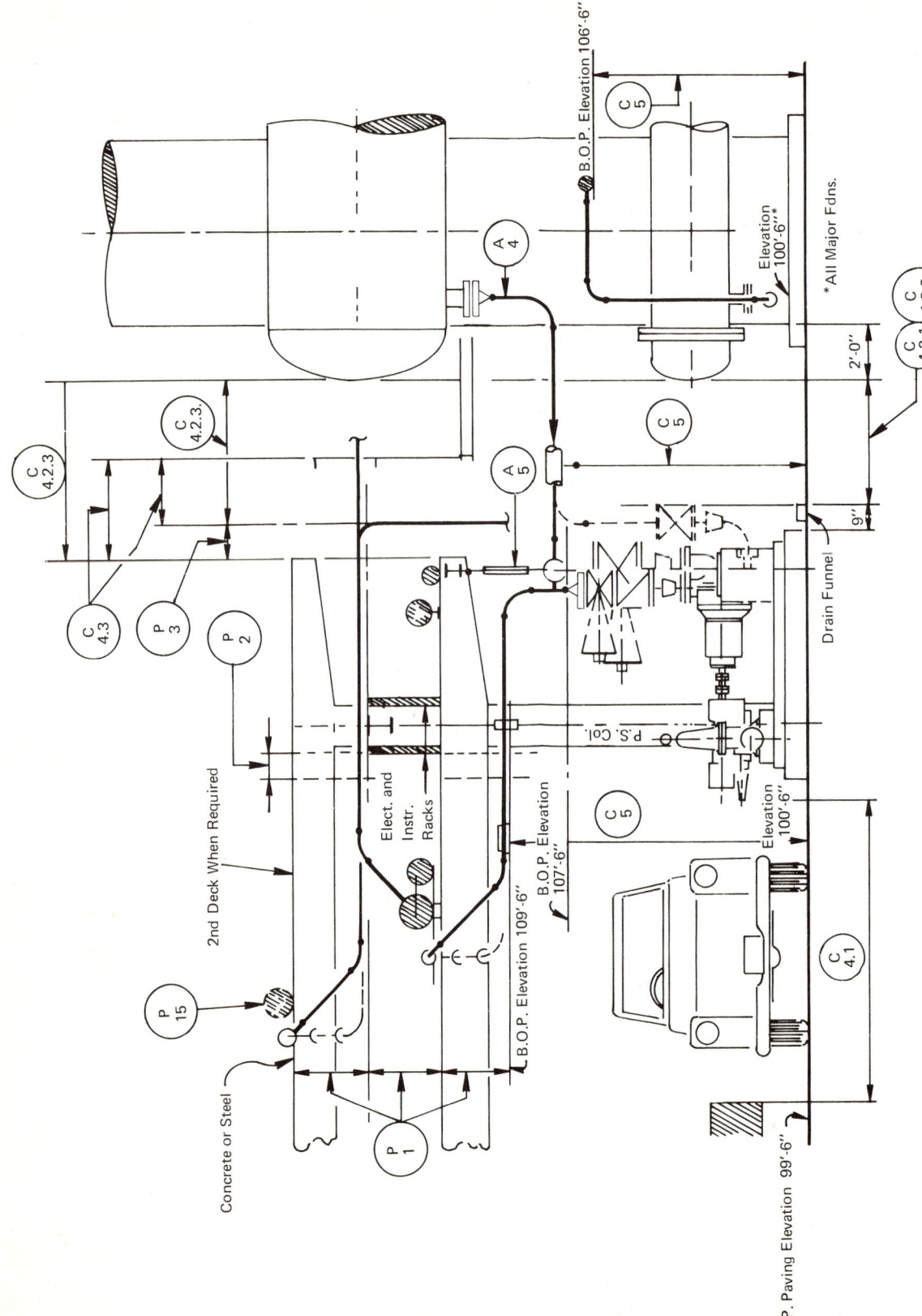

Figure 7-8. Typical section through pipeway. Courtesy of Fluor Corp.

Vessels

Manhole Orientation

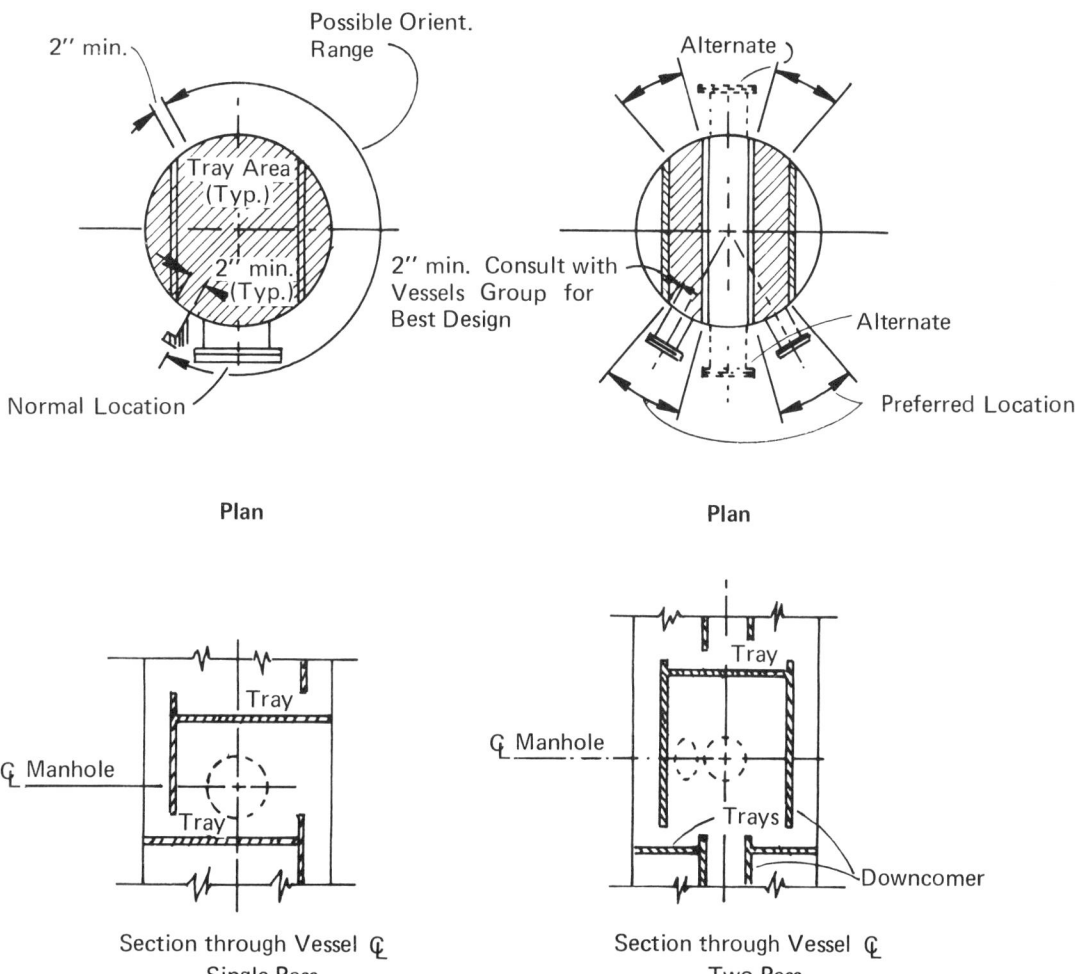

Manholes or handholes shall be provided for access and inspection of vessel internals, trays, internal piping, etc. Manhole elevations are normally dependent on tray arrangement. Orient manhole in rear half of vessel away from pipeway, with all manholes on or about the same centerline. Consideration shall be given to raising or lowering manholes one tray to provide the most economical ladder and platform arrangement.

Figure 7-9. Manhole orientation. Courtesy of Fluor Corp.

through. A bubble cap was connected to the standpipe which forced the vapor to make contact with the tray liquid.
2. *Valve trays* are in the most common use today. Stamped out by big presses, these trays come with small valves attached to them which allow vapor traffic. They are made in sections wide enough to permit passage through a manhole and are installed in these pieces, usually 17" wide. They rest on a tray support ring welded to the vessel wall. These trays are usually bolted to this support ring.
3. *Sieve trays* are perforated flat plates. They are inexpensive for small diameter vessels but large diameter towers must have extensive supports for these trays. Sieve trays are used for heavy hydrocarbon fractionation.

All trays have foam on top of the liquid. The height of the foam will vary with the process. Internal piping will interfere with the foam and should be avoided where possible. Foam may rise a foot or more above the tray liquid. Any horizontal pipe extending into it will serve as a dam. There are cases where the designer must have internal piping to properly distribute the incoming liquid on the tray. These will be shown later in this chapter.

To make fractionating towers work, the hot liquid must be retained in the bottom of the tower, drawn off, heated and vaporized and returned to the tower in a large vapor inlet nozzle. This vapor then starts its passage up the tower, making contact with the liquid on each tray, losing some temperature and some heavier hydrocarbons at each tray but vaporizing some lighter hydrocarbons from the liquid flowing down the tower.

Reflux Liquid

The liquid enters the vessel at the top tray and the nozzle must be oriented to assure that the liquid will flow evenly on the tray surface. This liquid is called "reflux." If the velocity is high an inlet baffle may be required (see Figure 7-11, Type "A"). The liquid is cool when it enters the top tray. As it flows down each tray it makes contact with the warmer rising vapor and some of the heat is retained. So each tray operates at a different temperature, getting a little hotter as the liquid comes down the tower. The liquid in the bottom of the tower may be 700°F, while the vapor leaving the top of the tower may be only 200°F.

Tower Temperature Gradient

Since tower temperature is different at each tray, the tower rate of expansion will vary depending on what part of the vessel you are considering. To determine this expansion, the piping designer will develop a tower temperature gradient and will calculate the expansion from grade to several points on the tower. This is the only way to determine if the piping to the tower is flexible enough to withstand the growth of the tower.

Figure 7-9 also shows the two pass tray. This type of tray has two downcomers which alternate from center downcomers to side downcomers. Note that the one pass tray has only side downcomers. Generally, the larger diameter towers will have two or more passes.

Tray types and the number of downcomers are determined by the tray manufacturer. Engineering contractors send the tray vendor a specification outlining the requirements of a particular tower. The tray vendor calculates the required service, defines the tray type and guarantees the performance of his product. Most tray suppliers have the right to approve the designer's nozzle orientation and any internal pipe arrangements.

Figure 7-9 shows possible location for manholes. The single pass tray offers great flexibility for manhole location. The manhole may be located on any tray with a wide orientation range. The two pass tray has a limited location. The manhole could not be located on a tray with center downcomers coming down from above. They would block entrance to the vessel since the downcomers would drop in front of the manhole. So location of manholes is limited (1) to every other tray and (2) to an area between the two side downcomers.

Manholes are used to gain entry to vessels for maintenance personnel and for replacement of vessel internal parts such as tray pieces, demister parts and internal piping. Access to the tower from grade, is usually from a road on the side opposite the pipeway, so manholes are to be located on the tower side opposite the main pipeway.

Tower Ladders and Platforms

Ladders and platforms on vertical fractionating towers must be layed out by the piping designer as he orients the nozzles. Platforms are to be provided at each manhole, usually three or four per tower. Other platforms will be needed for access to critical instruments or operating valves. Some operating companies require platforms to be located at each flanged nozzle so that they could easily tighten bolts should a leak occur. Others require no platforming in clean service; they believe the cost and maintenance is too high for the few times the platforms would be used. They feel it is cheaper to erect temporary scaffolding. Some feel they can elevate a workman with a crane to tighten bolts. Before trying to lay out platforming, the designer must thoroughly know his job specifications.

At a minimum, the author suggests a platform at each manhole. Instruments should be located for access from the ladder. Where this is not possible, platforms will be provided for instrumentation.

Ladders should be caged if they are over 20'-0" above grade. The maximum continuous ladder run should be 30'-0". This means the ladder should be offset every 30'-0", and the designer will supply an "offset platform" at these points.

The author also suggests that all ladders and platforms for vertical vessels should be galvanized. This costs about 5¢ per pound more but will save countless paintings over the years, and the ladders and platforms will always present a neat appearance.

Figure 7-10 shows typical ladder, cage and platform details for vertical vessels. Note that ladder rungs are spaced every 1'-0". A rung should be located at the top of each platform for easy entrance and exit. Cages are omitted for the first 8'-0" of the ladder from grade or one originating at a platform. Cages are also partially omitted at a side entry platform for 8'-0".

Platforms at towers should be located 2'-6" below the manhole centerline. They should be constructed of ¼" thick checkered floor plate except for areas of heavy snow load, where grating should be used. Grating should be avoided where possible as a workman's tools could fall through and might hit personnel below. Platforms for towers should be 3'-0" wide plus the 6" clearance from the vessel OD.

Process Nozzle Orientation

Figure 7-11 shows some basic orientation possibilities for some process nozzles, whether they are located in one or two pass trays. The reflux nozzle in the single pass tray has a baffle shown inside the vessel. This will reduce the incoming velocity and assure even distribution over the top tray. The type "B" reflux nozzle is used when the piping layout forces the orientation to be in the downcomer general area. Then the nozzle will be elevated higher and an internal distributor pipe is required to drop the liquid on the back part of the tray. If the service is corrosive, a pair of breakout flanges are added to the internal pipe to make replacement easy. This type of installation is costly and is to be avoided.

Orienting reflux liquid to the two pass tray requires a piping manifold internally above the top tray. Liquid must be allocated evenly to both sections should the tray have center downcomers as shown. Should it have side downcomers, the reflux could dump in the center and flow to the sides without any internal pipe.

Liquid feed nozzles are shown for single and two pass trays. Again, the object is to introduce liquid to the tray with the most even flow possible over the tray. Liquid feeds usually enter the tower at or above the middle tray. For a two-sized vessel, the liquid feed is located in the conical transition piece, giving added liquid to the larger diameter trays.

Instrument Connections

Location of instrumentation on towers is a critical item and requires great thought. The designer must understand the function of the instrument, its operation and how often the operator reads or services it. Figure 7-12 shows some instrument locations.

A temperature indicator (TI) may be needed to check the temperature of either the liquid or the vapor. Unless the designer knows the instrument's purpose, he cannot properly locate it. A sample connection (SC) may also be designed to take a vapor or liquid sample. Liquid sample and temperature points should always be located where there is flowing fluid. Since the greatest amount of flow occurs in the downcomer area, this is the place to locate these connections.

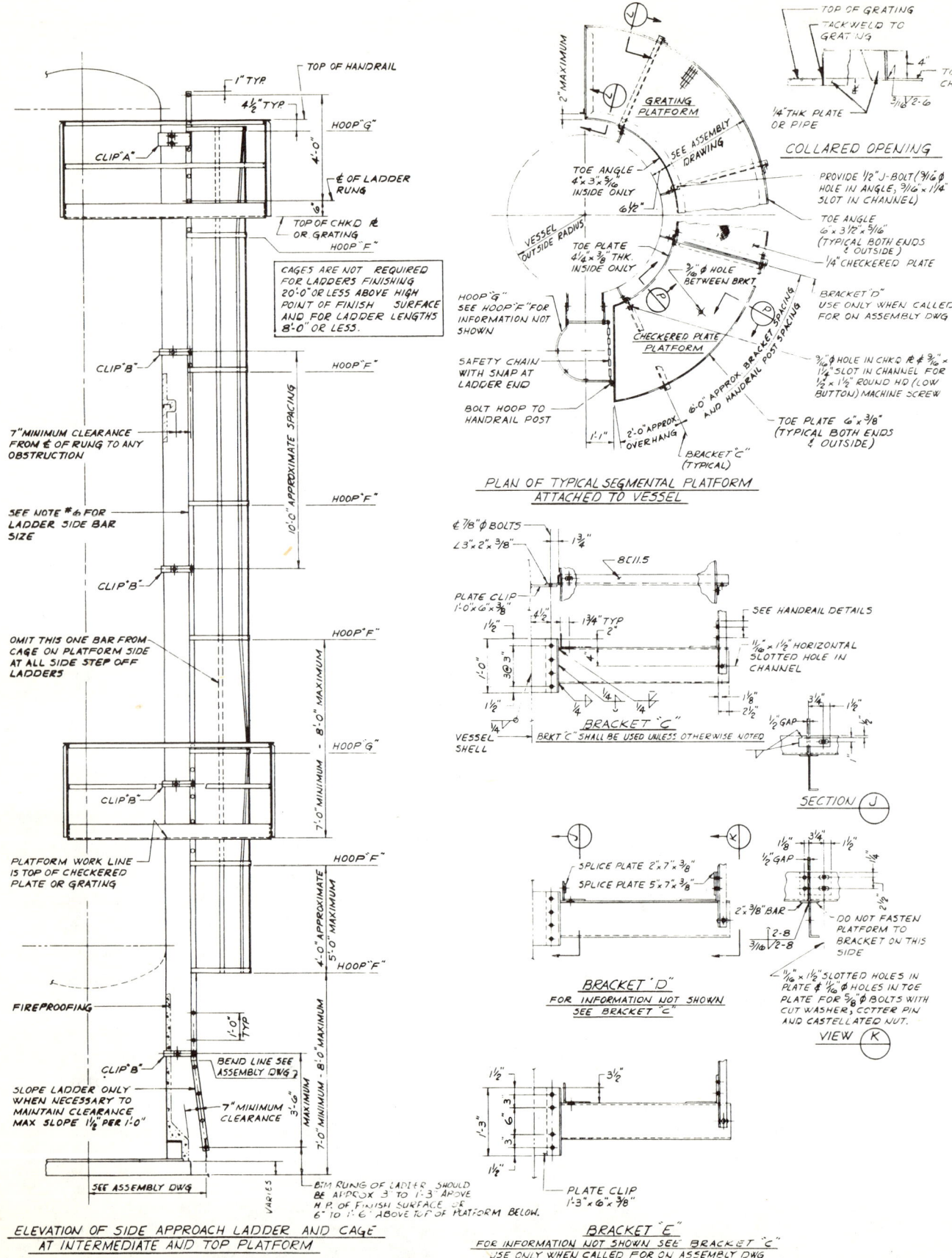

Figure 7-10. Typical ladder and platform details.

Vessels

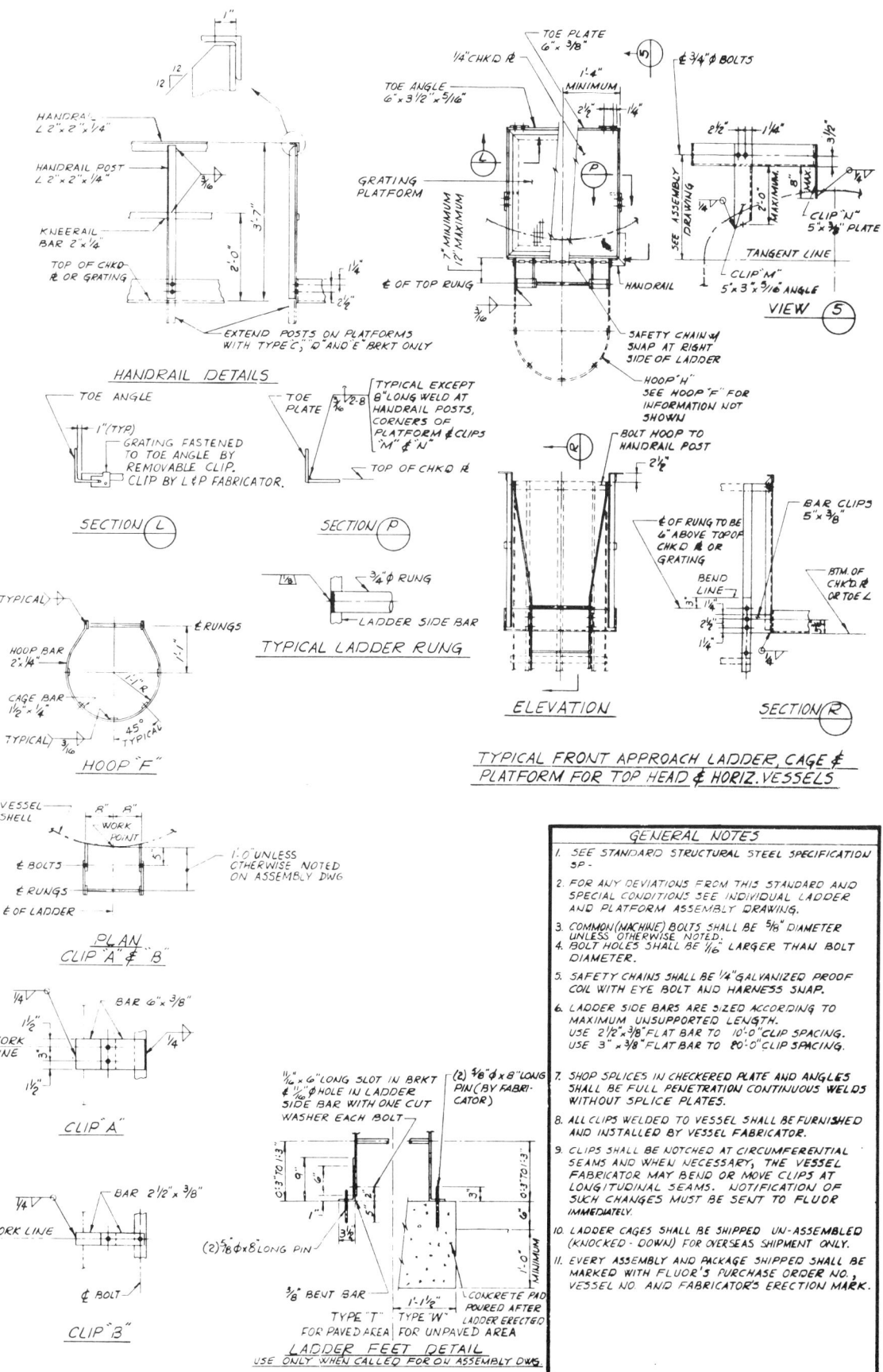

Figure 7-10. Typical ladder and platform details. Courtesy of Fluor Corp.

138 Process Piping Design

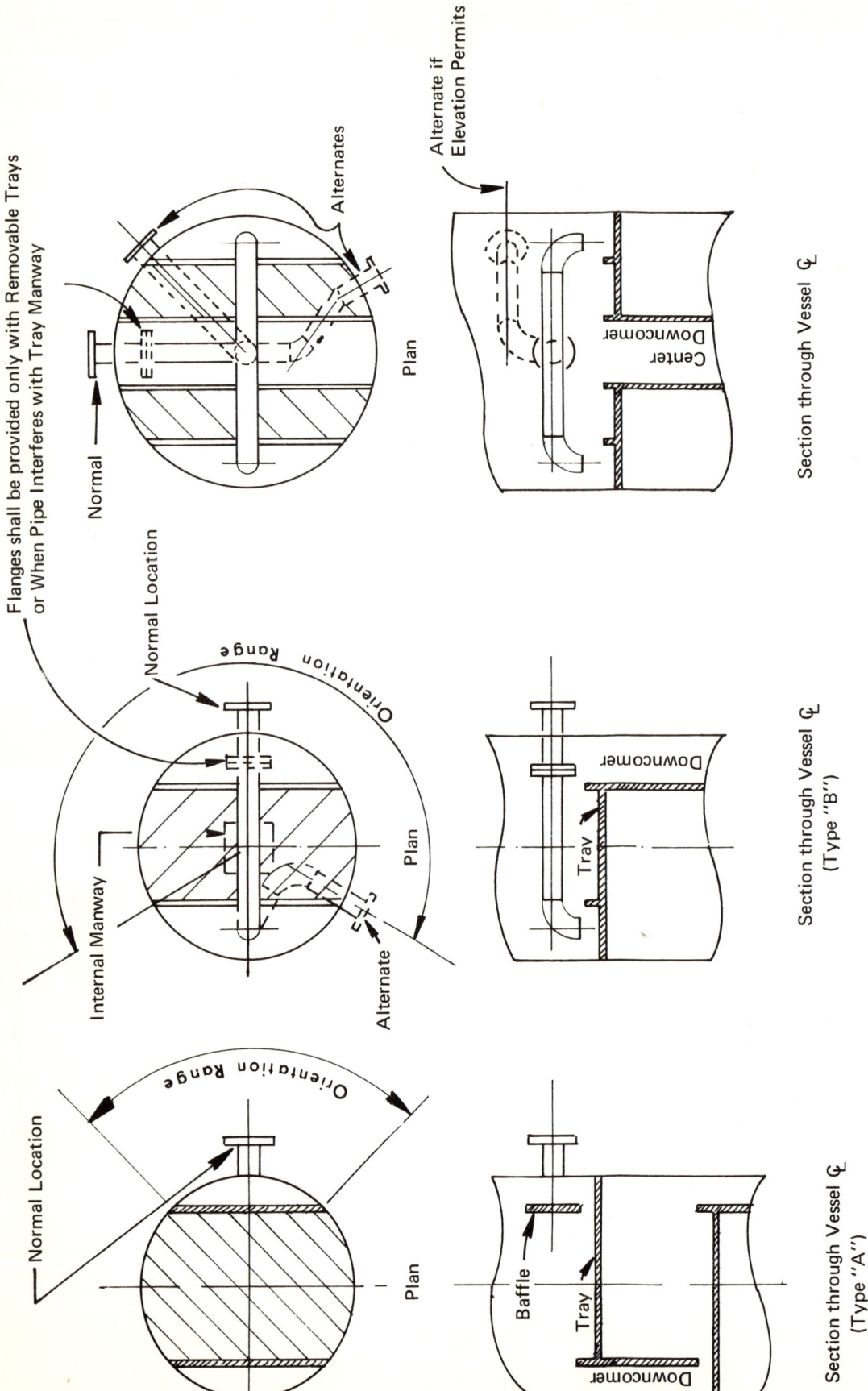

Figure 7-11. Process nozzle orientation. Courtesy of Fluor Corp.

Vessels

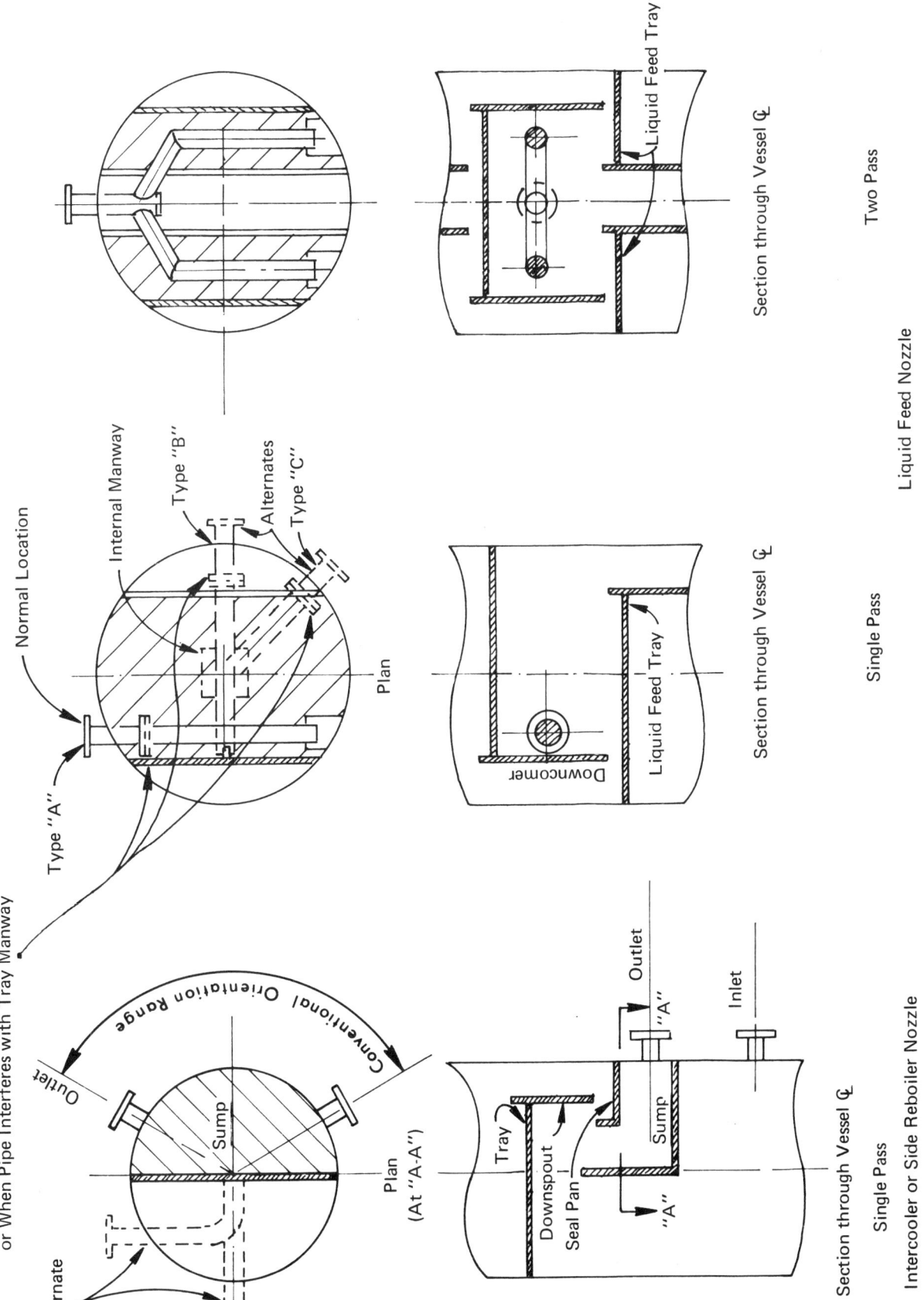

Figure 7-11. Process nozzle orientation. (continued)

140 Process Piping Design

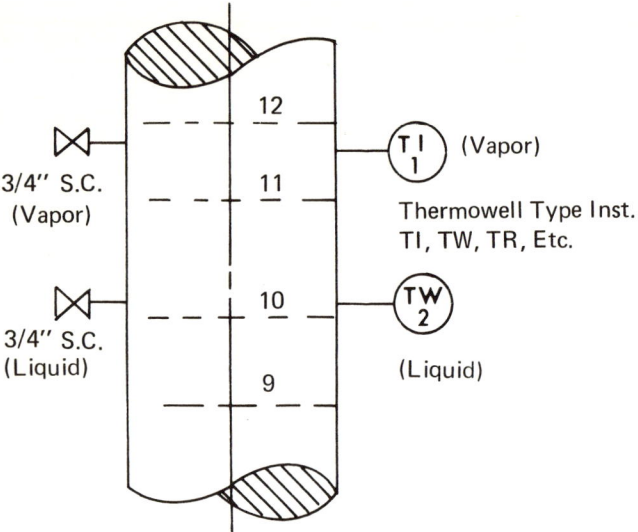

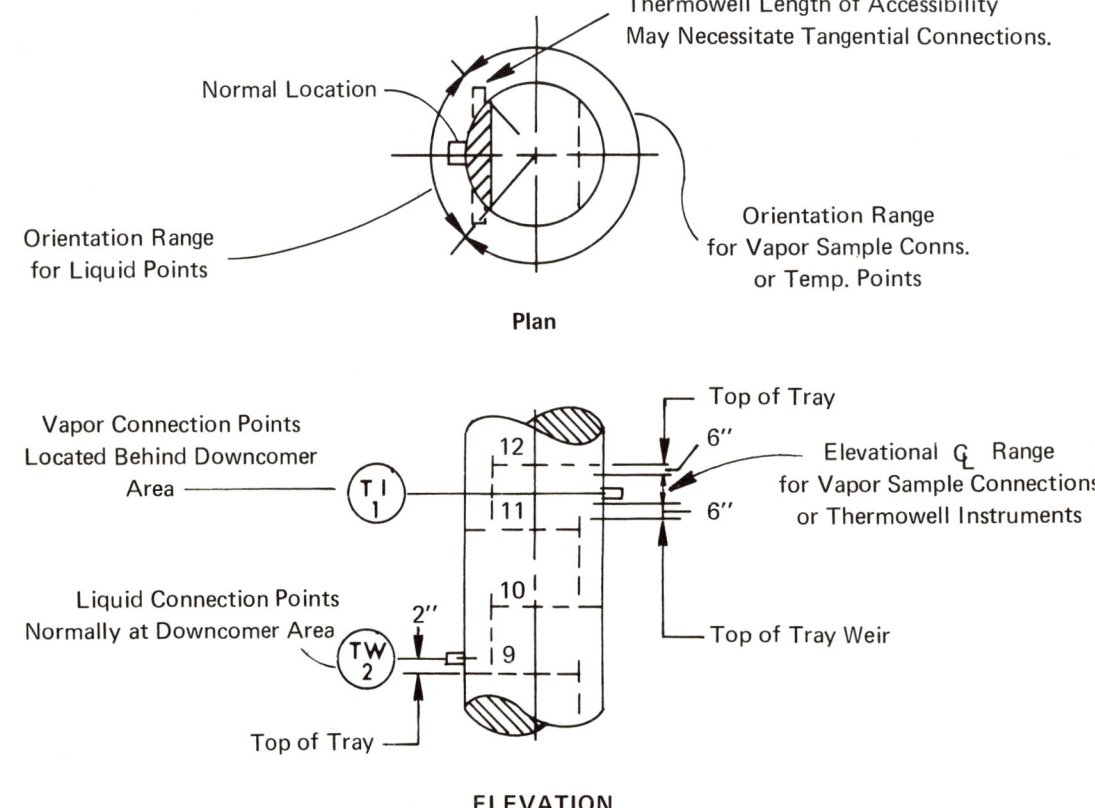

Figure 7-12. Instrumentation on vertical vessels. Courtesy of Fluor Corp.

The temperature connection has a thermowell that projects inside the vessel and the designer must know the projection dimension and the downcomer clearance dimension to locate the thermowell so that it will miss the downcomer. The liquid sample connection has no internal projection so it may be oriented any place in the downcomer area.

The thermowell is normally located 2" above the top of the tray. If the well is too long the location may be raised and the thermowell could be angled down about 45° to reduce the horizontal projection. When this is done, the bottom of the well should be set about 2" above the top of the tray.

Vapor sample and temperature points are to be located behind the downcomer in the vapor area of the tray.

Since all of these connections must be maintained or operated, they must have permanent access, either by permanent platform or ladder. Most designers will make them accessible from a ladder as platforms are expensive.

Tray quantity and location requirements will be shown on the mechanical flow diagram. The designer will orient as specified.

Other instrumentation which requires orientation may be the level controller (LC) the level gage (LG), the pressure indicator (PI) and the relief valve or pressure safety valve (PSV).

The LC and LG are to be located so that any incoming liquid or vapor will not disturb their function. A good way to accomplish this is to locate them next to the nozzle causing this turbulence. The velocity will then be directed toward the vessel center and away from the instruments.

The PI should be located in the vapor part of the tower so that any liquid head will not be reflected in its reading.

Other Nozzle Orientations

Probably the largest and most critical nozzle is the reboiler return. This connection will govern most nozzle and tray orientations. The reboiler return is feeding vapor and liquid to the tower. The liquid will drop and the vapor must go evenly up to the bottom tray to start its upward journey. For one pass trays, this nozzle should be located parallel to the downcomer and the top of the nozzle will be set about 1'-0" below the bottom tray. Never locate it perpendicular to the seal pan. The velocity coming in may blow the liquid out of the seal pan and cause loss of seal. Then the vapor could by-pass the bottom tray via the downcomer area. This would mean the bottom tray would be operating at only partial efficiency and it might not even make the product for which the tower was designed—all because one nozzle was improperly oriented.

If the tower has two pass trays, the bottom tray should have side downcomers. Then the reboiler return should be located parallel to both downcomers and equidistant between them. Should the bottom tray have center downcomers, locate the reboiler return parallel to and immediately below the downcomer seal pan. The vapor will flow evenly to both sides of the above tray and there is no way the seal could be blown out.

For a liquid and vapor flow (called two-phase flow) internal piping may be necessary to reduce velocity and direct the flow. For this case, the bottom half of the internal pipe will be cut out and an end plate fastened at the extreme end. This will direct the liquid down and the vapor will flow out and up to the bottom tray.

Liquid draw-off nozzles may be needed at one or more trays. These nozzles are used for taking various "cuts" or draws of a specific hydrocarbon. The best location for these draw-off nozzles is opposite the downcomer in a small draw-off pan located immediately below the tray for one pass trays. To locate this nozzle in two-pass trays, make it draw liquid from a tray that has side downcomers. Build a trough in the center of the tray and locate the nozzle to draw liquid from this trough. It is preferable to make this draw with the nozzle parallel to the trough. If the vessel is very large, two separate draw-off nozzles may be needed, one at each end of the trough. The piping connecting the two nozzles should be symmetrical to the point of connection to the one line going down the tower.

If the nozzle cannot be located parallel to the side downcomers, a draw may be made at the center of the vessel with internal pipe running from the trough to the draw-off nozzle.

Products

Products of fractionating towers are collected from the overhead vapor stream, the bottoms outlet and the side cut streams. The first fractionating tower in a refinery is usually the crude tower. Crude oil is heated to a temperature of 750°F or more and sent to the crude tower. It enters the tower "flash zone," the vapor area below the bottom tray, but above the normal liquid level.

The crude entering the tower has been partly vaporized due to the preheating. As it enters the flash zone it is both liquid and vapor. The liquid portion partially flashes, or vaporizes, and the heavier fractions remain in a liquid state and fall down to the liquid portion of the column. As the flashed vapor rises up through the trays, a temperature level is reached where the heavier fractions condense and become liquid. Lubricating oils and heavy oils collect in the lower trays. A few trays higher, the fuel oils are condensed out of the vapor.

Higher yet, kerosene is condensed. Near the top the gasoline components drop out of the vapor, depending on the design temperature of the top of the tower. The methanes, ethanes propanes and butanes remain as vapor and exit the crude tower at the top, going to the overhead condenser and accumulator.

To capture the other products a draw-off nozzle must be installed at the appropriate tray. These nozzles will drawoff the desired product and possibly send it to a side cut stripper for further fractionation.

Gasoline which is drawn off a crude tower is called "straight-run gasoline." For many years, gasoline was made this way, and with a few additives it was commercially marketed. Now, very few crude columns have a straight-run gasoline cut. This cut would be too low in octane rating for modern automobiles.

Also, it would not produce enough gasoline for each barrel of crude oil to be economical.

To meet these needs for economy and high performance, new and better processes were developed. To produce more gasoline per barrel of crude oil, cracking and polymerization were developed. *Cracking* is the making of gasoline out of heavier fractions by changing their molecular structures. *Polymerization* does the opposite, making gasoline out of lighter fractions by changing their molecular structure. *Reforming* is another process which actually makes a better gasoline, improving its antiknock rating.

Every year new methods for improving the marketable yields of crude oil are developed. Each new process must be studied by the modern piping designers, for each one offers a new challenge.

Vertical Vessel Davit

Figure 7-13 shows a vertical vessel davit. This is a device used to raise heavy objects from grade to the desired platform or to lower vessel parts to grade. The davit should be installed on towers which are over 30' tall if there are removable parts in the tower. Davits should be oriented so that they will drop their load into the "drop area," the clear open area at the back side of the column, away from the main rack. Very large diameter towers may require two davits. The swing of the davit must reach every platform which services a manhole to pick up a tray part and then must be swung around so that the part can be lowered to the drop area.

Vertical Vessel Pipe Supports

Figures 7-14 and 7-15 depict pipe supports and pipe guides for vertical vessels. Pipe attaching to the vessel nozzles is heavy and the nozzle is not designed to carry this dead load. Pipe supports are attached to the vessel shell, as near the nozzle as possible, to carry this load. Pipe guides are located below the pipe supports, at very wide spacing, to keep the pipe from swaying and putting moments on the vessel nozzle. Figure 7-15 defines this spacing.

Locating pipe supports is the piping designer's responsibility. As he designs the piping on a tower, he must define locations of pipe supports. Typical locations of pipe supports are shown in Figure 7-16.

The piping designer must also locate the pipe being supported in relation to the vessel. As shown in Figure 7-16, a minimum clearance of 1'-0" must be maintained between the vessel steel and the pipe steel. This dimension is the "L" dimension referred to in Figure 7-14. The Type $L1$ pipe support is to be used for sizes 2" through 10" for "L" dimensions of 12" to 20", etc. For large lines and long "L" dimensions, utilize the Type $L4$ support which requires special design.

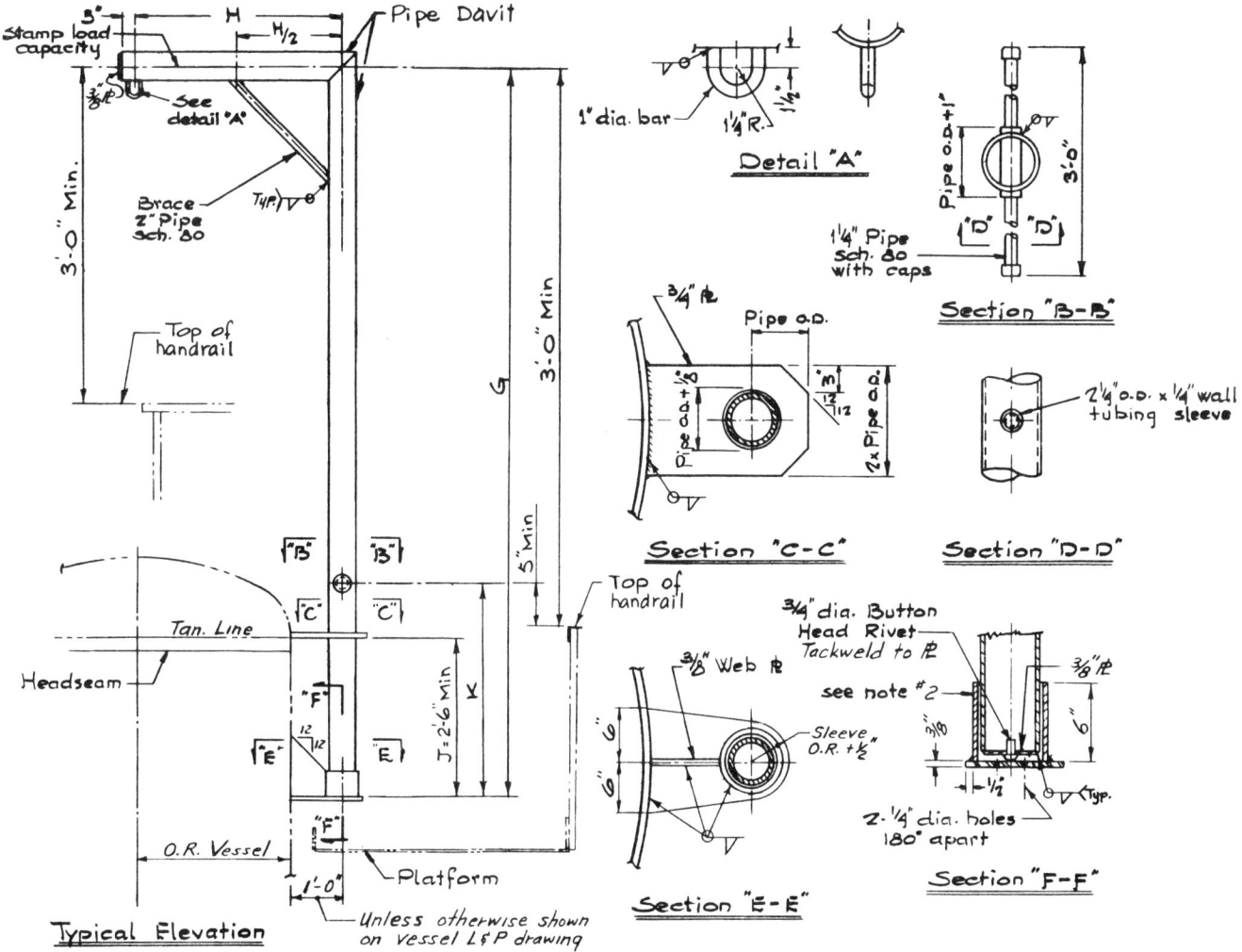

Figure 7-13. Davits for vertical vessels. Courtesy of Fluor Corp.

144 Process Piping Design

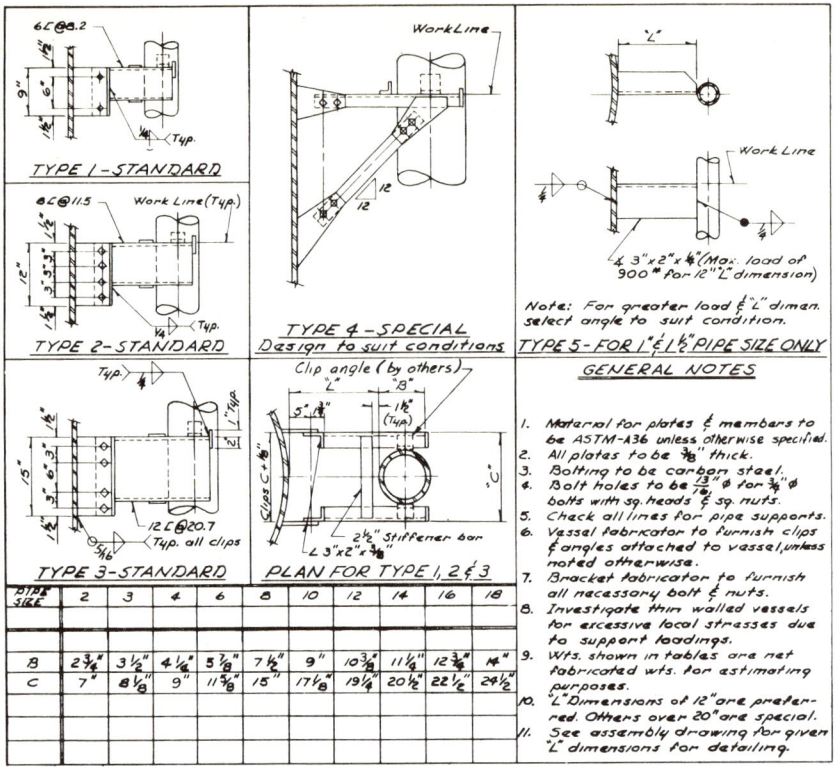

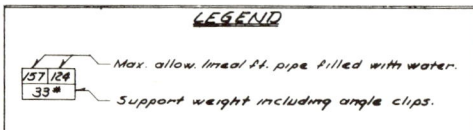

Figure 7-14. Vertical vessel pipe supports. Courtesy of Fluor Corp.

Vessels

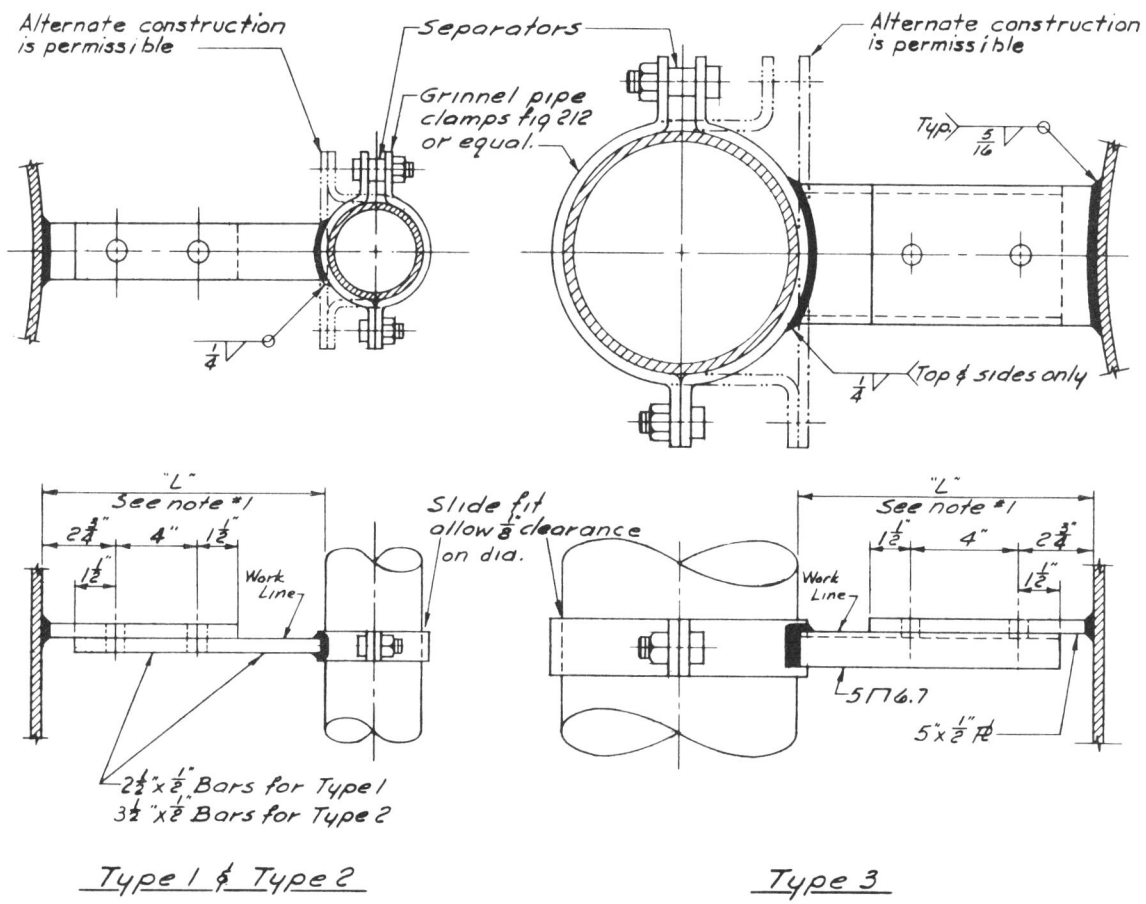

MAXIMUM GUIDE SPACING											
\	\	"L" DIMENSION — SEE NOTE #1									
TYPE OF GUIDE	LINE SIZE	L=12 GUIDE SPACING	L=14 GUIDE SPACING	L=16 GUIDE SPACING	L=18 GUIDE SPACING	L=20 GUIDE SPACING	L=22 GUIDE SPACING	L=24 GUIDE SPACING	L=26 GUIDE SPACING	L=28 GUIDE SPACING	L=30 GUIDE SPACING
TYPE 1	1"	25'	25'	25'	25'	25'					
	1½"	25'	25'	25'	25'	25'					
	2"	25'	25'	25'	25'	25'					
	3"	25'	25'	25'	25'	25'					
	4"	30'	30'	30'	30'	30'					
TYPE 2	6"	40'	40'	40'	40'	40'	40'	40'	40'	40'	40'
	8"	50'	50'	50'	50'	50'	50'	50'	50'	50'	50'
	10"	55'	55'	55'	55'	55'	55'	55'	55'	55'	55'
	12"	60'	60'	60'	60'	60'	55'	50'	50'	50'	45'
TYPE 3	14"	70'	70'	70'	60'	55'	50'	50'	45'	40'	40'
	16"	65'	65'	55'	55'	50'	45'	40'	40'	35'	35'
	18"	65'	55'	50'	45'	40'	40'	35'	35'	30'	30'

GENERAL NOTES

1. "L" Dimensions of 1'-0" are preferred.
2. See assembly drawing for given "L" dimensions for detailing.
3. Material for bars, plates & channels to be ASTM-A7 or better.
4. Bolt holes to be 13/16"ø for ¾"ø carbon steel bolts with sq. heads & sq. nuts.
5. Vessel fabricator to furnish all necessary bars & plates welded to vessel, unless noted otherwise.
6. Investigate thin walled vessels for distortion.
7. Care should be taken not to locate any guide too closely to a point where horizontal expansion is expected, otherwise distortion of guide & vessel may occur.
8. Computations based on 15# wind on projected area of insulated pipes with a maximum deflection of .003 × span, but not to exceed 1."

Figure 7-15. Vertical vessel pipe guides. Courtesy of Fluor Corp.

146 Process Piping Design

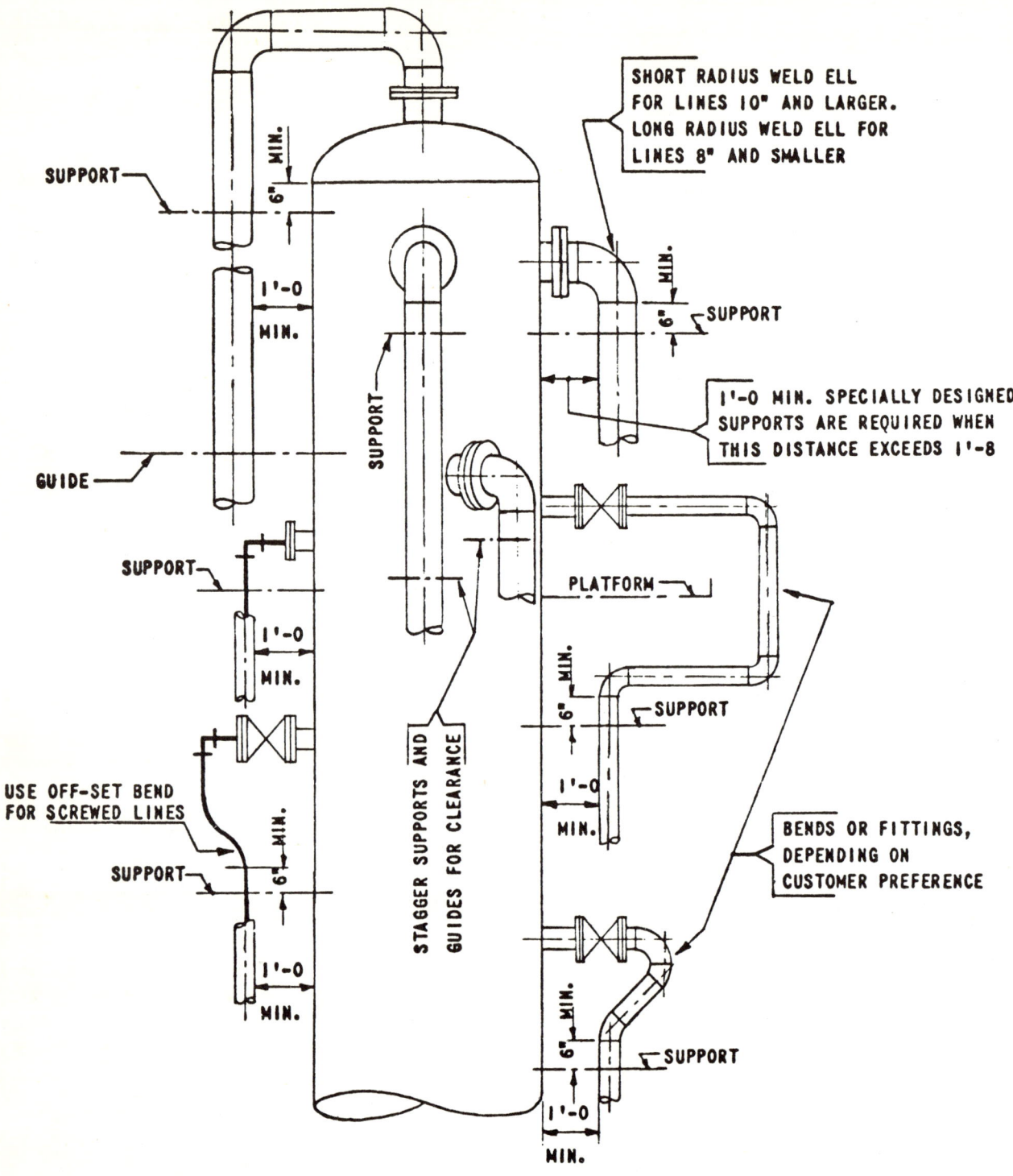

Figure 7-16. Location of pipe supports. Courtesy of Fluor Corp.

Chapter 7
Review Test

1. A chemical change occurs in what kind of vessel? _____

2. Separation occurs in what kind of vessel? _____

3. What is the purpose of a depentanizer? _____
_____

4. What is "reflux liquid?" _____
_____

5. What is a catalyst? _____
_____

6. Define "retention time" in a vessel. _____
_____

7. Horizontal vessels _____ feet or less above grade are usually not provided with permanent ladders and platforms.

8. Platforms weight about _____ pounds per square foot.

9. Ladders with cages weigh _____ pounds per linear foot.

10. When locating nozzles on horizontal vessels, the main rule to remember is _____
_____

11. What is entrained liquid? _____
_____

12. Where is the liquid level normally set for accumulators? _____
_____

13. What is the purpose of a horizontal vessel bootleg? _____
_____

14. What is a "hillside" connection? _____
_____

15. Define the following parts of a fractionation tower and give their use. _____

a. tray _____

b. downcomer _____

c. seal pan _____

d. pipe guide _____

e. manhole _____

f. skirt _____

g. reboiler return _____

h. liquid draw-off _____

8 Instrumentation

Instrumentation for piping designers is one of several important disciplines of piping, but most "pipers" have a very limited knowledge of instruments. This ignorance results in misapplied installations in almost every unit erected. These errors must be corrected in the field, which is costly.

In the last decade each year saw increasing use of instrumentation. Instruments give better control of process or utility units with less manpower. As manpower costs increase, instruments become more feasible. As an example, all control valves could be eliminated by supplying a man, 24 hours a day, to operate a globe valve. This would not only be expensive, but the job would be boring and undesirable. Without instruments in today's process units, costs of products would be several times more than they are.

Instrumentation is shown by a symbol on the flow diagram and piping drawings. A circle is used to depict the instrument type, function and identifying number. This circle is commonly called a "bubble" or a "balloon."

Instrument Types

There are four major instrument types: pressure, temperature, flow and liquid level. Each of the types may be locally mounted, which means they are read at the instrument, or they may be board mounted, which means they will have a transmitter sending impulse signals to an instrument located at the control board. The board's receiver converts this signal to a readable digit and this reading or recording is noted on the control board. Locally read instruments are identified by an open circle while board-mounted ones are shown with a circle split in the middle with a horizontal line.

Instrument Functions

There are also four major functions for each instrument type. The functions are to control, to indicate, to record and to alarm. Each instrument type may perform one or more of these functions. For example, a PI is a pressure indicator, a pressure-type instrument performing only one function, indicating. But a PIC is a pressure indicating controller which performs two functions, indicating and controlling.

Dual Instruments

Dual instruments are indicated on flow diagrams by touching circles. This means that two instruments are housed together, probably indicating and recording on the same chart. Dual instruments may be board mounted, locally mounted or on a local control panel.

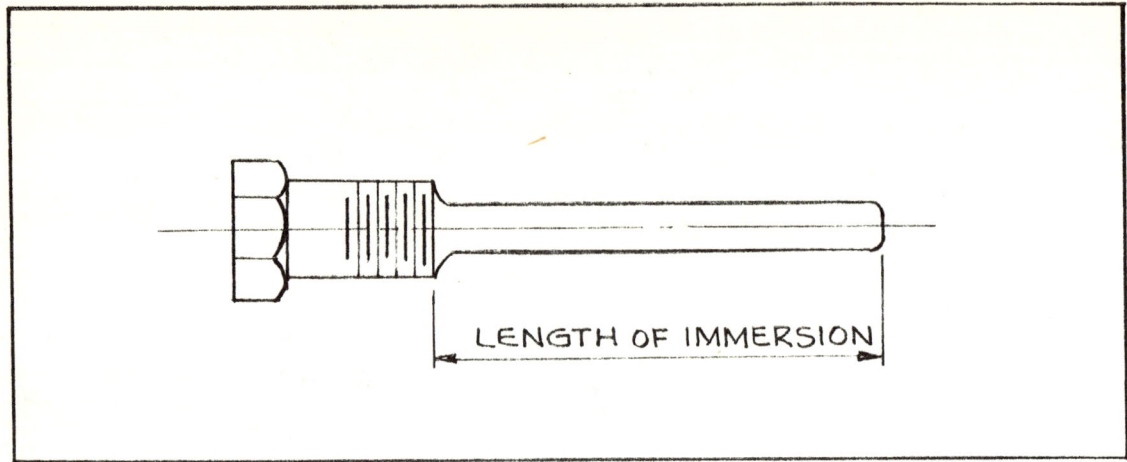

Figure 8-1. Thermowell detail.

Transmitters

Transmitters are shown on the flow diagram as a circle. If the transmitter is board mounted the horizontal line will split the circle. Local transmitters may be mounted on the pipe, grade or platform mounted on a pipe pedestal or on a platform's handrail. Unthinking designers often wrongly locate transmitters on platforms subject to vibration. If the transmitters are left there, vibration will eventually make the joints leak and the instrument will be inoperable. Permanent damage may occur.

Thermowells

Thermowells are the simplest instrument item. There are no moving parts. The thermowell is merely a protective case for another instrument item. Figure 8-1 shows a typical thermowell. The length of immersion is custom designed to fit the pipe into which it is being inserted. Normally thermowells project to the center of pipe. A 1″ coupling is welded to the pipe. The thermowell's outside threads will screw into the 1″ coupling. The temperature instrument then screws into the thermowell's inside threads and the temperature impulse point extends to the end of the immersion point. Instruments screwed into a thermowell are local temperature indicators, thermocouples for board-located temperature indicators, recorders and controllers or capillary tubing for local control of a temperature-regulating control valve.

Thermowells are often installed without any instrument. These are designated TW on the flow diagram. The plant operator uses these thermowells as check points and inserts a thermometer into them when he needs a temperature check. Quite often he just lays a simple glass tube type thermometer in the thermowell, which is called a "yellowback" because of its color. Thermowells should always be located in a spot convenient to the operator.

Thermowells are often located adjacent to a board-mounted temperature instrument which has its own thermowell. This TW would be the operator's check point. Now the designer has two thermowells projecting inside the pipe at least to the pipe's center, probably beyond. The unthinking designer has located them on the same centerline. The 1″ couplings will weld on the pipe, but after the pipe is erected the two thermowells cannot be installed because they hit each other inside the pipe. This requires one coupling to be plugged and a new coupling to be installed. Always stagger TW locations.

Piping designers are guilty of another error in locating thermowells. They forget that something must be inserted or removed from them and locate their coupling where there is a close obstruction.

Instrumentation

Always leave 2' clear from a thermowell to allow insertion of the instrument. A thermocouple in a large line may require more distance. If the designer is presented with this problem, he should check the thermocouple length and leave ample room.

This brings up the most prevalant error committed by piping people. They quite often orient thermowells for thermocouples so that they point horizontally into a walkway or platform area. The piper considers only the coupling, and he wants to make the connection accessible. It looks good on paper. Then the instrument installer comes along and inserts a thermocouple that sticks out 18" into a 3' aisleway or platform. After saying a few words about the piper's background, the field crew has to plug another coupling and install a new one to parallel the aisleway or platform.

Thermowells for local temperature indicators are often mislocated by pipers. Again pipers must consider what will be inserted into the thermowell. To properly locate the TW coupling, the designer must know the temperature indicator type. For the backmounted or "T" type indicator, the designer should never locate the coupling pointing above the horizontal. This would force the thermowell's glass to be pointed up, which will accumulate water on its face and eventually rust out the face. In common piping language, "never point them toward the moon."

Other types of temperature indicators are the bottom-mounted type and one type that rotates on a head. The rotating type costs a little more but has many advantages.

Other Temperature Instruments

A board-mounted temperature indicator will employ a thermocouple. Electrical impulse wire(s) will transmit the signal to the control room instrument panel. The receiving instrument converts this signal to a readable indication which is either visually observed (an indicator) or is recorded with pen and ink on a chart (a recorder). Recorders and indicators may also be located near the pipe or equipment.

Capillary tubing is used to transmit signals to many local temperature instruments. It is a factory-cut hollow tube filled with inert gas or mercury and sealed. Capillary tubing is ordered to a certain length and cannot be cut or lengthened in the field. The piping designer must locate the temperature connection and receiving instrument, sometimes a control valve, to be compatible with the capillary tubing length. This length is specified on the instrument specification sheet. There is a maximum economical length of tubing, which will vary from time to time, but is approximately 30-35'. For distances greater than this, employ a transmitter with a signal to the receiving instrument.

Pressure Instruments

The pressure indicator shown as PI on flow diagrams, is the most common pressure instrument. Pressure recorders and controllers are also utilized. Piping designers supply a 1/2" or 3/4" coupling, nipple and valve for these connections. The instrument designer usually starts with the valve and supplies the instrument hook-up. The first nipple and valve should always conform to the piping specification for material and rating. After the valve, a smaller instrument valve and tubing is quite often used.

Piping designers must locate pressure connections with great care, considering their function and accessibility, and orienting them toward their companion instrument when there is one. A pressure controller tap will connect to a control valve. This tap (or coupling, nipple and valve) should be located near and oriented toward the control valve to keep the piping as short and compact as possible.

Figure 8-2 shows a pressure recording controller portion from a mechanical flow diagram. The impulse point is on the 6" line. The A-3" line is mixed with the A-4" line which comes from a higher pressure line, the C-3". The control valve reduces the pressure from the C (300 pound) flange rating to an A (150 pound) flange rating. This could cause a change in velocity or some surging in the line. The impulse point is located downstream of the junction to allow the flow to stabilize. The 10' shown can be reduced for smaller line sizes, but it should be 30 pipe diameters minimum.

The pressure recording controller is shown as a board-mounted recorder. The impulse line is routed to the control board. From the control valve an impulse line goes to a transmitter which transmits to the recording controller on the main control panel.

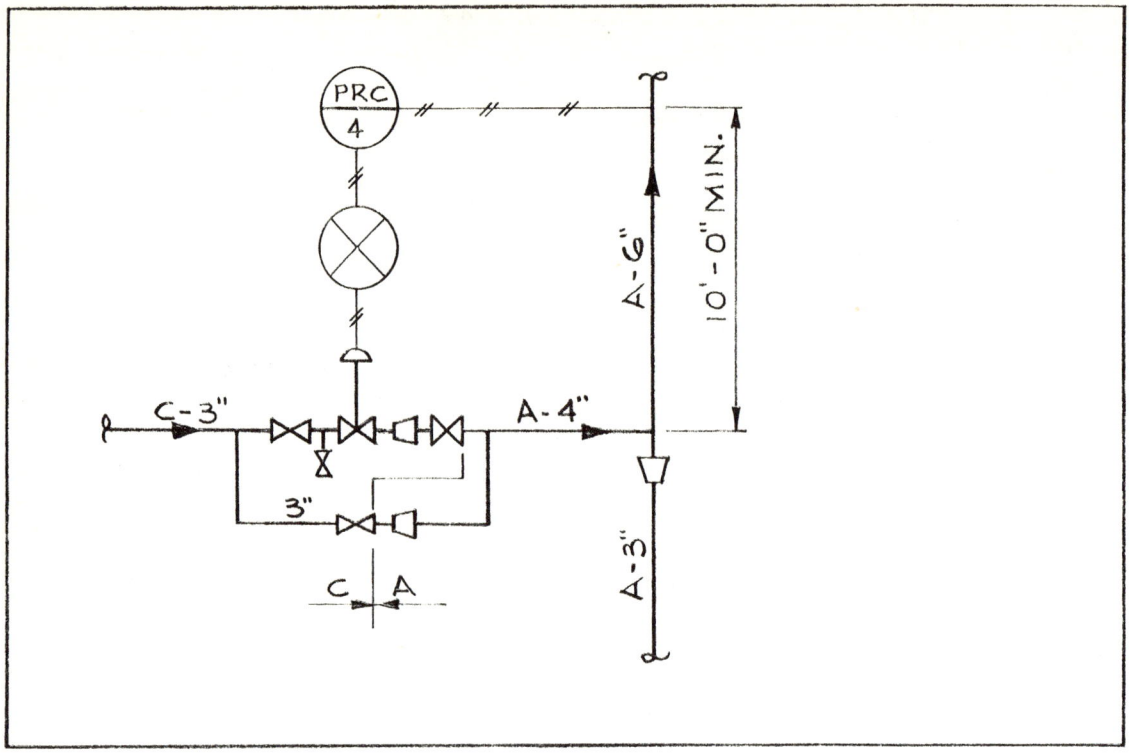

Figure 8-2. Pressure recording controller diagram.

At the control station, the specification break from *C* to *A* is carried through the downstream block valve. Should the control valve stick in the open position and the 4" downstream block valve should be closed it would have to withstand the pressure in the *C* line. Specification breaks at control stations should be downstream of the block valve and the by-pass as shown.

Flow Instruments

Flow instruments encompass flow meters, rotameters, sight glasses and orifice meters among others. Orifice flanges cause piping designers the greatest problems, especially in the larger sizes. Many orifice flanges, or meter runs as they are commonly called, are critical to a plant's operation and are designated "accounting meter" on flow diagrams. Flow accuracy is critical in them and all possible upstream straight-run piping should be provided to ensure this accuracy. Most meter runs, such as a FRC which records and controls flow, are not so critical. At the FRC meter run the flow is set by hand to flow a certain volume. This volume is recorded and if the proper amount is not going through the line the flow is adjusted manually to correct it.

Meter runs containing orifice flanges have lengths established by the American Gas Association (AGA) based on the configuration and diameter ratio between the orifice plate hole diameter and the pipe size. Piping designers assume a minimum diameter ratio of 0.7 or 0.75 to determine minimum upstream straight run of pipe. This assumption is conservative in most cases and results in a straight-run upstream length greater than necessary. The AGA curve might specify 27 pipe diameters upstream with a 0.75 diameter ratio. With a 10" line size, this would require 270" or 22'-6" of straight-run pipe upstream of the orifice flanges. When this length is not practical, consult the instrument engineer. The actual diameter ratio can be calculated, which may be 0.56. Using this ratio, the AGA curve will reduce the upstream requirement by several feet.

Instrumentation

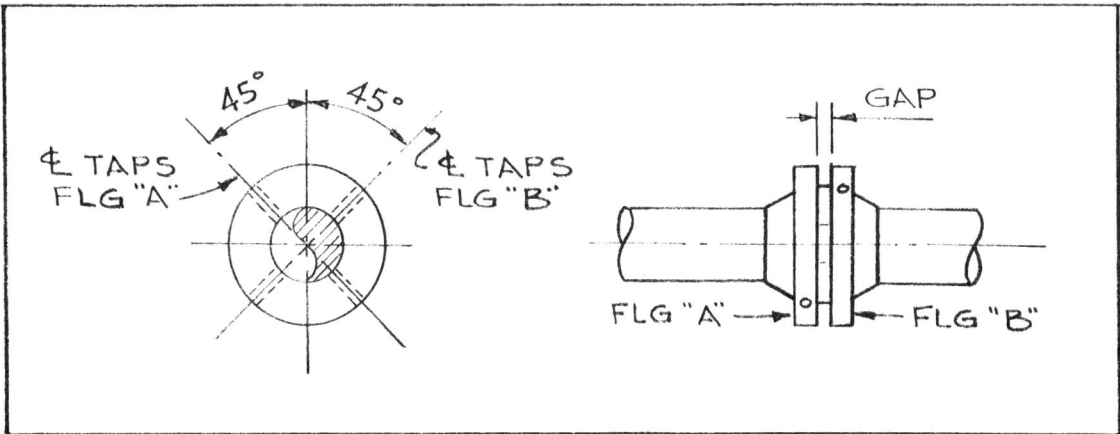

Figure 8-3. Detail of staggered 45° orifice taps.

For sizes 16" and larger, orifice runs get to be a piper's nightmare. One solution is to replace orifice flanges with a venturi tube which needs only a few feet of straight-run pipe upstream. Another solution is straightening vanes installed in the line upstream of the orifice flanges. When meter run lengths become excessive and expensive to provide, consult the instrument engineer for an alternative method.

Six diameters of straight-run pipe are to be provided downstream of orifice flanges. In some cases this may be reduced to four diameters but six should be the design basis.

Orifice flanges may be installed in a vertical pipe run for liquid flowing up or down and for vapor flowing down. Horizontal meter runs are preferred and should be provided where feasible. For many years, only horizontal runs were allowed. Today modern instrument engineers realize that piping configuration costs often outweigh the slight advantage of the horizontal meter run. All accounting meter runs are to be horizontal.

Orifice Flange Taps

It is preferable to locate orifice flange taps horizontally for liquid flow and vertically for vapor flow. Steam service falls between the two and each company has ideas for tap location, so no orientation is standard.

Many companies have settled on taps located 45° off the vertical for all services. No error can be made with this plan. Orifice flanges have two taps in each flange that are 180° apart. A further improvement is to locate taps 45° from the vertical but staggered 90° apart. See Figure 8-3 for details of staggered 45° taps. For vapor service the two taps above horizontal would be valved and the two taps below horizontal would be plugged. With liquid service the two lower taps would be valved while the upper taps are plugged. This is highly advantageous in critical services where orifice taps must employ flanged valves. When the two taps are located side by side, flanged valves will need offset piping to clear.

The staggered tap approach will not work with an orifice valve manifold, a manifold prefabricated with valves especially designed for orifice tap service.

Orifice flange rating is 300 pound minimum because a wide flange thickness is needed to allow taps to be drilled. Flanges are erected, bolted and hydrotested prior to insertion of the orifice plate. When the plate is inserted a gasket is placed on both sides and the flanges are then rebolted. This results in a gap, shown in Figure 8-3, caused by the plate and two gaskets. Thick flanges for the taps and this gap result in a need for longer bolts, a fact often overlooked by piping people.

Pipers must also leave about 15" clear from orifice taps for valves and piping.

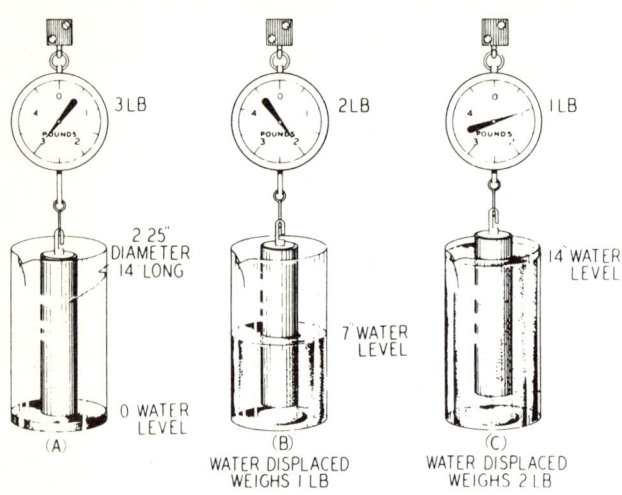

Figure 8-4. Theory of displacement type controller. Courtesy of Masoneilan.

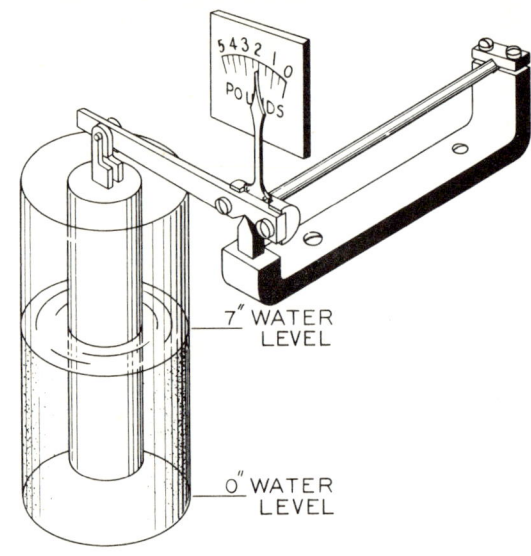

Figure 8-5. Torque type picture. Courtesy of Masoneilan.

Level Instruments

Level controllers are manufactured in several types. Some, such as the ball type, are located inside the equipment and some are located outside the equipment in a float cage. The ball type needs only one connection to project the ball float inside and flange up the controller. However, if maintenance is needed the equipment must be shut down so that the controller can be removed.

Float cage controllers are located outside the equipment with their own cage and two valved connections to the equipment, allowing maintenance with equipment under pressure. For this reason, float cage controllers are most often selected by designers.

The external level controller is made in displacement, constant displacement, variable displacement and interface types. The most common one is the variable displacement type. In this type, the displacer always weighs more than the liquid it displaces at full immersion. Although the displacer rises and falls with level changes, its movement is substantially less than the actual level movement.

The difference in movement is dependent on the cross-sectional area of the displacer, the liquid's specific gravity and the stiffness of the supporting spring or torque tube.

When a body is immersed or partly immersed in any liquid, it loses weight equal to the weight of the liquid being displaced. Figure 8-4, theory of displacement type controller, illustrates a method of using this principle for level measurements in vessels open to the atmosphere. The dimensions, volumetric displacement, weight and change in weight of the displacer in this example are the actual values used in the 14" range controller. In each picture a 2.25" diameter, 14" long displacer is shown.

In example *A* the displacer is suspended by a spring scale having a range of 0-5 pound, and the liquid level is even with the bottom of the displacer. The full weight of the displacer is supported by the spring scale and is 3 pounds. In example *B* the water level has been raised to 7". The displacer now loses weight equal to the weight of the liquid displaced (1 pound) and the net weight shown on the spring scale is 2 pounds. While the liquid level has risen 7", note that the displacer has risen very little.

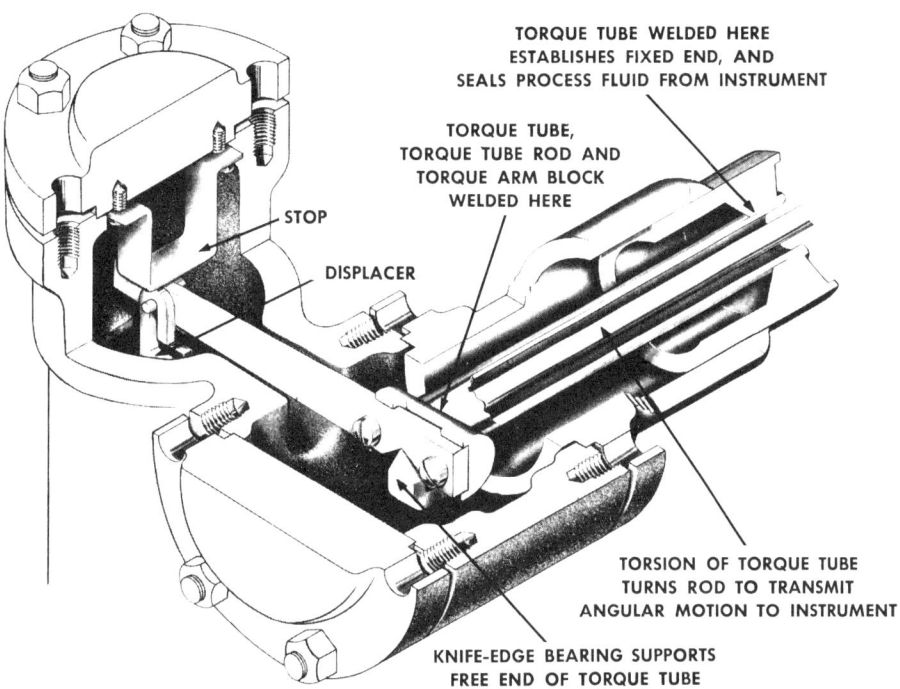

Figure 8-6. Torque tube subassembly. Courtesy of Masoneilan.

In example C the liquid level has increased to 14″ and the scale weight is reduced to 1 pound. The 14″ increase in liquid level has decreased the displacer weight by 2 pounds.

As the net weight of the displacer is decreased, the net load on the spring is reduced in direct proportion to the water level increase. In this case the water volume displaced by the 14″ increase in level is equal to approximately 56 cubic inches, a weight of 2 pounds.

Using an accurate spring scale, the scale could be calibrated in terms of level, thus providing a simple and accurate level indicator for liquids of known specific gravity. This type is limited to equipment open to the atmosphere but forms the basis for all level controller design. For accurate operation in pressurized containers, a frictionless seal was developed, which is now commonly called the torque tube.

Figure 8-5 depicts the torque tube design. The torsion spring or torque tube replaces the spring scale shown in Figure 8-4. The torsion spring can be designed to indicate net weight or level as shown previously.

Figure 8-6, a torque tube subassembly, is an exaggerated sectional view of an actual liquid level controller. The angular motion in a typical design is from 4° to 5°. This angular motion is used to actuate instruments which transmit air signals proportional to level changes for pneumatic indication, recording or controlling.

Figure 8-7, level controller types, shows the various types available from Masoneilan International, a world leader of control instruments. Controllers are available with flanged ends also.

Figure 8-8, level controller installation diagrams, shows installations for reboilers and crude towers and an application for flow recording controlling.

Figure 8-9, level controller installation diagrams, sheet 2, shows installations for three other applications.

Figure 8-10, level controllers—screwed connections, shows models available for screwed piping.

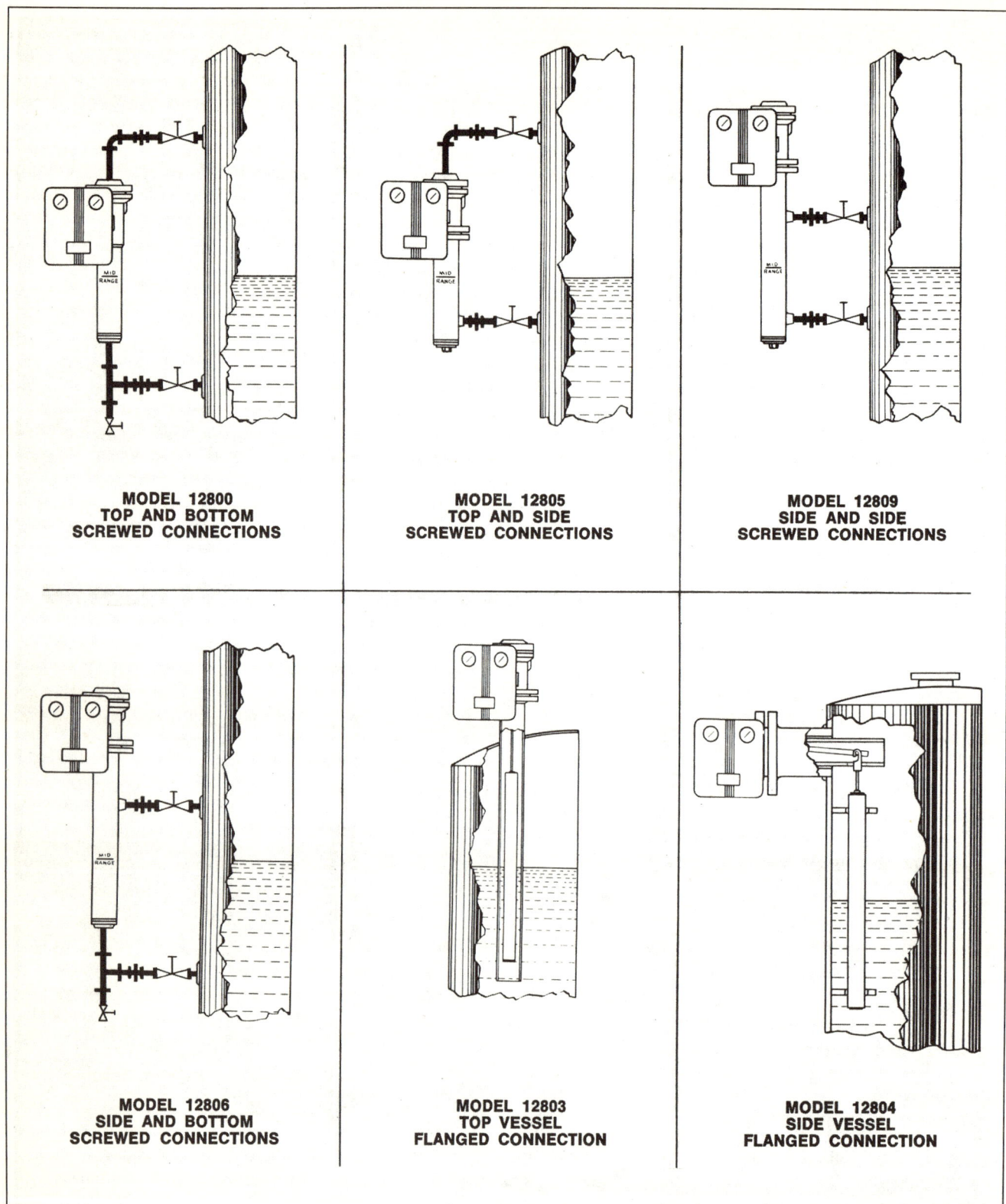

Figure 8-7. Level controller types. Courtesy of Masoneilan.

Instrumentation

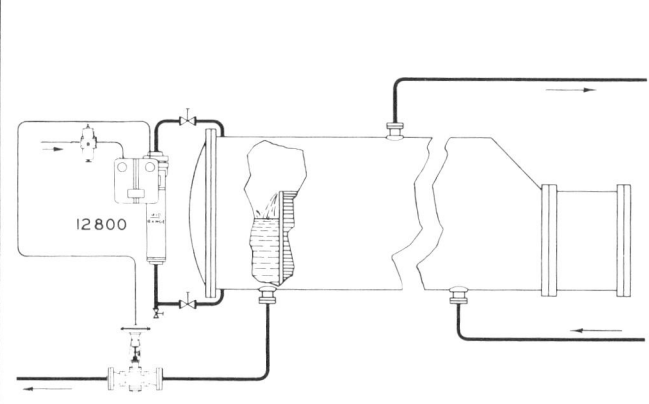

Control of product level within a fractionating tower reboiler

A weir in this kettle type reboiler maintains a constant level around the heating tubes. A 12800 level controller operates a control valve to maintain the rate of draw-off of the bottom product in accordance with the level changes in the downstream side of the weir.

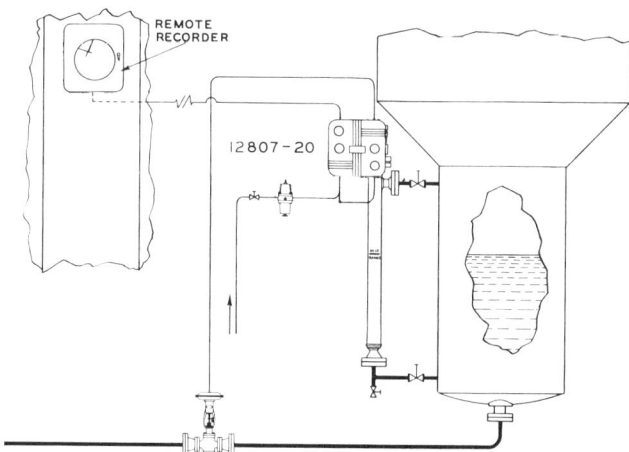

Control and transmission of crude tower level

A 12807-20 side-and-bottom connected controller-transmitter is used on this high temperature application. It controls the valve in the draw-off line from the bottom of the crude fractionating column in accordance with the level changes in the column. An independent pneumatic transmission system measures the level in the vessel which is transmitted to a recorder located at a central panel.

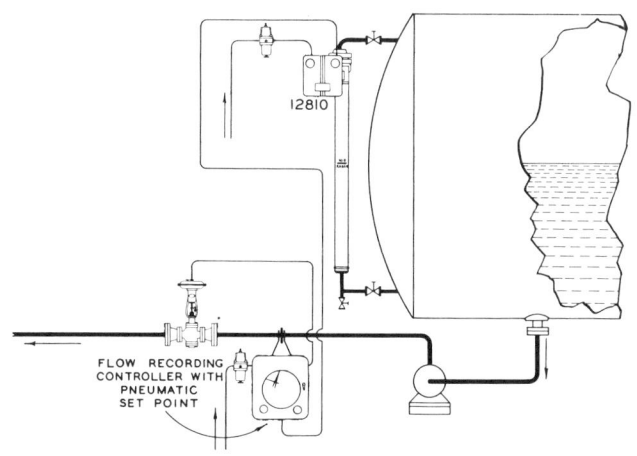

Control of feed rate by pneumatically setting the index of a flow controller

The feed rate through the heating units is stabilized by a flow controller. The 12810 level controller (with reset) averages the level changes within the feed accumulator by pneumatically adjusting the set point of a flow controller.

Figure 8-8. Level controller installation diagrams. Courtesy of Masoneilan.

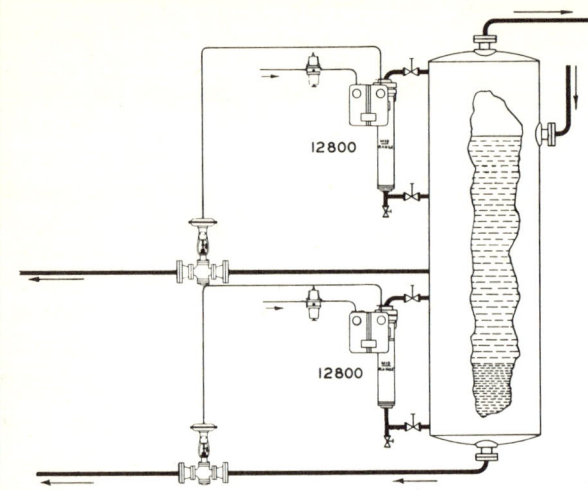

Control of distillate and water levels in a gas and water separator

On applications where three fluid phases exist, such as in a gas-water separator, two level controllers are used: one to control the interface level between the two liquids; the other to maintain the level of the lighter liquid.

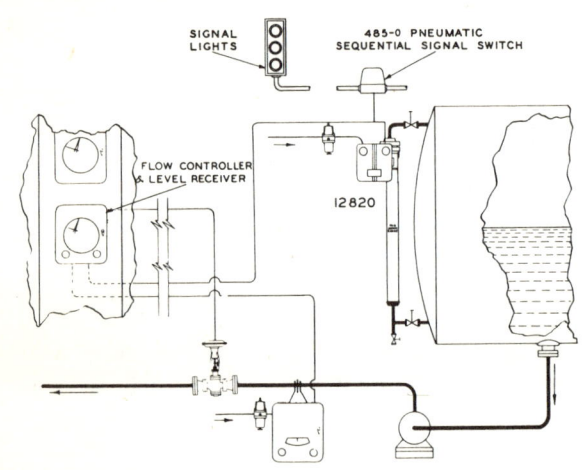

Transmission of level to remote recorder and signal lights

Where the flow to a heater must be maintained at a relatively constant rate (within the limits of the storage capacity available), a 12820 level transmitter is used to transmit the level changes of a feed accumulator to a central panel. A flow controller records on a single chart both the rate of flow which it receives from a flow transmitter located in the discharge line to the heater, and the level. This provides the operator with a continuous record of level and flow which permits him to adjust the flow rate to the existing conditions. A 485-0 sequential switch, which is connected to three signal lights, gives five-position indication of level which can be seen at some distance.

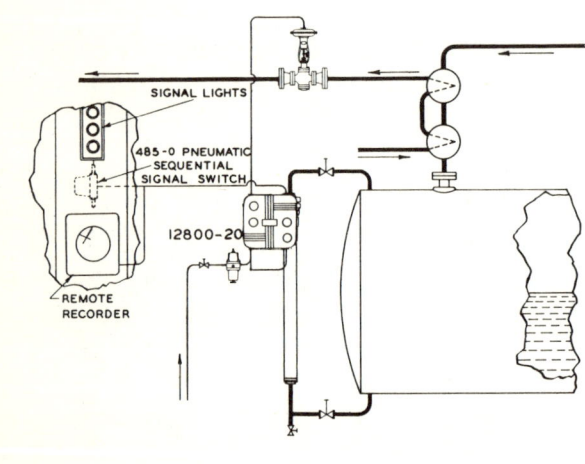

Level transmission and control of a reflux accumulator

Where the exchangers that condense the overhead product from a fractionating tower are operated to produce a reflux supply but no overhead liquid product, the 12800-20 controller-transmitter maintains the level in the reflux accumulator by controlling the flow of water through the condensers. The independent transmission system provides remote indication of level at the control station where it may be correlated with changes in process requirements.

Figure 8-9. Level controller installation diagrams, sheet 2. Courtesy of Masoneilan.

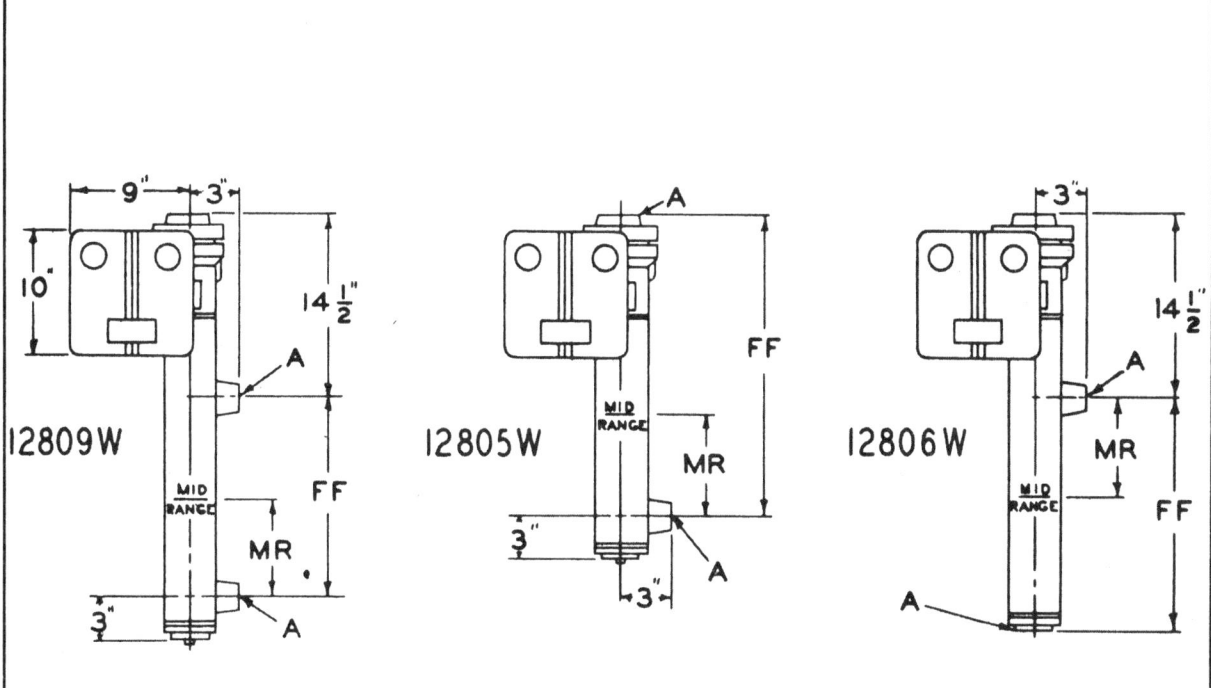

Dimensions (inches)

RANGE	12809W		12805W		12806W	
	FF	MR	FF	MR	FF	MR
14	14	7	20	7	16	7
32	32	16	38	16	34	16
48	48	24	54	24	50	24
60	60	30	66	30	62	30
72	72	36	78	36	74	36
84	84	42	90	42	86	42
96	96	48	102	48	98	48
120	120	60	126	60	122	60

Screwed Connections A = 1½" — 2" NPT — 600 lb. ASA rating

Figure 8-10. Level controllers — screwed connnections. Courtesy of Masoneilan.

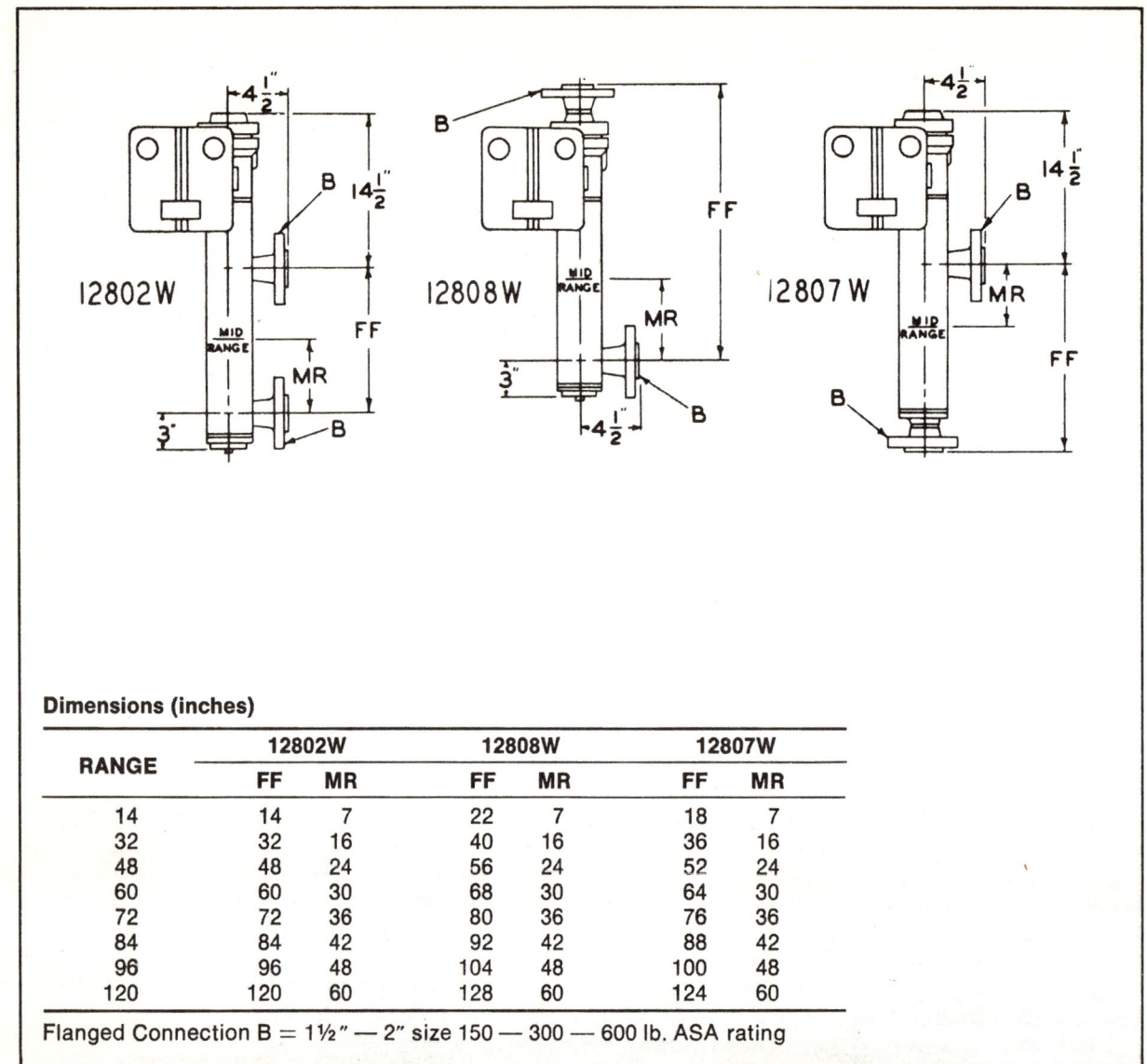

Figure 8-11. Level controllers — flanged connections. Courtesy of Masoneilan.

Figure 8-11, level controllers—flanged connections, shows the flanged models. Both are available in 1½″ or 2″ connection sizes. The 1½″ size is usually specified to keep piping costs to a minimum. Figure 8-12, level controller head orientation, gives allowable orientations for both models. Orientation of control arms is the piping designer's responsibility and access to the control box must be provided.

Level Gage

Level gages (also spelled gauges) show the liquid level in equipment. They come in several styles and various lengths. Figure 8-13, liquid level gage types, shows five common types. Everytime a level controller is installed a level gage must be installed to cover the float range of the controller. Figure 8-14, level gage dimensions, supplies dimensions for level gages. Piping designers are concerned with the center-to-center dimension of valves which locate their connections on equipment. Note the difference between "visible glass" and "distance between valve centers." For a 60″ float range level controller, visible glass of 60″ minimum must be provided. This may mean multiple gages, as most companies limit the distance between valve centers to 60″. For the standard gage this limits the viewing area to 55″, not enough to cover the 60″ float range. So two gage glasses are needed.

Instrumentation

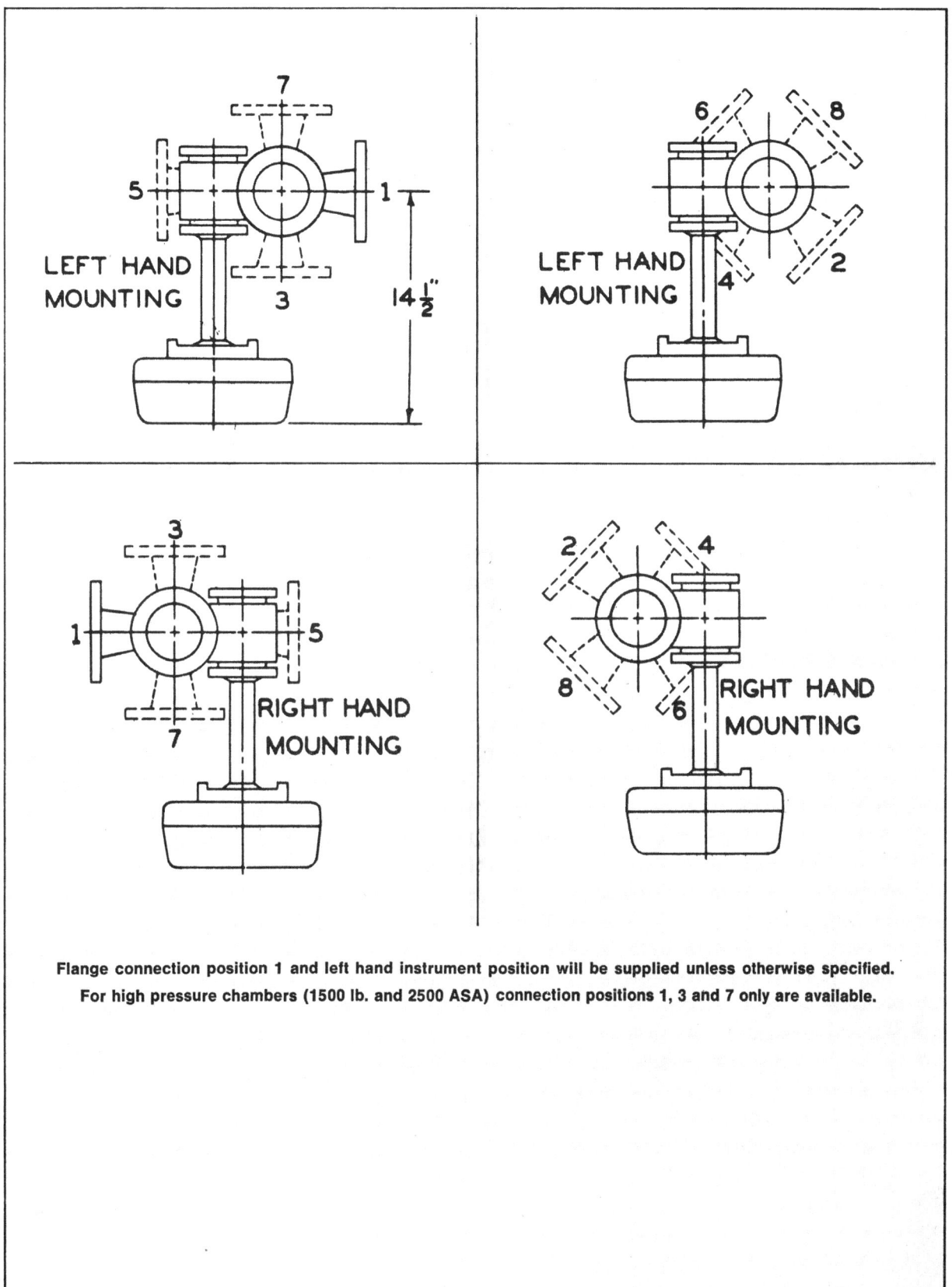

Figure 8-12. Level controller head orientation. Courtesy of Masoneilan.

Direct Reading LIQUID LEVEL GAGES
ITEM: Special Service Heating and Cooling and/or Design Gages and Valves

For accurate level measurements of liquids whose viscosity tends to vary, or volatile liquids which tend to boil under existing conditions, heating or cooling type gages are available. Gage and valves are traced with heating or cooling medium, either externally or internally as shown and described.

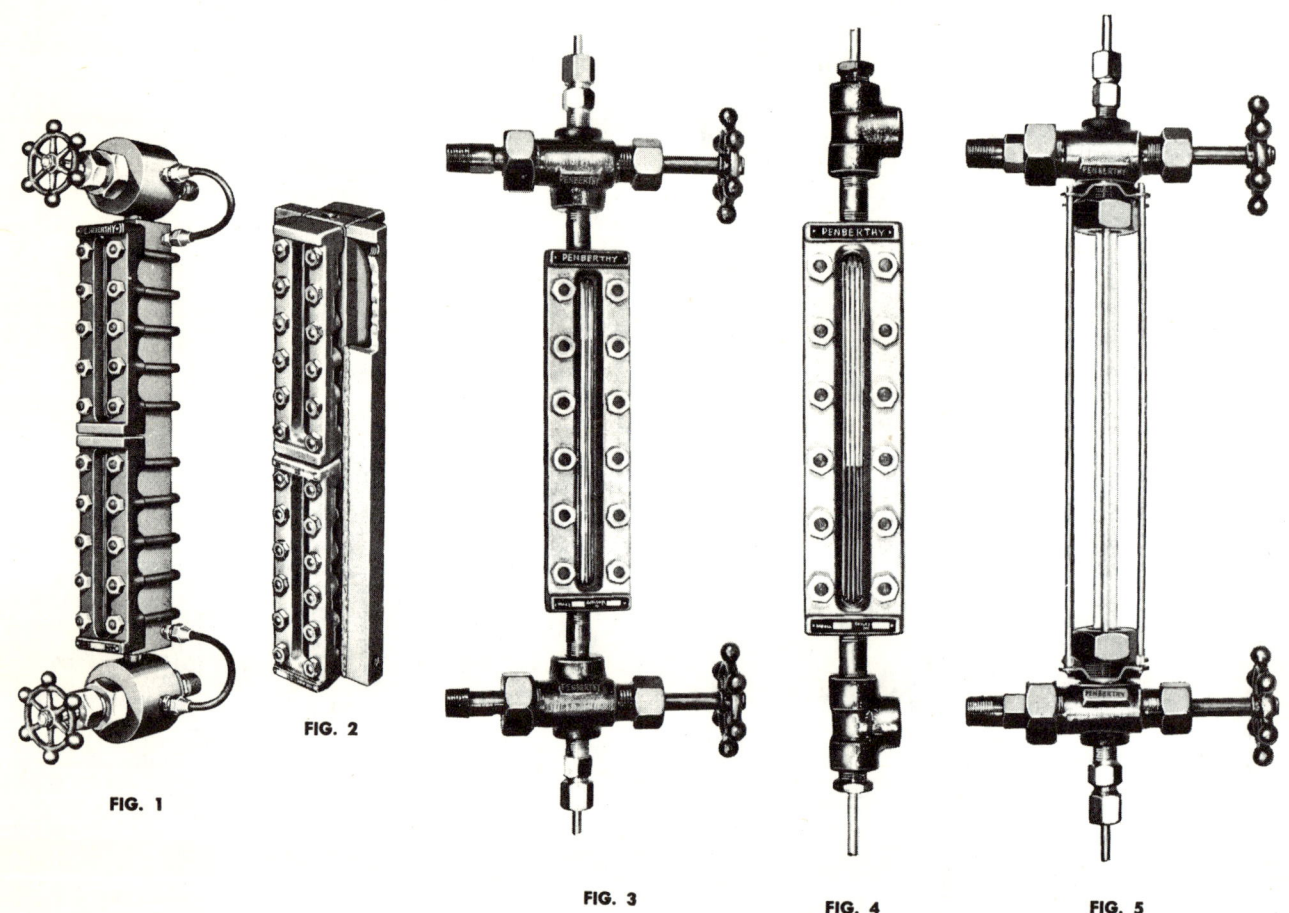

Fig. 1—External type Reflex gage with integral heating or cooling chamber at the back of liquid chamber.

Fig. 2—External type transparent gage with welded heating or cooling chambers on each side of liquid chamber.

Fig. 3—Standard reflex or transparent type gages and valves with internal heating or cooling stainless steel tube passing through liquid chamber and valves. Tube being held in place with packing adapters in the vent and drain connections of the valves.

Fig. 4—Packing block tees replace valves of Fig. 3 which enables side connection installation.

Fig. 5—Standard ¾" dia. tubular glass gages equipped with heating or cooling tube and packing adapters.

All Penberthy reflex series gages are available with externally heated or cooled feature—transparent type is only available in the ST Series.

When Ordering: For example "S-1108HC" is an S-1108 with external heating or cooling chamber.

Figure 8-13. Liquid level gage types. Courtesy of Penberthy.

Direct Reading LIQUID LEVEL GAGES
ITEM: Technical Data: Valve Centers

FOR PENBERTHY REFLEX AND TRANSPARENT LIQUID LEVEL GAGE SETS
Threaded Gage Connection Types

GAGE TYPE AND MODEL NUMBER[1]						NUMBER OF SECTIONS	VISIBLE GLASS (inches)	DISTANCE BETWEEN VALVE CENTERS IN INCHES [2]		
REFLEX			TRANSPARENT					STANDARD NO. 305 OR NO. 405 VALVES [3] WITH NIPPLES	UNION CONNECTION NO. 325 OR NO. 425 VALVES [3] WITH NIPPLES	CLOSE HOOKUP NOS. 305, 325 OR NOS. 405, 425 VALVES [3 & 4]
X	S	W	XT	ST	WT					
X-500	S-1100	—	XT-600	ST-1200	—	1	3¾	8½	11¼	3¾
X-501	S-1101	—	XT-601	ST-1201	—	1	4¾	9½	12¼	4¾
X-502	S-1102	—	XT-602	ST-1202	—	1	5¾	10½	13¼	5¾
X-503	S-1103	W-13	XT-603	ST-1203	WT-13	1	6¾	11½	14¼	6¾
X-504	S-1104	W-14	XT-604	ST-1204	WT-14	1	7⅞	12⅝	15⅜	7⅞
X-505	S-1105	W-15	XT-605	ST-1205	WT-15	1	9⅛	13⅞	16⅝	9⅛
X-506	S-1106	W-16	XT-606	ST-1206	WT-16	1	10¼	15	17¾	10¼
X-507	S-1107	W-17	XT-607	ST-1207	WT-17	1	11⅞	16⅝	19⅜	11⅞
X-508	S-1108	W-18	XT-608	ST-1208	WT-18	1	12⅝	17⅜	20⅛	12⅝
X-523	S-1123	W-23	XT-623	ST-1223	WT-23	2	15	19¾	22½	15
X-524	S-1124	W-24	XT-624	ST-1224	WT-24	2	17¼	22	24¾	17¼
X-525	S-1125	W-25	XT-625	ST-1225	WT-25	2	19¾	24½	27¼	19¾
X-526	S-1126	W-26	XT-626	ST-1226	WT-26	2	22	26¾	29½	22
X-527	S-1127	W-27	XT-627	ST-1227	WT-27	2	25¼	30	32¾	25¼
X-528	S-1128	W-28	XT-628	ST-1228	WT-28	2	26¾	31½	34¼	26¾
X-545	S-1145	W-35	XT-645	ST-1245	WT-35	3	30⅜	35⅛	37⅞	30⅜
X-546	S-1146	W-36	XT-646	ST-1246	WT-36	3	33¾	38½	41¼	33¾
X-547	S-1147	W-37	XT-647	ST-1247	WT-37	3	38⅝	43⅜	46⅛	38⅝
X-548	S-1148	W-38	XT-648	ST-1248	WT-38	3	40⅞	45⅝	48⅜	40⅞
X-566	S-1166	W-46	XT-666	ST-1266	WT-46	4	45½	50¼	53	45½
X-567	S-1167	W-47	XT-667	ST-1267	WT-47	4	52	56¾	59½	52
X-568	S-1168	W-48	XT-668	ST-1268	WT-48	4	55	59¾	62½	55
X-56	S-56	W-56	XT-56	ST-56	WT-56	5	57¼	62	64¾	57¼
X-57	S-57	W-57	XT-57	ST-57	WT-57	5	65⅜	70⅛	72⅞	65⅜
X-58	S-58	W-58	XT-58	ST-58	WT-58	5	69⅛	73⅞	76⅝	69⅛
X-67	S-67	W-67	XT-67	ST-67	WT-67	6	78¾	83½	86¼	78¾
X-68	S-68	W-68	XT-68	ST-68	WT-68	6	83¼	88	90¾	83¼
X-77	S-77	W-77	XT-77	ST-77	WT-77	7	92⅛	96⅞	99⅝	92⅛
X-78	S-78	W-78	XT-78	ST-78	WT-78	7	97⅜	102⅛	104⅞	97⅜
X-87	S-87	W-87	XT-87	ST-87	WT-87	8	105½	110¼	113	105½
X-88	S-88	W-88	XT-88	ST-88	WT-88	8	111½	116¼	119	111½
X-97	S-97	W-97	XT-97	ST-97	WT-97	9	118⅞	123⅝	126⅜	118⅞
X-98	S-98	W-98	XT-98	ST-98	WT-98	9	125⅝	130⅜	133⅛	125⅝
X-107	S-107	W-107	XT-107	ST-107	WT-107	10	132¼	137	139¾	132¼
X-108	S-108	W-108	XT-108	ST-108	WT-108	10	139¾	144½	147¼	139¾

(1) Select correct gage by referring to pressure-temperature rating.

(2) The center-to-center dimension given in table is the minimum distance obtainable between valve centers for the corresponding gage and valves. Large Chamber Gauge End connected add ⅞" to figures shown. For other requirements, refer to dimensional data of gage valves.

(3) Valve centers given in table also apply when gages are used with valves of other sizes in the corresponding Series.

For gage valve pressure-temperature rating and dimensions.

(4) Minimum Valve centers for ¾" Close Hook-up Gage is ¾" longer than shown in table for respective gage.

On Transparent gages, requiring several short sections to make up the total visibility desired, less interference in level reading will result if the gage is selected from higher pressure groups. This, however, does not apply to the Reflex type gages.

Figure 8-14. Level gage dimensions. Courtesy of Penberthy.

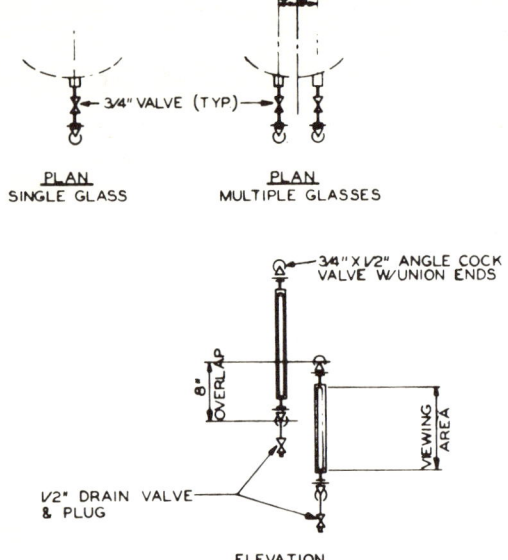

Figure 8-15. Multiple level gage installation. Courtesy of Fluor Corp.

Figure 8-15, multiple level gage installation, shows how gages are overlapped 8" to overlap viewing area 1" minimum. This viewing area overlap will vary with the manufacturer.

Control Valves

Control valves are designed with a body for throttling. Most control valves employ a globe-type body, but ball, butterfly and other types are also used. Control valves regulate flow throughout the unit by opening or closing fractional amounts letting more or less flow pass as the impulse signal commands them to do. The impulse signal may be pneumatic, electronic, hydraulic or electro-hydraulic. A few control valves are operated manually, but the great majority are pneumatically operated.

The control valve is comprised of two basic components, the actuator or motor operator and the valve body. Figure 8-16, a control valve actuator, gives details of actuator and terminology of diaphram actuator parts. Figure 8-17, a control valve body, gives details and terminology for body parts. A double-port body is shown but many parts are common for all control valves.

For fluid temperatures of 450°F or higher, radiating cooling fins are installed on top of the body but under the actuator to dispell heat to the atmosphere, protecting the actuator. This makes the overall height of the control valve greater and piping designers often forget to consider cooling fins. This results in actuators and by-passes vying to occupy the same spot and this cannot happen.

Figure 8-18, typical control stations, shows how to design control valve installations incorporating block valves on either side of the control valve. A globe valve by-pass is shown but gate valves can be used more economically. Dimensions from centerline of by-pass to control valve are a guide for students and will be sufficient in most cases. Each control valve height must be checked to ensure 9" minimum clearance above the actuator. Actuator diaphram widths vary and must also be checked to ensure clearance from the vertical pipe. When station types CS-1 or CS-2 are used, diaphrams are often wide enough to interfere with the block valves' flanges.

Control valves are sometimes purchased with handwheels, commonly called handjacks, mounted on the actuator. The handwheel allows manual operation, opening or closing the valve, which makes manual fluid control possible if the automatic control becomes inoperative. Manual control will continue until repairs are completed and automatic control is resumed. Block and by-pass valves are not supplied when manual handwheels are specified. Handwheels are usually mounted on the actuator's side; however, a reverse-acting control valve may have the handwheel on top.

Control valves must be located near the equipment they serve and must be accessible to the operator. A level control valve controls the level in a vessel by receiving impulse signals from a level controller attached to the vessel. Locating the control valve near the level controller reduces length of impulse line. When operating the control valve manually or when operating on by-pass, the operator should be able to see the vessel's gage glass.

Instrumentation

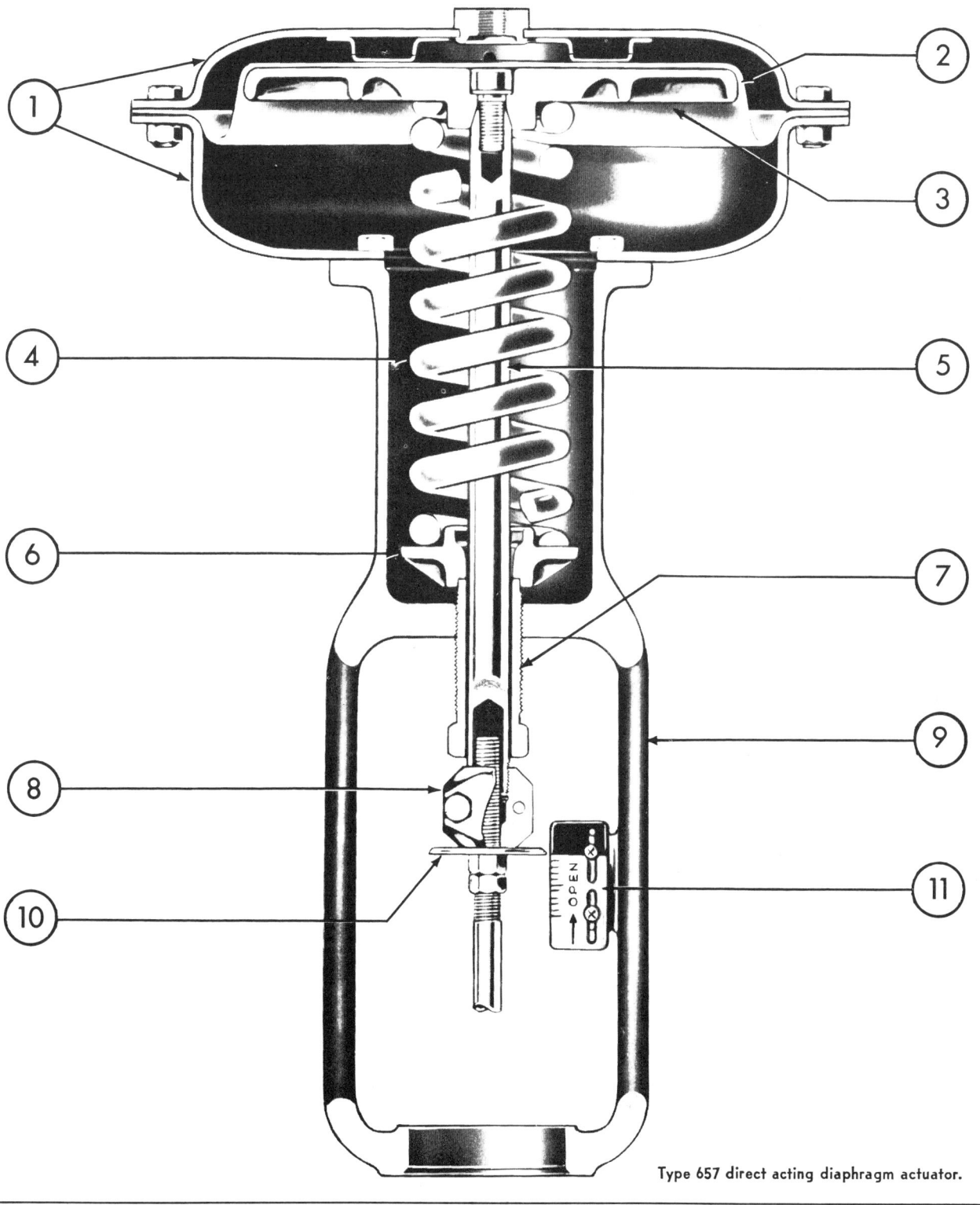

Type 657 direct acting diaphragm actuator.

1 Diaphragm Case	4 Actuator Spring	7 Spring Adjustor	10 Travel Indicator
2 Diaphragm	5 Actuator Stem	8 Stem Connector	11 Travel Indicator Scale
3 Diaphragm Plate	6 Spring Seat	9 Yoke	

Figure 8-16. Control valve actuator. Courtesy of Fisher Controls Co.

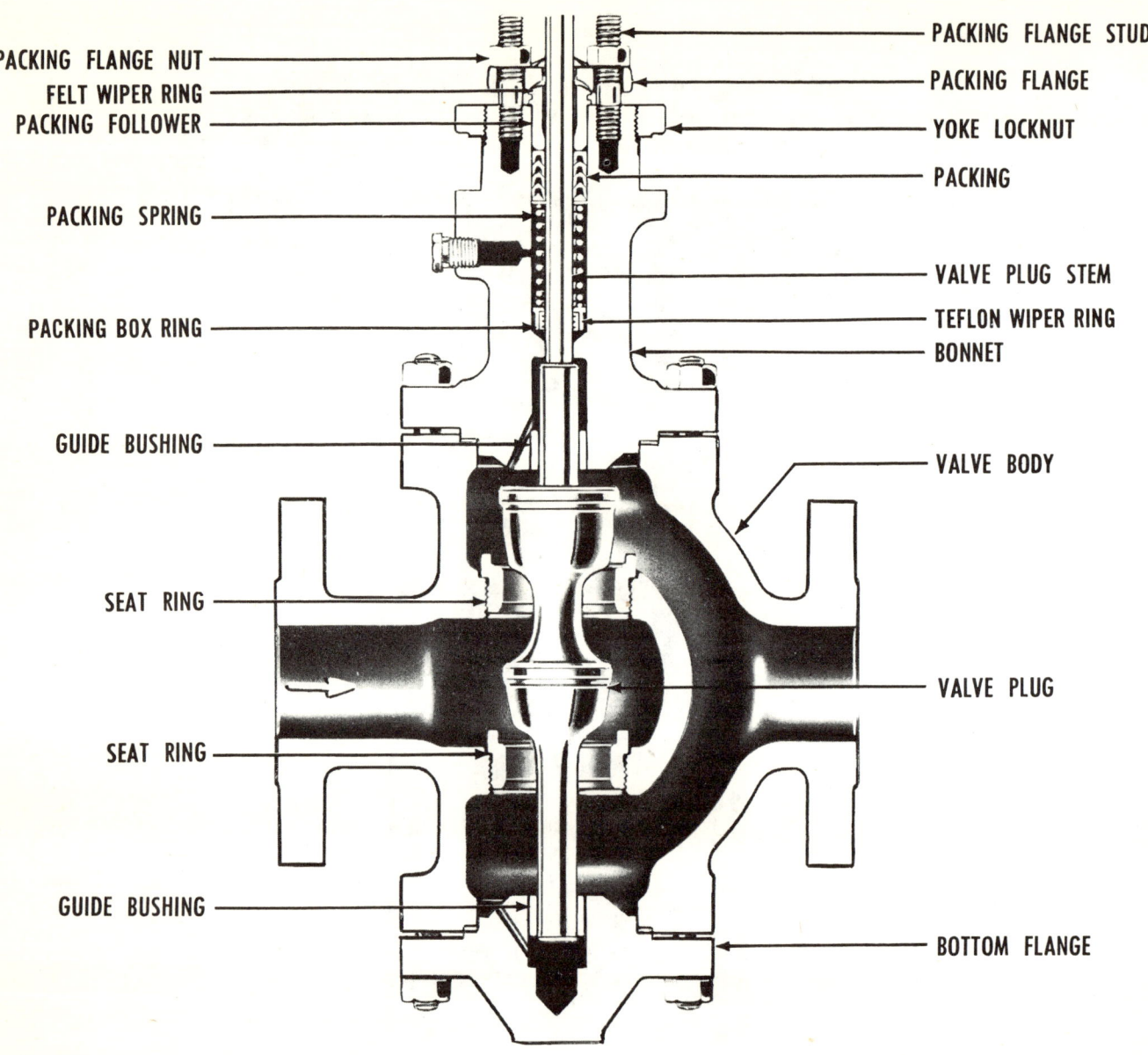

Figure 8-17. Control valve body. Courtesy of J.E. Lonergan Co.

Instrumentation

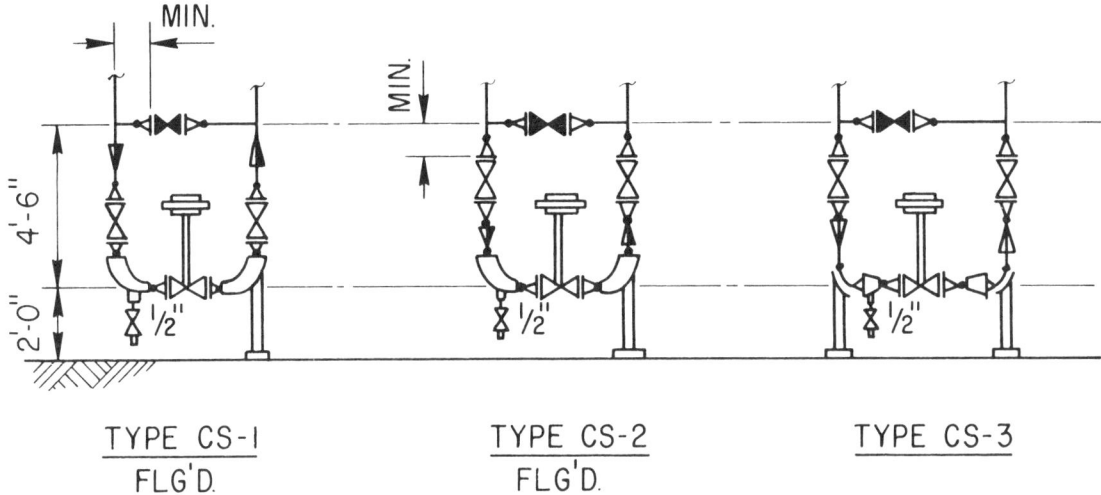

USE REDUCING ELLS WHEN BLOCK VALVES ARE 8" & UNDER (TYPE CS-1 & CS-2), FOR LARGER SIZES USE CONCENTRIC REDUCERS (TYPE CS-3)

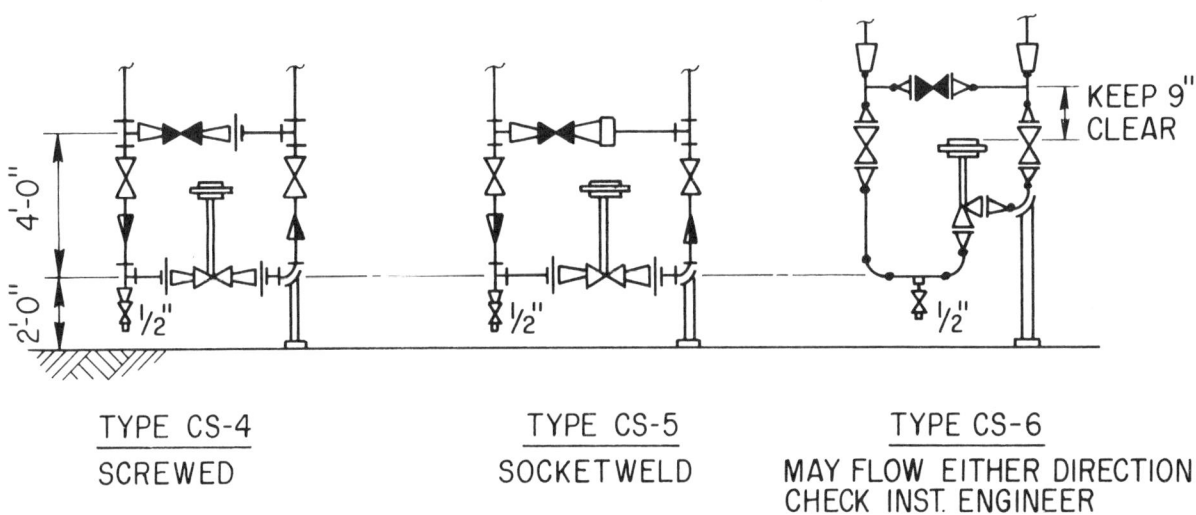

NOTES:
1. CHECK HEIGHTS & WIDTHS OF CONTROL VALVE OPERATORS FOR CLEARANCE.
2. CONTROL VALVE COOLING FINS REQUIRED AT 450°F. & ABOVE.
3. ON PULSATING PIPING LOWER BY-PASS AS MUCH AS POSSIBLE. MAKE THE INSTALLATION COMPACT TO REDUCE VIBRATION.
4. DIMENSION 2'-0" IS FROM HIGH POINT OF GRADE OR PAVING.

Figure 8-18. Typical control stations.

LONERGAN Safety-Relief Valves
  D Series
  Conventional

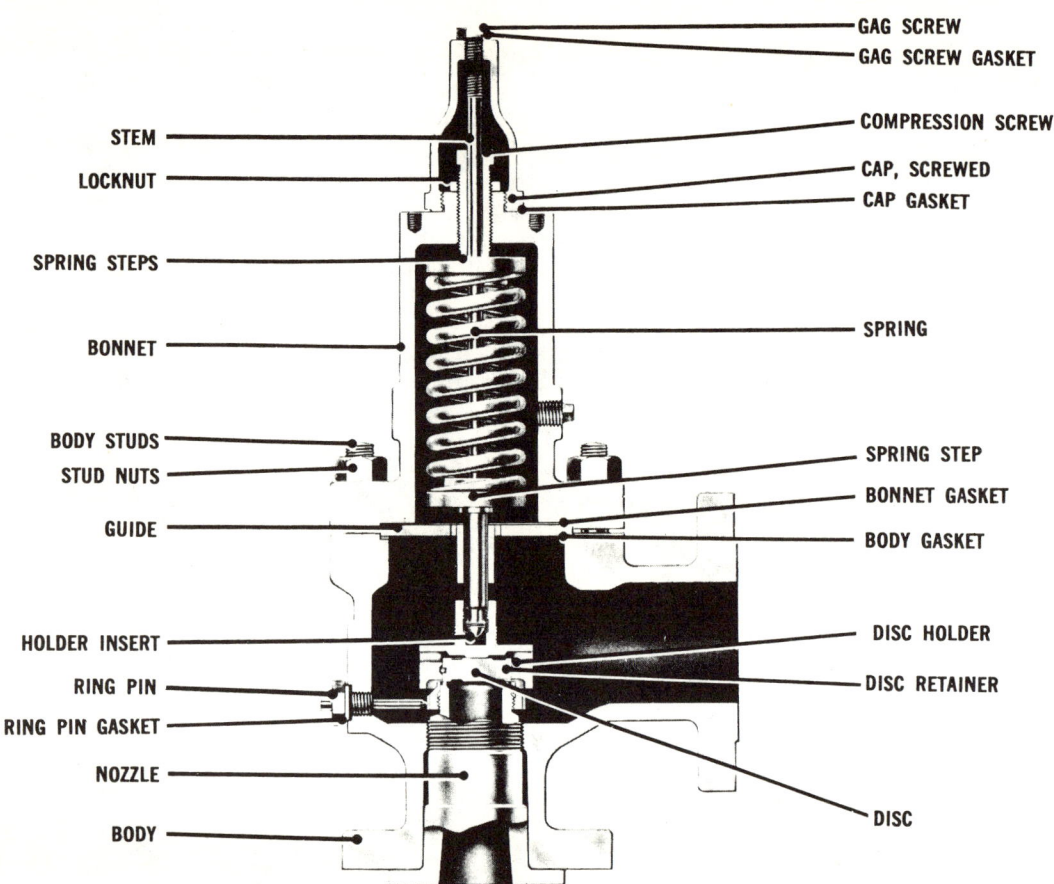

NAME OF PART	MATERIAL
Gag Screw	AISI 416 Stainless Steel
Gag Screw Gasket	Corrugated Soft Iron
Stem	AISI 416 Stainless Steel
Compression Screw	AISI 416 Stainless Steel
Bonnet Plug	Carbon Steel
Locknut	AISI 416 Stainless Steel
Cap Gasket	Corrugated Soft Iron
Spring Steps	AISI C-1117 Carbon Steel
Body Studs	ASTM A-193 GR B7, Alloy Steel
Stud Nuts	ASTM A-194 GR 2H, Alloy Steel
Bonnet Gasket	Corrugated Soft Iron
Body Gasket	Corrugated Soft Iron
Guide	ASTM A-351 GR CF8, Stainless Steel
Holder Insert	AISI 416 Stainless Steel, Hardened
Ring Pin Gasket	Corrugated Soft Iron
Ring Pin	AISI 416 Stainless Steel
Disc Holder	ASTM A-351 GR CF8, Stainless Steel
Disc Retainer	AISI 302 Stainless Steel
Disc	17-4 PH. Stainless Steel, Hardened
Adjusting Ring	AISI 302 Stainless Steel
Nozzle	ASTM A-351 GR CF8, Stainless Steel
Spring (Note 1)	Carbon Steel
Spring (Note 2)	Tungsten Steel (8.75-9.75)
Cap, Screwed	ASTM A-216, GR WCB, Carbon Steel
Bonnet (Note 3)	ASTM A-216, GR WCB, Carbon Steel
Bonnet (Note 4)	ASTM A-217, GR WC6, Alloy Steel
Body (Note 3)	ASTM A-216, GR WCB, Carbon Steel
Body (Note 4)	ASTM A-217, GR WC6, Alloy Steel

Figure 8-19. Conventional flanged relief valve. Courtesy of J.E. Lonergan Co.

Instrumentation

Control valve actuators should be installed in the vertical. In some rare cases this is not practical, as with body sizes of 10″ and larger with actuators utilizing cooling fins resulting in diaphram heights of 7-8′. Accessibility demands that these diaphrams be lowered, which results in horizontal installation, particularly on angle control valves.

With a horizontal valve plug stem or rod, two special problems arise. Supporting the diaphram is a mechanical problem best handled by cable from above. This allows free expansion. Many times the valve manufacturer can offer support solutions. The other problem is lubrication of the stem or rod. With the horizontal installation, gravity will compel the lubricant to seek and lubricate the lower half of the stem. The manufacturer must be notified if horizontal installation is decided so that he can alter the lubrication design to make it effective for horizontal installation.

Angle control valves are specified for large pressure drop applications. Flow is normally in the side and out the bottom when the diaphram is vertical. However, this varies with the service. Before issuing the final piping drawing, the manufacturer's certified outline drawing must be checked to determine flow pattern.

Control valves must be removable for maintenance. With flanged valves the flanges can be unbolted and removal is simple. If the flanges are ring joint, removal is much easier if one of the block valves is located in the vertical so a flanged elbow can be removed. For screwed or socketwelded control valves, provide a union on both sides of the control valve.

Relief Valves

Figure 8-19, conventional flanged relief valve, lists the relief valve's various parts. Except for small thermal relief valves, springs should always be installed upright. The inlet flange is part of the body while the nozzle forms the flange face or seating surface. This nozzle is ½ to 1″ thick, resulting in longer than standard bolting. When tabulating material needed for a relief valve always check the nozzle for added bolt length. As a guide, for relief valve flanged inlets up to 1″ size, add ½″ to normal bolting. For sizes 1½ to 4″ add ¾″ and for 6″ and larger add 1″

Chapter 8
Review Test

1. Define the four major instrument types. _____

   _____

2. Give the four major functions of the types. _____

   _____

3. What is a dual instrument? _____

4. What does "board mounted" mean? _____

5. Which instrument item has no moving parts? _____

6. Define AGA. _____

7. Orifice flanges are ____ pound minimum rating.

8. Level gages are usually limited to ____ inches length between valve centers.

9. Multiple gage glasses shall have a viewing area overlap of _____.

10. Most control valves are of the ____ type body.

11. Cooling fins are added to control valves for fluid temperatures of ____ °F or greater.

12. What is a control valve handjack? _____

    _____

13. For angle control valves, flow normally goes in the _____ and out the _____.

14. Relief valve springs should normally be installed _____.

15. Because of the inlet nozzle, relief valve inlet flanges require longer _____.

Appendix

In any international operation, measurement conversion is common. Engineers, designers and draftsmen are constantly seeking conversion tables and charts and usually find every one but the one they need. In this chapter conversion factors are assembled in alphabetical order. Following the factors are conversion tables useful to engineers and designers.

A

Acre	x 10	= Square chain (Gunters)
Acre	x 160	= Rods
Acre	x 1×10^5	= Square links (Gunters)
Acre	x 0.4047	= Hectare or square hectometer
Acres	x 43,560	= Square feet
Acres	x 4,047	= Square meters
Acres	x 1.562×10^{-3}	= Square miles
Acres	x 4,840	= Square yards
Acre-feet	x 43,560	= Cubic feet
Acre-feet	x 3.259×10^5	= Gallons
Amperes/square centimeters	x 6.452	= Amperes/square inch
Amperes/square centimeters	x 10^4	= Amperes/ square meter
Amperes/square inch	x 0.1550	= Amperes/square centimeters
Amperes/ square inches	x 1,550	= Amperes/square meter
Amperes/square meter	x 10^{-4}	= Amperes/square centimeter
Amperes/square meter	x 6.452×10^{-4}	= Amperes/square inch
Are	x 0.02471	= Acre (USA)
Ares	x 119.60	= Square yards
Ares	x 100	= Square meters
Atmospheres	x 14.7	= Pounds/square inch
Atmospheres	x 1.058	= Tons/square foot

Conversion factor A (continued on following page)

Conversion factor A (concluded)

Atmospheres	x 29.92	= Inches of mercury at 0°C
Atmospheres	x 76	= Centimeters of mercury
Atmospheres	x 33.90	= Feet of water at 4°C
Atmospheres	x 1.033	= Kilograms/square centimeter
Atmospheres	x 10,332	= Kilograms/square meter
Atmospheres	x 1,013.2	= Millibar
Atmospheres	x 760	= Millimeters of mercury
Atmospheres	x 10.332	= Meters of water at 4°C

B

Barrels (USA, dry)	x 7,056	= Cubic inches
Barrels (USA, dry)	x 105	= Quarts (dry)
Barrels (USA, liquid)	x 31.5	= Gallons
Barrels (oil)	x 42	= Gallons (oil)
Barrels (oil)	x 0.159	= Cubic meters
Barrels (oil)	x 159	= Liters
Barrels/day	x 6.6245×10^{-3}	= Cubic meters/hour
Barrels (oil)	x 5.6154	= Cubic feet
Barrels/day	x 29.167×10^{-3}	= Gallons per minute
Bars	x 0.9869	= Atmospheres
Bars	x 10^6	= Dynes/square centimeter
Bars	x 1.020×10^4	= Kilograms/square meter
Bars	x 2,089	= Pounds/square foot
Bars	x 14.50	= Pounds/square inch
Btu	x 1.055×10^{10}	= Ergs
Btu	x 778.3	= Foot-pounds
Btu	x 252	= Gram-calories
Btu	x 3.931×10^{-4}	= Horsepower-hours
Btu	x 1,054.8	= Joules
Btu	x 0.252	= Kilogram-calories
Btu	x 107.5	= Kilogram-meters
Btu	x 2.298×10^{-4}	= Kilowatt-hours
Btu/hour	x 0.2162	= Foot-pounds/second
Btu/hour	x 0.070	= Gram-calorie/second
Btu/hour	x 3.929×10^{-4}	= Horsepower-hours (British)
Btu/hour	x 0.2931	= Watts
Btu/hour foot °F	x 1.4882	= Kilocalorie/meter hour °C
Btu/hour foot²	x 2.7125	= Kilocalorie/meter² hour
Btu/pound	x 0.5556	= Kilocalorie/kilograms
Btu/pound °F	x 1	= Kilocalorie/kilograms °C
Btu inches/hour foot² °F	x 0.12402	= Kilocalorie/meter hour °C
Btu/minute	x 12.96	= Foot-pounds/second
Btu/minute	x 0.02356	= Horsepower
Btu/minute	x 0.01757	= Kilowatts
Btu/minute	x 17.57	= Watts
Btu/square foot/minute	x 0.1221	= Watts/square inch

Conversion factor B (continued on facing page)

Conversion factor B (concluded)

Bucket (British-dry)	x 1.818×10^4	= Cubic centimeters
Bushels	x 1.2445	= Cubic feet
Bushels	x 2,150.4	= Cubic inches
Bushels	x 0.03524	= cubic meters
Bushels	x 35.24	= Liters
Bushels	x 4	= Pecks
Bushels	x 64	= Pints (dry)
Bushels	x 32	= Quarts (dry)

C

Candle/square centimeter	x 3.142	= Lamberts
Candle/square inch	x 0.487	= Lamberts
Centares	x 1.0	= Square meters
Centigrade	x 9/5 + 32	= Fahrenheit
Centiliter	x 0.3382	= Fluid ounce (USA)
Centiliter	x 0.6103	= Cubic inch
Centiliter	x 2.705	= Drams
Centimeters	x 3.281×10^{-2}	= Feet
Centimeters	x 0.3937	= Inches
Centimeters	x 1.094×10^{-2}	= Yards
Centimeters of mercury	x 0.01316	= Atmospheres
Centimeters of mercury	x 0.4461	= Feet of water
Centimeters of mercury	x 136	= Kilograms/square meter
Centimeters of mercury	x 27.85	= Pounds/square foot
Centimeters of mercury	x 0.1934	= Pounds/square inch
Centimeters/second	x 1.1969	= Feet/minute
Centimeters/second	x 0.03281	= Feet/second
Centimeters/second	x 0.036	= Kilometers/hour
Centimeters/second	x 0.1943	= Knots
Centimeters/second	x 0.6	= Meters/minute
Centimeters/second	x 0.02237	= Miles/hour
Chain	x 792	= Inches
Chain	x 20.12	= Meters
Cords	x 8	= Cord feet
Cord feet	x 16	= Cubic feet
Coulomb	x 2.998×10^9	= Statcoulombs
Coulombs	x 1.036×10^{-5}	= Faradays
Coulombs/square centimeter	x 64.52	= Coulombs/square inch
Coulombs/square centimeter	x 10^4	= Coulombs/square meter
Coulombs/square inch	x 0.155	= Coulombs/square centimeter
Coulombs/square inch	x 1,550	= Coulombs/square meter
Cubic centimeters	x 3.531×10^{-5}	= Cubic feet
Cubic centimeters	x 0.06102	= Cubic inches
Cubic centimeters	x 10^{-6}	= Cubic meters
Cubic centimeters	x 1.308×10^{-6}	= Cubic yards
Cubic centimeters	x 2.624×10^{-4}	= Gallons (USA liquid)
Cubic centimeters	x 0.001	= Liters

Conversion factor C (continued on following page)

Conversion factor C (continued)

Cubic centimeters	x 2.113 x 10^{-3}	= Pints (USA liquid)
Cubic centimeters	x 1.057 x 10^{-3}	= Quarts (USA liquid)
Cubic feet	x 0.8036	= Bushels (dry)
Cubic feet	x 28,320	= Cubic centimeters
Cubic feet	x 1,728	= Cubic inches
Cubic feet	x 0.02832	= Cubic meters
Cubic feet	x 0.03704	= Cubic yards
Cubic feet	x 7.48052	= Gallons (USA liquid)
Cubic feet	x 28.32	= Liters
Cubic feet	x 59.84	= Pints (USA liquid)
Cubic feet	x 29.92	= Quarts (USA liquid)
Cubic feet	x 0.1781	= Barrels (oil, USA)
Cubic feet/minute	x 472	= Cubic centimeters/second
Cubic feet/minute	x 0.1247	= Gallons/second
Cubic feet/minute	x 0.4720	= Liters/second
Cubic feet/minute	x 62.43	= Pounds of water/minute
Cubic feet/minute	x 1.6989	= Cubic meters/hour
Cubic feet/second	x 0.646317	= Million gallons/day
Cubic feet/second	x 448.831	= Gallons/minute
Cubic feet/second	x 101.94	= Cubic meters/hour
Cubic inches	x 16.39	= Cubic centimeters
Cubic inches	x 5.787 x 10^{-4}	= Cubic feet
Cubic inches	x 1.639 x 10^{-5}	= Cubic meters
Cubic inches	x 2.143 x 10^{-5}	= Cubic yards
Cubic inches	x 4.329 x 10^{-3}	= Gallons
Cubic inches	x 0.01639	= Liters
Cubic inches	x 1.061 x 10^{5}	= Mil-feet
Cubic inches	x 0.03463	= Pints (USA liquid)
Cubic inches	x 0.01732	= Quarts (USA liquid)
Cubic meters	x 6.290	= Barrels (USA oil)
Cubic meters	x 28.38	= Bushels (dry)
Cubic meters	x 10^{6}	= Cubic centimeters
Cubic meters	x 35.314	= Cubic feet
Cubic meters	x 61,023	= Cubic inches
Cubic meters	x 1,308	= Cubic yards
Cubic meters	x 264.17	= Gallons (USA liquid)
Cubic meters	x 1,000	= Liters
Cubic meters	x 2,113	= Pints (USA liquid)
Cubic meters	x 1,057	= Quarts (USA liquid)
Cubic meters/hour	x 9.810 x 10^{-3}	= Cubic feet/second
Cubic meters/hour	x 0.5886	= Cubic feet/minute
Cubic meters/hour	x 4.4033	= Gallons/minute (USA)
Cubic meters/hour	x 150.95	= Barrels/day
Cubic meters/hour	x 3.6651	= Imperial gallons/minute
Cubic meters/hour	x 35.31	= Cubic feet/hour
Cubic meters/hour	x 277.8	= Cubic centimeters/second
Cubic yards	x 7.646 x 10^{5}	= Cubic centimeters

Conversion factor C (Continued on facing page)

Conversion factor C (concluded)

Cubic yards	x 27	= Cubic feet
Cubic yards	x 46,656	= Cubic inches
Cubic yards	x 0.7646	= Cubic meters
Cubic yards	x 202	= Gallons (USA liquid)
Cubic yards	x 764.6	= Liters
Cubic yards	x 1,615.9	= Pints (USA liquid)
Cubic yards	x 807.9	= Quarts (USA liquid)
Cubic yards/minute	x 0.45	= Cubic feet/second
Cubic yards/minute	x 3.367	= Gallons/second
Cubic yards/minute	x 12.74	= Liters/second
Cubit, Bible	x 21.8	= Inch
Cup	x 0.5	= Pint
Cup	x 16	= Tablespoon

D

Dalton	x 1.650×10^{-24}	= Gram
Days	x 1,440	= Minutes
Days	x 86,400	= Seconds
Decigrams	x 0.1	= Grams
Deciliters	x 0.1	= Liters
Decimeters	x 0.1	= Meters
Degrees (Angle)	x 60	= Minutes
Degrees (Angle)	x 3600	= Seconds
Degrees (Angle)	x 0.01745	= Radians
Degrees (Angle)	x 0.01111	= Quadrants
Degrees/second	x 0.1667	= Revolutions/minute
Degrees/second	x 2.778×10^{-3}	= Revolutions/second
Dekagrams	x 10	= Grams
Dekaliters	x 10	= Liters
Dekameters	x 10	= Meters
Drams (apothecaries or troy)	x 0.1371429	= Ounces (avoirdupois)
Drams (apothecaries or troy)	x 0.125	= Ounces (troy)
Drams	x 27.34375	= Grains
Drams	x 1.771845	= Grams
Drams	x 0.0625	= Ounces
Dyne/centimeters	x 0.01	= Erg/square millimeters
Dyne/square centimeters	x 9.869×10^{-7}	= Atmospheres
Dyne/square centimeters	x 2.953×10^{-5}	= Inch of mercury at 0°C
Dyne/square centimeters	x 4.015×10^{-4}	= Inch of water at 4°C
Dynes	x 1.020×10^{-3}	= Grams
Dynes	x 10^{-7}	= Joules/centimeters
Dynes	x 10^{-5}	= Joules/meters (newtons)
Dynes	x 1.020×10^{-6}	= Kilograms
Dynes	x 7.233×10^{-5}	= Poundals
Dynes	x 2.248×10^{-6}	= Pounds
Dynes/square centimeters	x 10^{-6}	= Bars

E

Ell	x 114.30	= Centimeters
Ell	x 45	= Inches
Em, pica	x 0.167	= Inch
Em, pica	x 0.4233	= Centimeter
Erg/second	x 1	= Dyne-centimeter/second
Erg	x 9.480×10^{-11}	= Btu
Erg	x 1	= Dyne-centimeter
Erg	x 7.367×10^{-8}	= Foot-pounds
Expansion coefficient, °F	x 1.8	= Expansion coefficient, °C

F

Fahrenheit −32	x .555	= Centigrade
Famm	x 5.8455	= Foot, USA
Famm	x 1.7814	= Meter
Faradays	x 26.8	= Ampere-hour
Fathom, British	x 6.08	= Feet
Fathom, British	x 1.8532	= Meter
Fathom, British	x 0.001	= Nautical mile, British
Fathom, USA	x 6	= Feet
Fathom, USA	x 1.8288	= Meter
Fathom, USA	x 2	= Yard
Feet, USA	x 12	= Inches
Feet, USA	x 0.3048	= Meters
Feet, USA	x 0.3333	= Yards
Feet, USA	x 0.18939×10^{-3}	= Miles, USA statute
Feet, USA	x 1.2×10^4	= Mils
Feet, USA	x 0.0606	= Rod
Feet of water	x 0.0295	= Atmospheres
Feet of water	x 0.8826	= Inches of mercury
Feet of water	x 0.03048	= Kilograms/square centimeters
Feet of water	x 304.8	= Kilograms/square meter
Feet of water	x 62.43	= Pounds/square foot
Feet of water	x 0.4335	= Pounds/square inch
Feet/hour	x 0.01666	= Feet/minute
Feet/hour	x 0.2777×10^{-3}	= Feet/second
Feet/hour	x 0.1894×10^{-3}	= Miles/hour
Feet/minute	x 0.5080	= Centimeter/second
Feet/minute	x 0.01666	= Feet/second
Feet/minute	x 0.18288	= Kilometer/hour
Feet/minute	x 0.009868	= Knot
Feet/minute	x 0.3048	= Meter/minute
Feet/minute	x 0.00508	= Meter/second
Feet/minute	x 0.01136	= Mile/hour

Conversion factor F (continued on facing page)

Conversion factor F (continued)

Feet/minute	x 0.1894 x 10^{-3}	= Mile/minute
Feet/second	x 30.48	= Centimeters/second
Feet/second	x 1.097	= Kilometers/hour
Feet/second	x 0.5921	= Knots
Feet/second	x 18.29	= Meters/minute
Feet/second	x 0.681818	= Miles/hour
Feet/second	x 0.0113636	= Miles/minute
Feet/second	x 3600	= Feet/hour
Feet/second	x 60	= Feet/minute
Feet/second/second	x 30.48	= Centimeters/second/second
Feet/second/second	x 1.097	= Kilometers/hour/second
Feet/second/second	x 0.3048	= Meters/second/second
Feet/second/second	x 0.6818	= Miles/hour/second
Feet/100 feet	x 1	= Percent grade
Firkin	x 9	= Gallon, liquid, USA
Firkin	x 34.06798	= Liter
Foot-candle	x 10.764	= Lumen/square meter
Foot-candle	x 1	= Lumen/square foot
Foot-candle	x 10.764	= Lux
Foot-candle	x 1.076	= Milliphot
Foot-candle	x 0.001076	= Phot
Foot-candle	x distance in feet^2	= Candlepower
Foot-Lambert	x 0.3425 x 10^{-3}	= Candle/square centimeters
Foot-Lambert	x 0.3183	= Candle/square foot
Foot-Lambert	x 0.00221	= Candle/square inch
Foot-Lambert	x 0.001076	= Lambert
Foot-Lambert	x square foot Area	= Lumen
Foot-Lambert	x 1.076	= Millilambert
Foot-Lambert	x 0.342 x 10^{-3}	= Stilb
Foot-pound	x 1.2853 x 10^{-3}	= Btu
Foot-pound	x 1.356 x 10^7	= Ergs
Foot-pound	x 0.32389	= Gram-calorie
Foot-pound	x 5.0505 x 10^{-7}	= Horsepower-hours, USA
Foot-pound	x 5.12 x 10^{-7}	= Horsepower-hours, metric
Foot-pound	x 12	= Inch-pound
Foot-pound	x 1.35582	= Joule absolute
Foot-pound	x 1.3554	= Joule international
Foot-pound	x 3.238 x 10^{-4}	= Kilogram-calories
Foot-pound	x 0.1383	= Kilogram-meters
Foot-pound	x 3.766 x 10^{-7}	= Kilowatt-hours
Foot-pound	x 0.001356	= Kilowatt-second
Foot-pound	x 0.01338	= Liter-atmosphere
Foot-pound	x 0.3766 x 10^{-3}	= Watt-hour
Foot-pound	x 1.356	= Watt-second
Foot-pound/minute	x 0.077118	= Btu/hour
Foot-pound/minute	x 1.286 x 10^{-3}	= But/minute

Conversion factor F (concluded)

Foot-pound/minute	x 2.259 x 10^5	= Erg/second
Foot-pound/minute	x 0.01666	= Foot-pound/second
Foot-pound/minute	x 3.066 x 10^{-5}	= Horsepower, metric
Foot-pound/minute	x 3.0303 x 10^{-5}	= Horsepower, USA
Foot-pound/minute	x 2.2597 x 10^{-5}	= Kilowatt
Foot-pound/minute	x 0.022597	= Watt
Foot-pound/second	x 0.0771	= Btu/minute
Foot-pound/second	x 4.6263	= Btu/hour
Foot-pound/second	x 1.843 x 10^{-3}	= Horsepower, metric
Foot-pound/second	x 1.818 x 10^{-3}	= Horsepower, USA
Foot-pound/second	x 1.356	= Joule
Foot-pound/second	x 1.3558 x 10^{-3}	= Kilowatts
Foot-pound/second	x 1.3558	= Watt
Fot	x 0.974	= Foot, USA
Fot	x 100	= Lines
Fot	x 0.2969	= Meter
Fot	x 10	= Turn
Foute	x 1	= Foot, USA
Furlong	x 6.6	= Chain, engineer
Furlong	x 10	= Chain, Gunter
Furlong	x 660	= Feet
Furlong	x 201.168	= Meters
Furlong	x 0.125	= Mile, statue, USA
Furlong	x 220	= Yards
Furlong	x 40	= Rods
Fuss	x 0.9842	= Foot, USA
Fuss	x 0.300	= Meter

G

Gallon, British, Imperial Liquid	x 0.125	= Bushel, dry, British
Gallon, British, Imperial Liquid	x 4546	= Cubic centimeter
Gallon, British, Imperial Liquid	x 0.16046	= Cubic foot
Gallon, British, Imperial Liquid	x 0.0045	= Cubic meter
Gallon, British, Imperial Liquid	x 1.032	= Gallon, dry, USA
Gallon, British, Imperial Liquid	x 1.2009	= Gallon, liquid, USA
Gallon, British, Imperial Liquid	x 4.54596	= Kilogram
Gallon, British, Imperial Liquid	x 10	= Pound, water, 62°F
Gallon, dry, USA	x 0.125	= Bushel, USA
Gallon, dry, USA	x 4404.92	= Cubic centimeter
Gallon, dry, USA	x 0.155555	= Cubic foot
Gallon dry, USA	x 268.803	= Cubic inch
Gallon, dry, USA	x 1.16365	= Gallon, liquid, USA
Gallon, dry, USA	x 4.4049	= Liter
Gallon, dry, USA	x 0.05	= Peck
Gallon, dry, USA	x 8	= Pint
Gallon, dry, USA	x 4.6546	= Quart, liquid, USA

Conversion factor G (continued on facing page)

Conversion factor G (continued)

Gallon, liquid, USA	x 0.0238	= Barrel, oil
Gallon, liquid, USA	x 3785.434	= Cubic centimeter
Gallon, liquid, USA	x 3.785434	= Cubic decimeter
Gallon, liquid, USA	x 0.13368	= Cubic foot
Gallon, liquid, USA	x 231	= Cubic inch, water, 62°F
Gallon, liquid, USA	x 3.7854×10^{-3}	= Cubic meter
Gallon, liquid, USA	x 4.951×10^{-3}	= Cubic yard
Gallon, liquid, USA	x 0.859365	= Gallon, dry, USA
Gallon, liquid, USA	x 0.832673	= Gallon, liquid, British
Gallon, liquid, USA	x 3.7853	= Liter
Gallon, liquid, USA	x 8	= Pint, liquid, USA
Gallon, liquid, USA	x 4	= Quart, liquid, USA
Gallon, liquid, USA	x 8.3453	= Pounds, water
Gallons/hour, USA	x 0.1337	= Cubic feet/hour
Gallons/hour, USA	x 2.228×10^{-3}	= Cubic feet/minute
Gallons/hour, USA	x 0.01666	= Gallons/minute
Gallons/hour, USA	x 2.777×10^{-4}	= Gallons/second
Gallons/minute, USA	x 34.2857	= Barrels/day, oil
Gallons/minute, USA	x 1.42857	= Barrels/hour, oil
Gallons/minute, USA	x 0.023809	= Barrels/minute, oil
Gallons/minute, USA	x 192.49999	= Cubic feet/day
Gallons/minute, USA	x 8.021	= Cubic feet/hour
Gallons/minute, USA	x 0.13368	= Cubic feet/minute
Gallons/minute, USA	x 2.228×10^{-3}	= Cubic feet/second
Gallons/minute, USA	x 0.2271	= Cubic meters/hour
Gallons/minute, USA	x 1440	= Gallons/day
Gallons/minute, USA	x 60	= Gallons/hour
Gallons/minute, USA	x 0.01666	= Gallons/second
Gallons/minute, USA	x 5.35565	= Tons, long, water, 62°F/day
Gallons/minute, USA	x 5.99839	= Tons, short, water, 62°F/day
Gallons/minute, USA	x 0.06308	= Liters/second
Gallons/second, USA	x 481	= Cubic feet/hour
Gallons/second, USA	x 8.02	= Cubic feet/minute
Gallons/second, USA	x 0.1337	= Cubic feet/second
Gallons/second, USA	x 60	= Gallons/minute, USA
Gills, British	x 142.07	= Cubic centimeter
Gills, British	x 0.1183	= Liters
Gills, British	x 0.25	= Pints, liquid
Grade	x 0.0025	= Circle
Grade	x 9,000	= Degree
Grade	x 54	= Minute
Grade	x 0.01571	= Radian
Grain	x 0.01666	= Dram, apothecary
Grain	x 0.03657	= Dram, avoirdupois
Grain	x 1	= Grain, troy
Grain	x 0.0648	= Grams
Grain	x 2.0833×10^{-3}	= Ounces

Conversion factor G (continued on following page)

Conversion factor G (concluded)

Gram	x 5	= Carat
Gram	x 3.858	= Carat, metric
Gram	x 100	= Centigram
Gram	x 0.2572	= Dram, apothecary
Gram	x 0.56438	= Dram, avoirdupois
Gram	x 980.665	= Dyne
Gram	x 15.4324	= Grain
Gram	x 9.807×10^{-5}	= Joules/centimeters
Gram	x 9.807×10^{-3}	= Joules/meter (newtons)
Gram	x 0.001	= Kilograms
Gram	x 1000	= Milligrams
Gram	x 0.03527	= Ounces, avoirdupois
Gram	x 0.03215	= Ounces, troy
Gram	x 0.07093	= Poundals
Gram	x 2.205×10^{-3}	= Pounds
Grams/centimeter	x 5.6×10^{-3}	= Pounds/inch
Grams/cubic centimeter	x 62.43	= Pounds/cubic foot
Grams/cubic centimeter	x 0.03613	= Pounds/cubic foot
Grams/liter	x 58.417	= Grains/gallon, USA
Grams/liter	x 8.345	= Pounds/1,000 gallons
Grams/liter	x 0.062427	= Pounds/cubic foot
Grams/liter	x 1,000	= Parts/million
Grams/square centimeter	x 2.0481	= Pounds/square foot
Gram-calories	x 3.968×10^{-3}	= Btu
Gram-calories	x 4.1868×10^{7}	= Ergs
Gram-calories	x 3.088	= Foot-pounds
Gram-calories	x 1.55856×10^{-6}	= Horsepower/hours
Gram-calories	x 1.163×10^{-6}	= Kilowatt-hours
Gram-calories	x 1.163×10^{-3}	= Watt-hours
Gram/calories/second	x 14.286	= Btu/hour
Gram-centimeters	x 9.29658×10^{-8}	= Btu
Gram-centimeters	x 980.7	= Ergs
Gram-centimeters	x 9.807×10^{-5}	= Joules
Gross	x 12	= Dozen
Gross, great	x 144	= Dozen
Gross, great	x 12	= Gross

H

Hand	x 10.16	= Centimeter
Hand	x 4	= Inch
Hand	x 48	= Foot
Hand	x 1,016	= Meter
Head, feet elevation, water	x 0.433	= Pounds/square inch
Hectare	x 2.471	= Acre
Hectare	x 100	= Are
Hectare	x 1.07639×10^{5}	= Square feet

Conversion factor H (continued on facing page)

Conversion factor H (continued)

Hectare	x 0.01	=	Square kilometer
Hectare	x 10,000	=	Square meters
Hectare	x 3.861×10^{-3}	=	Square miles
Hectare	x 11,960	=	Square yard
Hectogram	x 100	=	Gram
Hectoliter	x 3.532	=	Cubic feet
Hectoliter	x 0.1	=	Cubic meter
Hectoliter	x 0.1308	=	Cubic yard
Hectoliter	x 26.42	=	Gallon, USA
Hectoliter	x 100	=	Liter
Hectometer	x 328.089	=	Feet
Hectometer	x 100	=	Meter
Hectometer	x 0.06214	=	Mile, statute, USA
Hectometer	x 109.36	=	Yard
Hectowatts	x 100	=	Watts
Henries	x 1,000	=	Millihenries
Hogsheads, British	x 10.114	=	Cubic feet
Hogsheads, USA	x 8.42184	=	Cubic feet
Hogsheads, USA	x 63	=	Gallons, USA
Hogsheads, USA	x 238.476	=	Liter
Hogsheads, USA	x 504	=	Pint
Hogsheads, USA	x 252	=	Quart
Horsepower, USA	x 42.44	=	Btu/minute
Horsepower, USA	x 33,000	=	Foot-pounds/minute
Horsepower, USA	x 550	=	Foot-pounds/second
Horsepower, USA	x 0.7457	=	Kilowatts
Horsepower, boiler	x 33,479	=	Btu/hour
Horsepower, boiler	x 34.5	=	Pounds water/hour
Horsepower, boiler	x 9.803	=	Kilowatts
Horsepower, electric	x 0.7072	=	Btu/second
Horsepower, electric	x 746	=	Joule/second
Horsepower, electric	x 0.746	=	Kilowatts
Horsepower, electric	x 746	=	Watts
Horsepower, hours, USA	x 2,547	=	Btu
Horsepower, hours, USA	x 2.6845×10^{13}	=	Ergs
Horsepower, hours, USA	x 1.98×10^6	=	Foot-pounds
Horsepower, hours, USA	x 641,190	=	Gram-calories
Horsepower, hours, USA	x 1.01387	=	Horsepower-hour, metric
Horsepower, hours, USA	x $2,376 \times 10^4$	=	Inch-pound
Horsepower, hours, USA	x 26.8453×10^5	=	Joule
Horsepower, hours, USA	x 0.7457	=	Kilowatt-hour
Horsepower-hours, metric	x 2509.83	=	Btu
Horsepower-hours, metric	x 1.9529×10^6	=	Foot-pounds
Horsepower-hours, metric	x 0.98632	=	Horsepower-hour, USA
Horsepower-hours, metric	x 632,467	=	Gram-calories
Horsepower-hours, metric	x 26.4761	=	Joule

Conversion factor H (concluded)

Horsepower-hours, metric	x 0.73545	= Kilowatt-hour
Hours	x 0.0417	= Day
Hours	x 60	= Minute
Hours	x 0.00137	= Month
Hours	x 0.1142×10^{-3}	= Year
Hours	x 5.952×10^{-3}	= Week
Hundredweight (long)	x 112	= Pounds
Hundredweight (long)	x 0.05	= Tons, long
Hundredweight (short)	x 1.8	= Cubic foot
Hundredweight (short)	x 45.36	= Kilograms
Hundredweight (short)	x 100	= Pounds
Hundredweight (short)	x 0.05	= Ton, short
Hundredweight (short)	x 0.04536	= Tons, metric
Hundredweight (short)	x 0.044643	= Tons, long

I

Inch	x 254×10^6	= Angstrom
Inch	x 2.54	= Centimeter
Inch	x 0.833×10^{-3}	= Chain, engineer
Inch	x 1.2626×10^{-3}	= Chain, Gunter
Inch	x 0.254	= Decemeter
Inch	x 0.08333	= Foot, USA
Inch	x 0.0254	= Meter
Inch	x 1.578×10^{-5}	= Mile, statute, USA
Inch	x 25.4	= Millimeters
Inch	x 1,000	= Mils
Inch	x 5.05×10^{-3}	= Rods
Inch	x 0.02778	= Yards
Inch, mercury	x 0.03342	= Atmospheres
Inch, mercury	x 1.133	= Feet of water
Inch, mercury	x 13.61	= Inch height, water
Inch, mercury	x 70.73	= Pound/square foot
Inch, mercury	x 0.49116	= Pound/square inch
Inch, mercury	x 0.03453	= Kilograms/square centimeter
Inch-pound	x 1.07×10^{-4}	= Btu
Inch-pound	x 0.0833	= Foot-pound
Inch, water, 4°C	x 2.458×10^{-3}	= Atmospheres
Inch, water, 4°C	x 0.07355	= Inches of mercury
Inch, water, 4°C	x 2.54×10^{-3}	= Kilograms/square centimeter
Inch, water, 4°C	x 0.5781	= Ounces/square inch
Inch, water, 4°C	x 5.204	= Pounds/square foot
Inch, water, 4°C	x 0.03613	= Pounds/square inch

Appendix

J

Joule	x 9.48×10^{-4}	= Btu
Joule	x 10^7	= Erg
Joule	x 0.7376	= Foot-pound
Joule	x 2.389×10^{-4}	= Kilogram-calorie
Joule	x 1.0197×10^4	= Gram-centimeter
Joule	x 2.778×10^{-4}	= Watt-hours

K

Kilogram	x 1,000	= Grams
Kilogram	x 70.93	= Poundals
Kilogram	x 2.205	= Pounds
Kilogram	x 9.842×10^{-4}	= Tons, long
Kilogram	x 1.102×10^{-3}	= Tons, short
Kilogram	x 0.001	= Tons, metric
Kilograms/cubic meter	x 0.001	= Grams/cubic centimeter
Kilograms/cubic meter	x 0.06243	= Pounds/cubic foot
Kilograms/cubic meter	x 3.613×10^{-5}	= Pounds/cubic inch
Kilograms/cubic meter	x 3.405×10^{-10}	= Pounds/mil-foot
Kilograms/cubic meter	x 8.428×10^{-3}	= Ton, short/cubic yard
Kilograms/meter	x 10	= Gram/centimeter
Kilograms/meter	x 391.983	= Gram/inch
Kilograms/meter	x 0.672	= Pounds/foot
Kilograms/meter	x 0.056	= Pounds/inch
Kilograms/square centimeter	x 0.9678	= Atmospheres
Kilograms/square centimeter	x 32.81	= Feet of water
Kilograms/square centimeter	x 28.96	= Inch of mercury
Kilograms/square centimeter	x 2,048	= Pounds/square foot
Kilograms/square centimeter	x 14.22	= Pounds/square inch
Kilograms/square meter	x 9.678×10^{-5}	= Atmospheres
Kilograms/square meter	x 3.281×10^{-3}	= Feet of water
Kilograms/square meter	x 2.896×10^{-3}	= Inches of mercury
Kilograms/square meter	x 0.03937	= Inches of water
Kilograms/square meter	x 0.2048	= Pounds/square foot
Kilograms/square meter	x 1.422×10^{-3}	= Pounds/square inch
Kilometers	x 3281	= Feet
Kilometers	x 3.937×10^4	= Inches
Kilometers	x 1,000	= Meters
Kilometers	x 0.6214	= Miles
Kilometers	x 1,094	= Yards
Kilometers/hour	x 27.78	= Centimeters/second
Kilometers/hour	x 54.68	= Feet/minute

Conversion factor K (continued on following page)

Conversion factor K (concluded)

Kilometers/hour	x 0.9113	= Feet/second
Kilometers/hour	x 0.5396	= Knots
Kilometers/hour	x 16.67	= Meters/minute
Kilometers/hour	x 0.6214	= Miles/hour
Kilowatts	x 56.92	= Btu/minute
Kilowatts	x 1.35972	= Horsepower, metric
Kilowatts	x 1.341	= Horsepower, USA
Kilowatts	x 1,000	= Watts
Kilowatt-hours	x 3,413	= Btu
Kilowatt-hours	x 3.6×10^{13}	= Ergs
Kilowatt-hours	x 1.36	= Horsepower-hour, metric
Kilowatt-hours	x 1.34	= Horsepower-hour, USA
Kilowatt-hours	x 3.6×10^6	= Joules
Kilowatt-hours	x 860.5	= Kilogram-calories
Kilowatt-hours	x 3.671×10^5	= Kilogram-meters
Kilowatt-hours	x 3.53	= Pounds of water*
Kilowatt-hours	x 22.75	= Pounds of water†
Kip	x 1	= Kilopound
Kip	x 1,000	= Pound
Knott, USA	x 51.48	= Centimeter/second
Knott, USA	x 6080.2	= Feet/hour
Knott, USA	x 1.8532	= Kilometer/hour
Knott, USA	x 30.887	= Meter/minute
Knott, USA	x 1.15155	= Mile/hour
Knott, USA	x 2027	= Yards/hour

L

League, land	x 24	= Furlong
League, land	x 4.828	= Kilometer
League, land	x 3	= Mile
League, marine	x 5.56	= Kilometer
League, marine	x 3	= Mile, nautical
League, marine	x 3.45	= Mile, statute
Light year	x 5.9×10^{12}	= Miles
Light year	x 9.46091×10^{12}	= Kilometers
Links, engineers'	x 12	= Inches
Links, surveyors'	x 7.92	= Inches
Links, surveyors'	x 0.66	= Feet
Links, surveyors'	x 0.22	= Yard
Liters	x 0.02838	= Bushels, USA, dry
Liters	x 100	= Centiliters

*Evaporated from and at 212°F
†Raised from 62°F to 212°F

Conversion factor L (continued on facing page)

Appendix

Conversion factor L (concluded)

Liters	x 1,000	= Cubic centimeters
Liters	x 0.035316	= Cubic feet
Liters	x 6.291×10^{-3}	= Barrels, oil, USA
Liters	x 61.027	= Cubic inches
Liters	x 0.001	= Cubic meter
Liters	x 1.308×10^{-3}	= Cubic yard
Liters	x 0.2642	= Gallon, USA, liquid
Liters	x 1.7598	= Pint, USA, dry
Liters	x 2.1134	= Pint, USA, liquid
Liters	x 2.202	= Pounds of water
Liters/minute	x 5.886×10^{-4}	= Cubic foot/second
Liters/minute	x 4.403×10^{-3}	= Gallons/second
Liters/second	x 2.1186	= Cubic feet/minute
Lumen	x 0.07958	= Candlepower
Lumen	x 1.47×10^{-3}	= Watt
Lumens/square foot	x 1	= Foot-candles
Lux	x 0.0929	= Foot-candles

M

Maas	x 1.5	= Liter
Meter	x 100	= Centimeter
Meter	x 3.2808	= Feet, USA
Meter	x 0.01	= Hectometer
Meter	x 39.37	= Inch
Meter	x 0.001	= Kilometer
Meter	x 5.396×10^{-4}	= Miles, nautical
Meter	x 6.214×10^{-4}	= Miles, statute
Meter	x 1000	= Millimeters
Meter	x 1.094	= Yards
Meter	x 1.179	= Vara
Meters/minute	x 1.667	= Centimeters/second
Meters/minute	x 3.281	= Feet/minute
Meters/minute	x 0.05468	= Feet/second
Meters/minute	x 0.06	= Kilometers/hour
Meters/minute	x 0.03238	= Knots
Meters/minute	x 0.03728	= Miles/hour
Meters/second	x 196.8	= Feet/minute
Meters/second	x 3.281	= Feet/second
Meters/second	x 3.6	= Kilometers/hour
Meters/second	x 0.06	= Kilometers/minute
Meters/second	x 2.237	= Miles/hour
Meters/second	x 0.03728	= Miles/minute
Micron	x 0.0001	= Centimeter
Micron	x 1000	= Millimicron
Mile, USA, nautical	x 6,080.2	= Feet, USA
Mile, USA, nautical	x 6,080	= Feet, British

Conversion factor M (continued on following page)

Conversion factor M (concluded)

Mile, USA, nautical	x 72,962.5	= Inches
Mile, USA, nautical	x 1.853	= Kilometer
Mile, USA, nautical	x 0.333	= League
Mile, USA, nautical	x 1,853.248	= Meter
Mile, USA, nautical	x 1.15155	= Mile, USA, statute
Mile, USA, nautical	x 2,026.73	= Yard
Miles, USA, statute	x 5,280	= Feet, USA
Miles, USA, statute	x 8	= Furlongs
Miles, USA, statute	x 63,360	= Inches
Miles, USA, statute	x 1.60935	= Kilometer
Miles, USA, statute	x 8,000	= Link
Miles, USA, statute	x 1,609.35	= Meters
Miles, USA, statute	x 0.8684	= Mile, USA, nautical
Miles, USA, statute	x 1,900.8	= Vara
Miles, USA, statute	x 1,706	= Yard
Miles/hour	x 44.7	= Centimeters/second
Miles/hour	x 88	= Feet/minute
Miles/hour	x 1.467	= Feet/second
Miles/hour	x 1.609	= Kilometers/hour
Miles/hour	x 0.02682	= Kilometers/minute
Miles/hour	x 0.8684	= Knots
Miles/hour	x 26.82	= Meters/minute
Miles/hour	x 0.4470	= Meters/second
Miles/hour	x 0.01667	= Miles/minute
Miles/minute	x 5,280	= Feet/minute
Miles/minute	x 316,800	= Feet/hour
Miles/minute	x 88	= Feet/second
Miles/minute	x 60	= Mile/hour
Miles/minute	x 1.609	= Kilometers/minute
Miles/minute	x 0.8684	= Knots/minute
Millimeter	x 0.1	= Centimeter
Millimeter	x 3.281×10^{-3}	= Feet
Millimeter	x 0.03937	= Inches
Millimeter	x 10^{-6}	= Kilometers
Millimeter	x 0.001	= Meters
Millimeter	x 6.214×10^{-7}	= Miles
Millimeter	x 39.37	= Mils
Millimeter	x 1.094×10^{-3}	= Yards
Million gallons/day	x 1.54723	= Cubic feet/second
Mils	x 2.540×10^{-3}	= Centimeters
Mils	x 8.333×10^{-5}	= Feet
Mils	x 0.001	= Inches
Mils	x 2.540×10^{-8}	= Kilometers
Mils	x 2.778×10^{-5}	= Yards

N

Nail	x 2.5	=	Inch
Nepers	x 8.686	=	Decibels
Newton	x 1×10^5	=	Dynes

O

Ounces	x 16	=	Drams
Ounces	x 437.5	=	Grains
Ounces	x 28.349527	=	Grams
Ounces	x 0.0625	=	Pounds
Ounces	x 0.9115	=	Ounces, troy
Ounces	x 2.79×10^{-5}	=	Tons, long
Ounces	x 2.835×10^{-5}	=	Tons, metric
Ounces, fluid	x 1.805	=	Cubic inches
Ounces, fluid	x 0.02957	=	Liters
Ounces, troy	x 480	=	Grains
Ounces, troy	x 31.103481	=	Grams
Ounces, troy	x 1.09714	=	Ounces, avoirdupois
Ounces, troy	x 0.08333	=	Pounds, troy
Ounces/square inch	x 0.0625	=	Pounds/square inch

P

Parsec	x 19×10^{12}	=	Miles, USA, statute
Parsec	x 3.084×10^{13}	=	Kilometers
Parts/million	x 0.05833	=	Grains/gallon, USA
Parts/million	x 0.07016	=	Grains/gallon, British
Parts/million	x 8.345	=	Pounds/million gallons, USA
Peck, British	x 554.6	=	Cubic inches
Peck, British	x 2	=	Gallons, British
Peck, British	x 9.0919	=	Liters
Peck, USA	x 0.25	=	Bushels
Peck, USA	x 537.605	=	Cubic inches
Peck, USA	x 8.809582	=	Liters
Peck, USA	x 8	=	Quarts, dry
Peck, USA	x 9.3092	=	Quarts, liquid
Pennyweights, troy	x 24	=	Grains
Pennyweights, troy	x 0.05	=	Ounces, troy
Pennyweights, troy	x 1.55517	=	Grams
Pennyweights, troy	x 4.1667×10^{-3}	=	Pounds, troy
Pfund, Germany	x 500	=	Gram
Pint, USA, dry	x 0.015625	=	Bushel
Pint, USA, dry	x 550.6136	=	Cubic centimeter
Pint, USA, dry	x 0.01945	=	Cubic feet
Pint, USA, dry	x 33.6	=	Cubic inches

Conversion factor P (continued on following page)

Conversion factor P (continued)

Pint, USA, dry	x 2	= Cup
Pint, USA, dry	x 0.125	= Gallon, USA, dry
Pint, USA, dry	x 0.14545	= Gallon, USA, liquid
Pint, USA, dry	x 0.5506	= Liter
Pint, USA, dry	x 0.0625	= Peck
Pint, USA, dry	x 0.5	= Quart, USA, dry
Pint, USA, dry	x 0.58182	= Quart, USA, liquid
Pint, USA, liquid	x 437.2	= Cubic centimeters
Pint, USA, liquid	x 0.01671	= Cubic feet
Pint, USA, liquid	x 28.875	= Cubic inch
Pint, USA, liquid	x 2	= Cup
Pint, USA, liquid	x 0.1074	= Gallon, USA, dry
Pint, USA, liquid	x 0.125	= Gallon, USA, liquid
Pint, USA, liquid	x 4	= Gill
Pint, USA, liquid	x 0.4732	= Liters
Pint, USA, liquid	x 16	= Ounces
Pint, USA, liquid	x 0.5	= Quarts, USA, liquid
Pint, USA, liquid	x 0.42968	= Quarts, USA, dry
Pint, USA, liquid	x 128	= Dram, fluid
Poise	x 100	= Centipoise
Pole	x 16.5	= Feet
Pole	x 5.0292	= Meter
Pole	x 1	= Rod
Pole	x 5.5	= Yard
Ponce	x 2.71	= Centimeter
Pood	x 1,000	= Cubic inch
Pood	x 40	= Funt
Pood	x 4.32	= Gallon, USA
Pood	x 16.3805	= Kilogram
Poundals	x 13,826	= Dynes
Poundals	x 14.098	= Grams
Poundals	x 1.383×10^{-3}	= Joules/centimeter
Poundals	x 0.1383	= Joules/meter
Poundals	x 0.0141	= Kilograms
Poundals	x 0.1383	= Newton
Poundals	x 0.03108	= Pound-force
Pounds	x 2267.9616	= Carats
Pounds	x 256	= Drams
Pounds	x 7,000	= Grains
Pounds	x 453.5924	= Grams
Pounds	x 0.04448	= Joules/centimeters
Pounds	x 0.4536	= Kilograms
Pounds	x 16	= Ounces
Pounds	x 14.5833	= Ounces, troy
Pounds	x 32.174	= Poundals
Pounds	x 1.21528	= Pounds, troy
Pounds	x 4.464×10^{-4}	= Tons, long

Conversion factor P (continued on facing page)

Conversion factor P (concluded)

Pounds	x 4.536×10^{-4}	= Tons, metric
Pounds	x 5×10^{-4}	= Tons, short
Pounds, troy	x 5,760	= Grains
Pounds, troy	x 373.24177	= Grams
Pounds, troy	x 13.1657	= Ounces, avoirdupois
Pounds, troy	x 12	= Ounces, troy
Pounds, troy	x 240	= Pennyweights, troy
Pounds, troy	x 0.822857	= Pounds, avoirdupois
Pounds, troy	x 3.6735×10^{-4}	= Tons, long
Pounds, troy	x 3.7324×10^{-4}	= Tons, metric
Pounds, troy	x 4.1143×10^{-4}	= Tons, short
Pounds of water	x 0.01602	= Cubic feet
Pounds of water	x 27.68	= Cubic inches
Pounds of water	x 0.1198	= Gallons
Pounds of water/minute	x 2.67×10^{-4}	= cubic feet/second
Pounds/cubic foot	x 0.01602	= Grams/cubic centimeters
Pounds/cubic foot	x 16.02	= Kilograms/cubic meter
Pounds/cubic foot	x 5.787×10^{-4}	= Pounds/cubic inch
Pounds/cubic foot	x 27	= Pounds/cubic yard
Pounds/cubic inch	x 27.68	= Grams/cubic centimeter
Pounds/cubic inch	x 2.768×10^{4}	= Kilograms/cubic meter
Pounds/cubic inch	x 1,728	= Pounds/cubic foot
Pounds/cubic inch	x 46,656	= Pounds/cubic yard
Pounds/hour	x 10.714×10^{-3}	= Tons/day, long
Pounds/hour	x 12×10^{-3}	= Tons/day, short
Pounds/hour	x 10.886×10^{-3}	= Tons/day, metric
Pounds/hour	x 0.45359	= Kilograms/hour
Pounds/square foot	x 4.725×10^{-4}	= Atmospheres
Pounds/square foot	x 0.01602	= Feet of water
Pounds/square foot	x 0.01414	= Inches of mercury
Pounds/square foot	x 4.8824	= Kilograms/square meter
Pounds/square foot	x 0.1111	= Ounce/square inch
Pounds/square foot	x 0.107638	= Pound/square centimeter
Pounds/square foot	x 6.944×10^{-3}	= Pound/square inch
Pounds/square foot	x 10.76387	= Pound/square meter
Pounds/square inch	x 0.068046	= Atmospheres
Pounds/square inch	x 2.307	= Feet of water
Pounds/square inch	x 27.7	= Inch of water
Pounds/square inch	x 2.036	= Inch of mercury
Pounds/square inch	x 0.0703	= Kilogram/square centimeter
Pounds/square inch	x 703.1	= Kilogram/square meter
Pounds/square inch	x 51.714	= Millimeters of mercury
Pounds/square inch	x 2,304	= Ounce/square foot
Pounds/square inch	x 144	= Pound/square foot

Q

Quadrant	x 0.25	= Circumference
Quadrant	x 90	= Degrees
Quadrant	x 5,400	= Minutes
Quadrant	x 1.571	= Radians
Quarts, USA, dry	x 0.03125	= Bushel
Quarts, USA, dry	x 1,101.2	= Cubic centimeter
Quarts, USA, dry	x 0.03889	= Cubic foot
Quarts, USA, dry	x 67.20	= Cubic inches
Quarts, USA, dry	x 1.1012	= Liter
Quarts, USA, dry	x 1.16365	= Quart, USA, liquid
Quarts, USA, liquid	x 946.331	= Cubic centimeter
Quarts, USA, liquid	x 0.03342	= Cubic foot
Quarts, USA, liquid	x 57.75	= Cubic inches
Quarts, USA, liquid	x 9.464×10^{-4}	= Cubic meters
Quarts, USA, liquid	x 1.238×10^{-3}	= Cubic yard
Quarts, USA, liquid	x 4	= Cup
Quarts, USA, liquid	x 256	= Dram fluid
Quarts, USA, liquid	x 0.25	= Gallons
Quarts, USA, liquid	x 0.946331	= Liter
Quarts, USA, liquid	x 5.9523×10^{-3}	= Oil, barrel
Quarts, USA, liquid	x 32	= Ounces
Quarts, USA, liquid	x 2	= Pint
Quarts, USA, liquid	x 0.8594	= Quart, USA, dry

R

Radians	x 57.3	= Degrees
Radians	x 3,438	= Minutes
Radians	x 0.6366	= Quadrants
Radians	x 2.063×10^5	= Seconds
Rod	x 0.165	= Chain, engineer
Rod	x 0.25	= Chain, Gunters
Rod	x 16.5	= Foot
Rod	x 0.025	= Furlong
Rod	x 198	= Inch
Rod	x 25	= Link
Rod	x 5.029	= Meter
Rod	x 5.5	= Yard

S

Seconds, angle	x 2.778×10^{-4}	= Degrees
Seconds, angle	x 16.67×10^{-3}	= Minutes
Seconds, time	x 2.777×10^{-4}	= Hour
Seconds, time	x 0.1666	= Minutes

Conversion factor S (continued on facing page)

Conversion factor S (continued)

Snow, cubic foot	x 7.2	= Pounds, 32°F
Snow, inch deep	x 0.1	= Inch, water
Square centimeter	x 1.076×10^{-3}	= Square foot
Square centimeter	x 0.155	= Square inch
Square centimeter	x 0.0001	= Square meter
Square centimeter	x 3.861×10^{-11}	= Square miles
Square centimeter	x 100	= Square millimeters
Square centimeter	x 1.196×10^{-4}	= Square yards
Square feet, USA	x 2.296×10^{-5}	= Acre
Square feet, USA	x 9.29×10^{-4}	= Are
Square feet, USA	x 929.034	= Square centimeters
Square feet, USA	x 144	= Square inches
Square feet, USA	x 0.0929	= Square meter
Square feet, USA	x 3.587×10^{-8}	= Square miles
Square feet, USA	x 9.29×10^4	= Square millimeters
Square feet, USA	x 0.1111	= Square yards
Square inches	x 6.452	= Square centimeters
Square inches	x 6.944×10^{-3}	= Square feet
Square inches	x 645.2	= Square millimeters
Square inches	x 7.716×10^{-4}	= Square yard
Square kilometer	x 247.1	= Acre
Square kilometer	x 100	= Hectare
Square kilometer	x 10.76×10^6	= Square feet
Square kilometer	x 1.55×10^9	= Square inches
Square kilometer	x 10^6	= Square meters
Square kilometer	x 0.3861	= Square mile, USA
Square kilometer	x 1.196×10^6	= Square yards
Square meters	x 2.471×10^{-4}	= Acre
Square meters	x 0.01	= Are
Square meters	x 0.0001	= Hectare
Square meters	x 10,000	= Square centimeters
Square meters	x 10.764	= Square feet
Square meters	x 1,550	= Square inches
Square meters	x 3.861×10^{-7}	= Square miles
Square meters	x 1.196	= Square yards
Square miles	x 640	= Acre
Square miles	x 259	= Hectare
Square miles	x 27.88×10^6	= Square feet
Square miles	x 2.59	= Square kilometers
Square miles	x 2.59×10^6	= Square meters
Square miles	x 3.098×10^6	= Square yards
Square millimeters	x 0.01	= Square centimeters
Square millimeters	x 1.076×10^{-5}	= Square feet
Square millimeters	x 1.55×10^{-3}	= Square inches
Square rods	x 0.00625	= Acre
Square rods	x 272.25	= Square feet
Square rods	x 25.293	= Square meter

Conversion factor S (continued on following page)

Conversion factor S (concluded)

Square rods	x 30.25	= Square yard
Square vara	x 7.716	= Square feet
Square yard	x 2.066 x 10^{-4}	= Acres
Square yard	x 8361	= Square centimeter
Square yard	x 9	= Square feet
Square yard	x 1,296	= Square inches
Square yard	x 0.8361	= Square meters
Square yard	x 3.228 x 10^{-7}	= Square miles
Square yard	x 8.361 x 10^5	= Square millimeters
Square yard	x 0.03306	= Square rods
Stone	x 14	= Pound
Stone	x 6.35	= Kilogram

T

Tablespoon	x 0.0625	= Cup
Tablespoon	x 3	= Teaspoon
Teaspoon	x 0.0208	= Cup
Teaspoon	x 0.333	= Tablespoon
Temperature, °C + 17.78	x 1.8	= °F
Temperature, °F −32	x 0.555	= °C
Ton, long	x 1,016	= Kilogram
Ton, long	x 2,240	= Pounds
Ton, long	x 1.016	= Metric tons
Ton, long	x 1.12	= Short tons
Ton, metric	x 7.454	= Barrel, oil, 36API
Ton, metric	x 1,000	= Kilograms
Ton, metric	x 2,205	= Pounds
Ton, metric	x 0.9842	= Ton, long
Ton, metric	x 1.1023	= Ton, short
Ton, shipping, USA	x 40	= Cubic feet
Ton, shipping, USA	x 2.8317	= Cubic meter
Ton, shipping, USA	x 1.050	= Ton, shipping, British
Ton, short	x 40	= Cubic feet
Ton, short	x 268.8	= Gallons, USA, liquid
Ton, short	x 4	= Hogshead
Ton, short	x 907.18486	= Kilograms
Ton, short	x 1,000	= Liter
Ton, short	x 32,000	= Ounces
Ton, short	x 2,000	= Pounds
Ton, short	x 0.89286	= Tons, long
Ton, short	x 0.907	= Tons, metric
Tons, short/square foot	x 9,765	= Kilograms/square meter
Tons, short/square foot	x 2,000	= Pounds/square inch
Tons, short/day	x 83.333	= Pounds/hour
Tons, short/day	x 0.16643	= Gallons/minute

Conversion factor T (continued on facing page)

Appendix

Conversion factor T (concluded)

Tons, short/day	x 0.9072	= Tons, metric/day
Tons, short/day	x 0.8929	= Tons, long/day
Tons, short/day	x 37.8	= Kilograms/hour
Tons, metric/day	x 91.859	= Pounds/hour
Tons, metric/day	x 41.667	= Kilograms/hour
Tons, metric/day	x 0.9843	= Tons, long/day
Tons, metric/day	x 1.1023	= Tons, short/day
Tons, long/day	x 1.12	= Tons, short/day
Tons, long/day	x 1.016	= Tons, metric/day
Tons, long/day	x 93.333	= Pounds/hour
Tons, long/day	x 42.335	= Kilograms/hour

V

Vara	x 2.7777	= Feet
Vara	x 33.3333	= Inch
Vara	x 0.9259	= Yard
Volt/inch	x 0.3937	= Volt/centimeter

W

Water, 62°F	x 8.3311	= Pound
Water height in feet	x 0.4335	= Pound/square inch
Water height in feet	x 0.03048	= Kilograms/square centimeters
Water height in inches	x 0.03613	= Pound/square inch
Water height in inches	x 0.00254	= Kilograms/square centimeter
Water height in meters	x 1.42067	= Pound/square inch
Water height in meters	x 0.100	= Kilograms/square centimeters
Watts	x 3.4128	= Btu/hour
Watts	x 0.05688	= Btu/minute
Watts	x 107	= Ergs/second
Watts	x 44.27	= Foot-pounds/minute
Watts	x 0.7378	= Foot-pounds/second
Watts	x 1.341×10^{-3}	= Horsepower, USA
Watts	x 1.36×10^{-3}	= Horsepower, metric
Watts	x 0.001	= Kilowatt
Watt-hours	x 3.4128	= Btu
Watt-hours	x 3.60×10^{10}	= Ergs
Watt-hours	x 2,656	= Foot-pounds
Watt-hours	x 859.85	= Gram-calories
Watt-hours	x 1.341×10^{-3}	= Horsepower-hours, USA
Watt-hours	x 1.3596×10^{-3}	= Horsepower-hours, metric
Watt-hours	x 0.8605	= Kilogram-calories
Watt-hours	x 367.2	= Kilogram-meters
Watt-hours	x 0.001	- Kilowatt-hours

Y

Yard, USA	x 91.4402	= Centimeter
Yard, USA	x 3	= Feet
Yard, USA	x 36	= Inch
Yard, USA	x 9.144 x 10^{-4}	= Kilometer
Yard, USA	x 0.9144	= Meter
Yard, USA	x 4.934 x 10^{-4}	= Mile, nautical, USA
Yard, USA	x 5.682 x 10^{-4}	= Mile, statute, USA
Yard, USA	x 914.402	= Millimeters
Yard, USA	x 0.1818	= Rod
Year	x 8,765	= Hours
Year	x 525,948	= Minutes

Appendix

Table A-1
Conversion Table, Pounds per Square Inch to Kilograms per Square Centimeter
Based on 1 Inch = 25.4 Millimetres; 1 Pound = 0.45359243 Kilograms

lb./sq. in.	0 kg/cm^2	1 kg/cm^2	2 kg/cm^2	3 kg/cm^2	4 kg/cm^2	5 kg/cm^2	6 kg/cm^2	7 kg/cm^2	8 kg/cm^2	9 kg/cm^2
—	—	0.07031	0.14061	0.21092	0.28123	0.35153	0.42184	0.49215	0.56246	0.63276
10	0.70307	0.77338	0.84368	0.91399	0.98430	1.05460	1.12491	1.19522	1.26553	1.33583
20	1.40641	1.47645	1.54675	1.61706	1.68737	1.75767	1.82798	1.89829	1.96860	2.03890
30	2.10921	2.17952	2.24982	2.32013	2.39044	2.46074	2.53105	2.60136	2.67166	2.74197
40	2.81228	2.88259	2.95289	3.02320	3.09351	3.16381	3.23412	3.30443	3.37473	3.44504
50	3.51535	3.58566	3.65596	3.72627	3.79658	3.86688	3.93719	4.00750	4.07780	4.14811
60	4.21842	4.28873	4.35903	4.42934	4.49965	4.56995	4.64026	4.71057	4.78087	4.85118
70	4.92149	4.99179	5.0621	5.1324	5.2027	5.2730	5.3433	5.4136	5.4839	5.5543
80	5.6246	5.6949	5.7652	5.8355	5.9058	5.9761	6.0464	6.1167	6.1870	6.2573
90	6.3276	6.3979	6.4682	6.5385	6.6089	6.6792	6.7495	6.8198	6.8901	6.9604
100	7.0307	7.1010	7.1713	7.2416	7.3119	7.3822	7.4525	7.5228	7.5932	7.6635
10	7.7338	7.8041	7.8744	7.9447	8.0150	8.0853	8.1556	8.2259	8.2962	8.3665
20	8.4368	8.5071	8.5775	8.6478	8.7181	8.7884	8.8587	8.9290	8.9993	9.0696
30	9.1399	9.2102	9.2805	9.3508	9.4211	9.4914	9.5617	9.6321	9.7024	9.7727
40	9.8430	9.9133	9.9836	10.0539	10.1242	10.1945	10.2648	10.3351	10.4054	10.4757
50	10.5460	10.6164	10.6867	10.7570	10.8273	10.8976	10.9679	11.0382	11.1085	11.1788
60	11.2491	11.3194	11.3897	11.4600	11.5303	11.6006	11.6710	11.7413	11.8116	11.8819
70	11.9522	12.0225	12.0928	12.1631	12.2334	12.3037	12.3740	12.4443	12.5146	12.5849
80	12.6553	12.7256	12.7959	12.8662	12.9365	13.0068	13.0771	13.1474	13.2177	13.2880
90	13.3583	13.4286	13.4989	13.5692	13.6396	13.7099	13.7802	13.8505	13.9208	13.9911
200	14.0614	14.1317	14.2020	14.2723	14.3426	14.4129	14.4832	14.5535	14.6238	14.6942
10	14.7645	14.8348	14.9051	14.9754	15.0457	15.1160	15.1863	15.2566	15.3269	15.3972
20	15.4675	15.5378	15.6081	15.6785	15.7488	15.8191	15.8894	15.9597	16.0300	16.1003
30	16.1706	16.2409	16.3112	16.3815	16.4518	16.5221	16.5924	16.6628	16.7331	16.8034
40	16.8737	16.9440	17.0143	17.0846	17.1549	17.2252	17.2955	17.3658	17.4361	17.5064
50	17.5767	17.6470	17.7174	17.7877	17.8580	17.9283	17.9986	18.0689	18.1392	18.2095
60	18.2798	18.3501	18.4204	18.4907	18.5610	18.6313	18.7017	18.7720	18.8423	18.9126
70	18.9829	19.0532	19.1235	19.1938	19.2641	19.3344	19.4047	19.4750	19.5453	19.6156
80	19.6860	19.7563	19.8266	19.8969	19.9672	20.0375	20.1078	20.1781	20.2484	20.3187
90	20.3890	20.4593	20.5296	20.5999	20.6702	20.7406	20.8109	20.8812	20.9515	21.0218
300	21.0921	21.1624	21.2327	21.3030	21.3733	21.4436	21.5139	21.5842	21.6545	21.7249
10	21.7952	21.8655	21.9358	22.0061	22.0764	22.1467	22.2170	22.2873	22.3576	22.4279
20	22.4982	22.5685	22.6388	22.7092	22.7795	22.8498	22.9201	22.9904	23.0607	23.1310
30	23.2013	23.2716	23.3419	23.4122	23.4825	23.5528	23.6231	23.6934	23.7638	23.8341
40	23.9044	23.9747	24.0450	24.1153	24.1856	24.2559	24.3262	24.3965	24.4668	24.5371
50	24.6074	24.6777	24.7481	24.8184	24.8887	24.9590	25.0293	25.0996	25.1699	25.2402

Table A-1 (continued on following page)

Table A-1 (continued)

lb./sq. in.	0 kg/cm²	1 kg/cm²	2 kg/cm²	3 kg/cm²	4 kg/cm²	5 kg/cm²	6 kg/cm²	7 kg/cm²	8 kg/cm²	9 kg/cm²
60	25.3105	25.3808	25.4511	25.5214	25.5917	25.6620	25.7324	25.8027	25.8730	25.9433
70	26.0136	26.0839	26.1542	26.2245	26.2948	26.3651	26.4354	26.5057	26.5760	26.6463
80	26.7166	26.7870	26.8573	26.9276	26.9979	27.0682	27.1385	27.2088	27.2791	27.3494
90	27.4197	27.4900	27.5603	27.6306	27.7009	27.7713	27.8416	27.9119	27.9822	28.0525
400	28.1228	28.1931	28.2634	28.3337	28.4040	28.4743	28.5446	28.6149	28.6852	28.7555
10	28.8259	28.8962	28.9665	29.0368	29.1071	29.1774	29.2477	29.3180	29.3883	29.4586
20	29.5289	29.5992	29.6695	29.7398	29.8102	29.8805	29.9508	30.0211	30.0914	30.1617
30	30.2320	30.3023	30.3726	30.4429	30.5132	30.5835	30.6538	30.7241	30.7945	30.8648
40	30.9351	31.0054	31.0757	31.1460	31.2163	31.2866	31.3569	31.4272	31.4975	31.5678
50	31.6381	31.7084	31.7788	31.8491	31.9194	31.9897	32.0600	32.1303	32.2006	32.2709
60	32.3412	32.4115	32.4818	32.5521	32.6224	32.6927	32.7630	32.8334	32.9037	32.9740
70	33.0443	33.1146	33.1849	33.2552	33.3255	33.3958	33.4661	33.5364	33.6067	33.6770
80	33.7473	33.8177	33.8880	33.9583	34.0286	34.0989	34.1692	34.2395	34.3098	34.3801
90	34.4504	34.5207	34.5910	34.6613	34.7316	34.8019	34.8723	34.9426	35.0129	35.0832
500	35.1535	35.2238	35.2941	35.3644	35.4347	35.5050	35.5753	35.6456	35.7159	35.7862
10	35.8566	35.9269	35.9972	36.0675	36.1378	36.2081	36.2784	36.3487	36.4190	36.4893
20	36.5596	36.6299	36.7002	36.7705	36.8409	36.9112	36.9815	37.0518	37.1221	37.1924
30	37.2627	37.3330	37.4033	37.4736	37.5439	37.6142	37.6845	37.7548	37.8251	37.8955
40	37.9658	38.0361	38.1064	38.1767	38.2470	38.3173	38.3876	38.4579	38.5282	38.5985
50	38.6688	38.7391	38.8094	38.8798	38.9501	39.0204	39.0907	39.1610	39.2313	39.3016
60	39.3719	39.4422	39.5125	39.5828	39.6531	39.7234	39.7937	39.8641	39.9344	40.0047
70	40.0750	40.1453	40.2156	40.2859	40.3562	40.4265	40.4968	40.5671	40.6374	40.7077
80	40.7780	40.8483	40.9187	40.9890	41.0593	41.1296	41.1999	41.2702	41.3405	41.4108
90	41.4811	41.5514	41.6217	41.6920	41.7623	41.8326	41.9030	41.9733	42.0436	42.1139
600	42.1842	42.2545	42.3248	42.3951	42.4654	42.5357	42.6060	42.6763	42.7466	42.8169
10	42.8873	42.9576	43.0279	43.0982	43.1685	43.2388	43.3091	43.3794	43.4497	43.5200
20	43.5903	43.6606	43.7309	43.8012	43.8715	43.9419	44.0122	44.0825	44.1528	44.2231
30	44.2934	44.3637	44.4340	44.5043	44.5746	44.6449	44.7152	44.7855	44.8558	44.9262
40	44.9965	45.0668	45.1371	45.2074	45.2777	45.3480	45.4183	45.4886	45.5589	45.6292
50	45.6995	45.7698	45.8401	45.9104	45.9808	46.0511	46.1214	46.1917	46.2620	46.3323
60	46.4026	46.4729	46.5432	46.6135	46.6838	46.7541	46.8244	46.8947	46.9651	47.0354
70	47.1057	47.1760	47.2463	47.3166	47.3869	47.4572	47.5275	47.5978	47.6681	47.7384
80	47.8087	47.8790	47.9494	48.0197	48.0900	48.1603	48.2306	48.3009	48.3712	48.4415
90	48.5118	48.5821	48.6524	48.7227	48.7930	48.8633	48.9336	49.0040	49.0743	49.1446
700	49.2149	49.2852	49.3555	49.4258	49.4961	49.5664	49.6367	49.7070	49.7773	49.8476
10	49.9179	49.9883	50.059	50.129	50.199	50.269	50.340	50.410	50.480	50.551
20	50.621	50.691	50.762	50.832	50.902	50.973	51.043	51.113	51.183	51.254
30	51.324	51.394	51.465	51.535	51.605	51.676	51.746	51.816	51.887	51.957
40	52.027	52.097	52.168	52.238	52.308	52.379	52.449	52.519	52.590	52.660

Table A-1 (continued on facing page)

Table A-1 (concluded)

lb./sq. in.	0 kg/cm²	1 kg/cm²	2 kg/cm²	3 kg/cm²	4 kg/cm²	5 kg/cm²	6 kg/cm²	7 kg/cm²	8 kg/cm²	9 kg/cm²
50	52.730	52.801	52.871	52.941	53.011	53.082	53.152	53.222	53.293	53.363
60	53.433	53.504	53.574	53.644	53.715	53.785	53.855	53.925	53.996	54.066
70	54.136	54.207	54.277	54.347	54.418	54.488	54.558	54.629	54.699	54.769
80	54.839	54.910	54.980	55.050	55.121	55.191	55.261	55.332	55.402	55.472
90	55.543	55.613	55.683	55.753	55.824	55.894	55.964	56.035	56.105	56.175
800	56.246	56.316	56.386	56.456	56.527	56.597	56.667	56.738	56.808	56.878
10	56.949	57.019	57.089	57.160	57.230	57.300	57.370	57.441	57.511	57.581
20	57.652	57.722	57.792	57.863	57.933	58.003	58.074	58.144	58.214	58.284
30	58.355	58.425	58.495	58.566	58.636	58.706	58.777	58.847	58.917	58.988
40	59.058	59.128	59.198	59.269	59.339	59.409	59.480	59.550	59.620	59.691
50	59.761	59.831	59.902	59.972	60.042	60.112	60.183	60.253	60.323	60.394
60	60.464	60.534	60.605	60.675	60.745	60.816	60.886	60.956	61.026	61.097
70	61.167	61.237	61.308	61.378	61.448	61.519	61.589	61.659	61.730	61.800
80	61.870	61.940	62.011	62.081	62.151	62.222	62.292	62.362	62.433	62.503
90	62.573	62.644	62.714	62.784	62.854	62.925	62.995	63.065	63.136	63.206
900	63.276	63.347	63.417	63.487	63.557	63.628	63.698	63.768	63.839	63.909
10	63.979	64.050	64.120	64.190	64.261	64.331	64.401	64.471	64.542	64.612
20	64.682	64.753	64.823	64.893	64.964	65.034	65.104	65.175	65.245	65.315
30	65.385	65.456	65.526	65.596	65.667	65.737	65.807	65.878	65.948	66.018
40	66.089	66.159	66.229	66.299	66.370	66.440	66.510	66.581	66.651	66.721
50	66.792	66.862	66.932	67.003	67.073	67.143	67.213	67.284	67.354	67.424
60	67.495	67.565	67.635	67.706	67.776	67.846	67.917	67.987	68.057	68.127
70	68.198	68.268	68.338	68.409	68.479	68.549	68.620	68.690	68.760	68.831
80	68.901	68.971	69.041	69.112	69.182	69.252	69.323	69.393	69.463	69.534
90	69.604	69.674	69.745	69.815	69.885	69.955	70.026	70.096	70.166	70.237

Table A-2
Circumference, Area and Volume of Circles and Cylinders
Courtesy of Chicago Bridge and Iron Co.

Diam. in Feet	Circumference		Area of Circle		Volume of Cylinder Per Foot of Height			Diam. in Feet
	Feet	Meters	Sq. Feet	Sq. Meters	U. S. Gals.	Imperial Gals.	U.S. Bbls. (42 Gals.)	
1	3.14	0.9576	0.785	.0730	5.9	4.9	0.140	1
2	6.28	1.9151	3.142	.2919	23.5	19.6	0.560	2
3	9.42	2.8727	7.069	.6567	52.9	44.0	1.259	3
4	12.57	3.8302	12.566	1.1675	94.0	78.3	2.238	4
5	15.71	4.7878	19.635	1.8241	146.9	122.3	3.497	5
6	18.85	5.7454	28.274	2.6268	211.5	176.1	5.04	6
7	21.99	6.7029	38.485	3.5753	287.9	239.7	6.85	7
8	25.13	7.6605	50.266	4.6698	376.0	313.1	8.95	8
9	28.27	8.6180	63.617	5.9102	475.9	396.3	11.33	9
10	31.42	9.5756	78.540	7.2966	587.5	489.2	13.99	10
11	34.56	10.5331	95.033	8.8289	710.9	591.9	16.93	11
12	37.70	11.4907	113.097	10.5071	846.0	704.5	20.14	12
13	40.84	12.4482	132.732	12.3312	992.9	826.8	23.64	13
14	43.98	13.4058	153.938	14.3013	1,151.5	958.9	27.42	14
15	47.12	14.3634	176.715	16.4173	1,321.9	1,100.7	31.47	15
16	50.27	15.3209	201.062	18.6793	1,504.0	1,252.4	35.81	16
17	53.41	16.2785	226.980	21.0871	1,697.9	1,413.8	40.43	17
18	56.55	17.2360	254.469	23.6409	1,903.6	1,585.1	45.32	18
19	59.69	18.1936	283.529	26.3407	2,120.9	1,766.1	50.50	19
20	62.83	19.1511	314.159	29.1864	2,350.1	1,956.9	55.95	20
21	65.97	20.1087	346.361	32.1780	2,591.0	2,157.4	61.69	21
22	69.12	21.0663	380.133	35.3155	2,843.6	2,367.8	67.70	22
23	72.26	22.0238	415.476	38.5989	3,108.0	2,587.9	74.00	23
24	75.40	22.9814	452.389	42.0283	3,384.1	2,817.9	80.57	24
25	78.54	23.9389	490.874	45.6037	3,672.5	3,057.6	87.43	25
26	81.68	24.8965	530.929	49.3249	3,971.6	3,307.1	94.56	26
27	84.82	25.8541	572.555	53.1921	4,283.0	3,566.4	101.98	27
28	87.97	26.8116	615.752	57.2052	4,606.1	3,835.4	109.67	28
29	91.11	27.7692	660.520	61.3643	4,941.0	4,114.3	117.64	29
30	94.25	28.7267	706.858	65.6693	5,287.7	4,402.9	125.90	30
31	97.39	29.6843	754.768	70.1202	5,646.1	4,701.4	134.43	31
32	100.53	30.6418	804.248	74.7171	6,016.2	5,009.6	143.24	32
33	103.67	31.5994	855.299	79.4598	6,398.1	5,327.5	152.34	33
34	106.81	32.5570	907.920	84.3486	6,791.7	5,655.3	161.71	34
35	109.96	33.5145	962.113	89.3832	7,197.1	5,992.9	171.36	35
36	113.10	34.4721	1,017.88	94.5638	7,614.2	6,340.2	181.29	36
37	116.24	35.4296	1,075.21	99.8903	8,043.1	6,697.4	191.50	37
38	119.38	36.3872	1,134.11	105.3627	8,483.8	7,064.3	201.99	38
39	122.52	37.3447	1,194.59	110.9811	8,936.2	7,441.0	212.77	39
40	125.66	38.3023	1,256.64	116.7454	9,400.3	7,827.4	223.82	40
41	128.81	39.2599	1,320.25	122.6556	9,876.2	8,223.7	235.15	41
42	131.95	40.2174	1,385.44	128.7118	10,363.8	8,629.7	246.76	42
43	135.09	41.1750	1,452.20	134.9139	10,863.2	9,045.6	258.65	43
44	138.23	42.1325	1,520.53	141.2619	11,374.4	9,471.2	270.82	44
45	141.37	43.0901	1,590.43	147.7559	11,897.2	9,906.6	283.27	45
46	144.51	44.0476	1,661.90	154.3958	12,431.9	10,351.8	296.00	46
47	147.65	45.0052	1,734.94	161.1816	12,978.3	10,806.8	309.01	47
48	150.80	45.9628	1,809.56	168.1134	13,536.4	11,271.5	322.30	48
49	153.94	46.9203	1,885.74	175.1911	14,106.3	11,746.0	335.86	49
50	157.08	47.8779	1,963.50	182.4147	14,688.0	12,230.4	349.71	50
51	160.22	48.8354	2,042.82	189.7842	15,281.4	12,724.5	363.84	51
52	163.36	49.7930	2,123.72	197.2997	15,886.5	13,228.4	378.25	52
53	166.50	50.7505	2,206.18	204.9611	16,503.4	13,742.0	392.94	53
54	169.65	51.7081	2,290.22	212.7685	17,132.0	14,265.5	407.91	54
55	172.79	52.6657	2,375.83	220.7218	17,772.4	14,798.7	423.15	55
56	175.93	53.6232	2,463.01	228.8210	18,424.6	15,341.8	438.68	56
57	179.07	54.5808	2,551.76	237.0661	19,088.5	15,894.6	454.49	57
58	182.21	55.5383	2,642.08	245.4572	19,764.1	16,457.2	470.57	58
59	185.35	56.4959	2,733.97	253.9942	20,451.5	17,029.6	486.94	59
60	188.50	57.4534	2,827.43	262.6772	21,150.7	17,611.7	503.59	60
61	191.64	58.4110	2,922.47	271.5060	21,861.6	18,203.7	520.51	61
62	194.78	59.3686	3,019.07	280.4808	22,584.2	18,805.4	537.72	62
63	197.92	60.3261	3,117.25	289.6016	23,318.6	19,416.9	555.21	63
64	201.06	61.2837	3,216.99	298.8682	24,064.8	20,038.2	572.97	64
65	204.20	62.2412	3,318.31	308.2808	24,822.7	20,669.3	591.02	65

Table A-2 (continued on facing page)

Table A-2 (continued)

Diam. in Feet	Circumference		Area of Circle		Volume of Cylinder Per Foot of Height			Diam. in Feet
	Feet	Meters	Sq. Feet	Sq. Meters	U.S. Gals.	Imperial Gals.	U.S. Bbls. (42 Gals.)	
66	207.35	63.1988	3,421.19	317.8394	25,592.3	21,310.2	609.34	66
67	210.49	64.1563	3,525.65	327.5438	26,373.7	21,960.9	627.95	67
68	213.63	65.1139	3,631.68	337.3942	27,166.9	22,621.3	646.83	68
69	216.77	66.0715	3,739.28	347.3905	27,971.8	23,291.5	665.99	69
70	219.91	67.0290	3,848.45	357.5328	28,788.4	23,971.5	685.44	70
71	223.05	67.9866	3,959.19	367.8210	29,616.8	24,661.3	705.16	71
72	226.19	68.9441	4,071.50	378.2551	30,457.0	25,360.9	725.17	72
73	229.34	69.9017	4,185.39	388.8352	31,308.9	26,070.3	745.45	73
74	232.48	70.8593	4,300.84	399.5611	32,172.5	26,789.4	766.01	74
75	235.62	71.8168	4,417.86	410.4331	33,047.9	27,518.3	786.86	75
76	238.76	72.7744	4,536.46	421.4509	33,935.1	28,257.0	807.98	76
77	241.90	73.7319	4,656.63	432.6147	34,834.0	29,005.5	829.38	77
78	245.04	74.6895	4,778.36	443.9244	35,744.6	29,763.8	851.06	78
79	248.19	75.6470	4,901.67	455.3800	36,667.0	30,531.9	873.02	79
80	251.33	76.6046	5,026.55	466.9816	37,601.2	31,309.7	895.27	80
81	254.47	77.5622	5,153.00	478.7291	38,547.1	32,097.4	917.79	81
82	257.61	78.5197	5,281.02	490.6226	39,504.7	32,894.8	940.59	82
83	260.75	79.4773	5,410.61	502.6619	40,474.1	33,702.0	963.67	83
84	263.89	80.4348	5,541.77	514.8472	41,455.3	34,519.0	987.03	84
85	267.04	81.3924	5,674.50	527.1785	42,448.2	35,345.8	1,010.67	85
86	270.18	82.3499	5,808.80	539.6556	43,452.9	36,182.3	1,034.59	86
87	273.32	83.3075	5,944.68	552.2787	44,469.3	37,028.7	1,058.79	87
88	276.46	84.2651	6,082.12	565.0478	45,497.4	37,884.8	1,083.27	88
89	279.60	85.2226	6,221.14	577.9627	46,537.3	38,750.7	1,108.03	89
90	282.74	86.1802	6,361.73	591.0236	47,589.0	39,626.4	1,133.07	90
91	285.88	87.1377	6,503.88	604.2304	48,652.4	40,511.9	1,158.39	91
92	289.03	88.0953	6,647.61	617.5832	49,727.6	41,407.1	1,183.99	92
93	292.17	89.0528	6,792.91	631.0819	50,814.5	42,312.2	1,209.87	93
94	295.31	90.0104	6,939.78	644.7265	51,913.1	43,227.0	1,236.03	94
95	298.45	90.9680	7,088.22	658.5170	53,023.5	44,151.6	1,262.47	95
96	301.59	91.9255	7,238.23	672.4535	54,145.7	45,086.0	1,289.18	96
97	304.73	92.8831	7,389.81	686.5359	55,279.6	46,030.2	1,316.18	97
98	307.88	93.8406	7,542.96	700.7643	56,425.3	46,984.2	1,343.46	98
99	311.02	94.7982	7,697.69	715.1386	57,582.7	47,947.9	1,371.02	99
100	314.16	95.7557	7,853.98	729.6588	58,751.9	48,921.5	1,398.85	100
101	317.30	96.7133	8,011.85	744.3249	59,932.8	49,904.8	1,426.97	101
102	320.44	97.6709	8,171.28	759.1370	61,125.4	50,897.9	1,455.37	102
103	323.58	98.6284	8,332.29	774.0950	62,329.8	51,900.8	1,484.04	103
104	326.73	99.5860	8,494.87	789.1989	63,546.0	52,913.5	1,513.00	104
105	329.87	100.5435	8,659.01	804.4488	64,773.9	53,935.9	1,542.24	105
106	333.01	101.5011	8,824.73	819.8446	66,013.6	54,968.2	1,571.75	106
107	336.15	102.4586	8,992.02	835.3863	67,265.0	56,010.2	1,601.55	107
108	339.29	103.4162	9,160.88	851.0740	68,528.2	57,062.0	1,631.62	108
109	342.43	104.3738	9,331.32	866.9076	69,803.1	58,123.6	1,661.98	109
110	345.58	105.3313	9,503.32	882.8871	71,089.7	59,195.0	1,692.61	110
111	348.72	106.2889	9,676.89	899.0126	72,388.2	60,276.2	1,723.53	111
112	351.86	107.2464	9,852.05	915.2840	73,698.3	61,367.1	1,754.72	112
113	355.00	108.2040	10,028.75	931.7013	75,020.2	62,467.8	1,786.20	113
114	358.14	109.1615	10,207.03	948.2645	76,353.9	63,578.4	1,817.95	114
115	361.28	110.1191	10,386.89	964.9737	77,699.3	64,698.7	1,849.98	115
116	364.42	111.0767	10,568.32	981.8288	79,056.5	65,828.7	1,882.30	116
117	367.57	112.0342	10,751.31	998.8299	80,425.4	66,968.6	1,914.89	117
118	370.71	112.9918	10,935.88	1,015.9769	81,806.1	68,118.3	1,947.76	118
119	373.85	113.9493	11,122.02	1,033.2698	83,198.5	69,277.7	1,980.92	119
120	376.99	114.9069	11,309.73	1,050.7086	84,602.7	70,446.9	2,014.35	120
121	380.13	115.8645	11,499.01	1,068.2934	86,018.6	71,625.9	2,048.06	121
122	383.27	116.8220	11,689.86	1,086.0241	87,446.3	72,814.7	2,082.05	122
123	386.42	117.7796	11,882.29	1,103.9008	88,885.7	74,013.3	2,116.33	123
124	389.56	118.7371	12,076.28	1,121.9233	90,336.8	75,221.7	2,150.88	124
125	392.70	119.6947	12,271.84	1,140.0918	91,799.8	76,439.8	2,185.71	125
126	395.84	120.6522	12,468.98	1,158.4063	93,274.4	77,667.7	2,220.82	126
127	398.98	121.6098	12,667.68	1,176.8666	94,760.9	78,905.5	2,256.21	127
128	402.12	122.5674	12,867.96	1,195.4729	96,259.0	80,153.0	2,291.88	128
129	405.27	123.5249	13,069.81	1,214.2252	97,769.0	81,410.2	2,327.83	129
130	408.41	124.4825	13,273.23	1,233.1233	99,290.6	82,677.3	2,364.06	130
131	411.55	125.4400	13,478.22	1,252.1674	100,824.0	83,954.0	2,400.57	131
132	414.69	126.3976	13,684.77	1,271.3574	102,369.2	85,240.8	2,437.36	132
133	417.83	127.3551	13,892.91	1,290.6934	103,926.1	86,537.2	2,474.43	133
134	420.97	128.3127	14,102.61	1,310.1753	105,494.8	87,843.4	2,511.78	134
135	424.12	129.2703	14,313.88	1,329.8031	107,075.2	89,159.4	2,549.41	135

Table A-2 (continued on following page)

Table A-2 (concluded)

Diam. in Feet	Circumference		Area of Circle		Volume of Cylinder Per Foot of Height			Diam. in Feet
	Feet	Meters	Sq. Feet	Sq. Meters	U.S. Gals.	Imperial Gals.	U.S. Bbls. (42 Gals.)	
136	427.26	130.2278	14,526.72	1,349.5769	108,667.4	90,485.2	2,587.32	136
137	430.40	131.1854	14,741.14	1,369.4965	110,271.3	91,820.7	2,625.51	137
138	433.54	132.1429	14,957.12	1,389.5622	111,887.0	93,166.1	2,663.98	138
139	436.68	133.1005	15,174.67	1,409.7737	113,514.4	94,521.2	2,702.73	139
140	439.82	134.0580	15,393.80	1,430.1312	115,153.6	95,886.1	2,741.75	140
141	442.96	135.0156	15,614.50	1,450.6346	116,804.6	97,260.8	2,781.06	141
142	446.11	135.9732	15,836.77	1,471.2839	118,467.2	98,645.3	2,820.65	142
143	449.25	136.9307	16,060.60	1,492.0792	120,141.7	100,039.5	2,860.52	143
144	452.39	137.8883	16,286.01	1,513.0204	121,827.8	101,443.6	2,900.66	144
145	455.53	138.8458	16,512.99	1,534.1076	123,525.8	102,857.4	2,941.09	145
146	458.67	139.8034	16,741.54	1,555.3406	125,235.4	104,281.0	2,981.80	146
147	461.81	140.7609	16,971.67	1,576.7196	126,956.9	105,714.4	3,022.78	147
148	464.96	141.7185	17,203.36	1,598.2446	128,690.1	107,157.6	3,064.05	148
149	468.10	142.6761	17,436.62	1,619.9154	130,435.0	108,610.6	3,105.60	149
150	471.24	143.6336	17,671.46	1,641.7322	132,191.7	110,073.3	3,147.42	150
151	474.38	144.5912	17,907.86	1,663.6950	133,960.1	111,545.9	3,189.53	151
152	477.52	145.5487	18,145.84	1,685.8036	135,740.3	113,028.2	3,231.91	152
153	480.66	146.5063	18,385.38	1,708.0582	137,532.2	114,520.3	3,274.58	153
154	483.81	147.4638	18,626.50	1,730.4587	139,335.9	116,022.0	3,317.52	154
155	486.95	148.4214	18,869.19	1,753.0052	141,151.3	117,533.9	3,360.75	155
156	490.09	149.3790	19,113.45	1,775.6976	142,978.5	119,055.3	3,404.25	156
157	493.23	150.3365	19,359.28	1,798.5359	144,817.4	120,586.6	3,448.04	157
158	496.37	151.2941	19,606.68	1,821.5202	146,668.1	122,127.6	3,492.10	158
159	499.51	152.2516	19,855.65	1,844.6503	148,530.6	123,678.4	3,536.44	159
160	502.65	153.2092	20,106.19	1,867.9264	150,404.7	125,239.0	3,581.07	160
161	505.80	154.1667	20,358.30	1,891.3485	152,290.7	126,809.4	3,625.97	161
162	508.94	155.1243	20,611.99	1,914.9165	154,188.4	128,389.5	3,671.15	162
163	512.08	156.0819	20,867.24	1,938.6304	156,097.8	129,979.5	3,716.62	163
164	515.22	157.0394	21,124.06	1,962.4902	158,019.0	131,579.2	3,762.36	164
165	518.36	157.9970	21,382.46	1,986.4960	159,951.9	133,188.7	3,808.38	165
166	521.50	158.9545	21,642.43	2,010.6477	161,896.6	134,808.0	3,854.68	166
167	524.65	159.9121	21,903.96	2,034.9453	163,853.0	136,437.1	3,901.26	167
168	527.79	160.8696	22,167.07	2,059.3889	165,821.2	138,076.0	3,948.13	168
169	530.93	161.8272	22,431.75	2,083.9784	167,801.2	139,724.6	3,995.27	169
170	534.07	162.7848	22,698.00	2,108.7138	169,792.8	141,383.1	4,042.69	170
171	537.21	163.7423	22,965.82	2,133.5952	171,796.3	143,051.3	4,090.39	171
172	540.35	164.6999	23,235.21	2,158.6225	173,811.5	144,729.3	4,138.37	172
173	543.50	165.6574	23,506.18	2,183.7957	175,838.4	146,417.1	4,186.63	173
174	546.64	166.6150	23,778.71	2,209.1149	177,877.1	148,114.7	4,235.17	174
175	549.78	167.5726	24,052.81	2,234.5800	179,927.5	149,822.0	4,283.99	175
176	552.92	168.5301	24,328.49	2,260.1910	181,989.7	151,539.2	4,333.09	176
177	556.06	169.4877	24,605.73	2,285.9480	184,063.7	153,266.1	4,382.47	177
178	559.20	170.4452	24,884.55	2,311.8508	186,149.4	155,002.8	4,432.13	178
179	562.35	171.4028	25,164.94	2,337.8997	188,246.8	156,749.3	4,482.07	179
180	565.49	172.3603	25,446.90	2,364.0944	190,356.0	158,505.6	4,532.29	180
181	568.63	173.3179	25,730.42	2,390.4351	192,476.9	160,271.7	4,582.79	181
182	571.77	174.2755	26,015.52	2,416.9217	194,609.6	162,047.5	4,633.56	182
183	574.91	175.2330	26,302.19	2,443.5543	196,754.1	163,833.1	4,684.62	183
184	578.05	176.1906	26,590.43	2,470.3327	198,910.3	165,628.6	4,735.96	184
185	581.19	177.1481	26,880.25	2,497.2571	201,078.2	167,433.8	4,787.58	185
186	584.34	178.1057	27,171.63	2,524.3275	203,257.9	169,248.8	4,839.48	186
187	587.48	179.0632	27,464.58	2,551.5437	205,449.3	171,073.5	4,891.65	187
188	590.62	180.0208	27,759.11	2,578.9060	207,652.5	172,908.1	4,944.11	188
189	593.76	180.9784	28,055.20	2,606.4141	209,867.5	174,752.4	4,996.85	189
190	596.90	181.9359	28,352.87	2,634.0682	212,094.2	176,606.5	5,049.86	190
191	600.04	182.8935	28,652.10	2,661.8682	214,332.6	178,470.5	5,103.16	191
192	603.19	183.8510	28,952.91	2,689.8141	216,582.8	180,344.1	5,156.74	192
193	606.33	184.8086	29,255.29	2,717.9059	218,844.8	182,227.6	5,210.59	193
194	609.47	185.7661	29,559.24	2,746.1437	221,118.5	184,120.9	5,264.73	194
195	612.61	186.7237	29,864.76	2,774.5275	223,403.9	186,023.9	5,319.14	195
196	615.75	187.6813	30,171.85	2,803.0571	225,701.1	187,936.8	5,373.84	196
197	618.89	188.6388	30,480.51	2,831.7327	228,010.1	189,859.4	5,428.81	197
198	622.04	189.5964	30,790.74	2,860.5542	230,330.8	191,791.8	5,484.07	198
199	625.18	190.5539	31,102.55	2,889.5217	232,663.2	193,734.0	5,539.60	199
200	628.32	191.5115	31,415.93	2,918.6351	235,007.4	195,685.9	5,595.42	200

Notes:
1. If diameters are assumed as meters, values in columns "Circumference Feet" and "Area of Circle Square Feet" will represent circumference in meters and area of circle in square meters respectively.
2. If diameters are assumed as meters, values in column "Area of Circle Square Feet" will represent volume of cylinder in cubic meters per vertical meter of height.

Formula to determine capacity per foot of vertical height of cylinder.
D = Diameter in Feet.
$0.1398854 D^2$ = Barrels of 42 U.S. Gallons per vertical foot.
$5.875185 \ D^2$ = U.S. Gallons per vertical foot.
$4.892148 \ D^2$ = Imperial Gallons per vertical foot.
$0.022240 \ D^2$ = Cubic Meters per vertical foot.
$0.785398 \ D^2$ = (D = Diameter in Meters) = Cubic meters per vertical meter.

Table A-3
Surface and Volume of Spheres

Diam. in Ft.	Surface of Sphere in Sq. Ft.	Volume of Sphere			Diam. in Ft.	Surface of Sphere in Sq. Ft.	Volume of Sphere		
		Cu. Ft.	U.S. Gals.	U.S. Bbls.			Cu. Ft.	U.S. Gals.	U.S. Bbls.
1	3.14	0.52	3.92	.09	61	11,690	118,847	889,037	21,168
2	12.57	4.19	31.33	.75	62	12,076	124,788	933,481	22,226
3	28.27	14.14	105.75	2.52	63	12,469	130,924	979,382	23,319
4	50.27	33.51	250.67	5.97	64	12,868	137,258	1,026,764	24,447
5	78.54	65.45	489.60	11.66	65	13,273	143,793	1,075,649	25,611
6	113.10	113.10	846.03	20.14	66	13,685	150,533	1,126,062	26,811
7	153.94	179.59	1,343.46	31.99	67	14,103	157,479	1,178,026	28,048
8	201.06	268.08	2,005.40	47.75	68	14,527	164,636	1,231,565	29,323
9	254.47	381.70	2,855.34	67.98	69	14,957	172,007	1,286,701	30,636
10	314.16	523.60	3,916.79	93.26	70	15,394	179,594	1,343,460	31,987
11	380	697	5,213	124	71	15,837	187,402	1,401,863	33,378
12	452	905	6,768	161	72	16,286	195,432	1,461,935	34,808
13	531	1,150	8,605	205	73	16,742	203,689	1,523,699	36,279
14	616	1,437	10,748	256	74	17,203	212,175	1,587,178	37,790
15	707	1,767	13,219	315	75	17,671	220,893	1,652,397	39,343
16	804	2,145	16,043	382	76	18,146	229,847	1,719,378	40,938
17	908	2,572	19,243	458	77	18,627	239,040	1,788,145	42,575
18	1,018	3,054	22,843	544	78	19,113	248,475	1,858,721	44,255
19	1,134	3,591	26,865	640	79	19,607	258,155	1,931,131	45,979
20	1,257	4,189	31,334	746	80	20,106	268,083	2,005,398	47,748
21	1,385	4,849	36,273	864	81	20,612	278,262	2,081,544	49,561
22	1,521	5,575	41,706	993	82	21,124	288,696	2,159,594	51,419
23	1,662	6,371	47,656	1,135	83	21,642	299,387	2,239,571	53,323
24	1,810	7,238	54,146	1,289	84	22,167	310,339	2,321,498	55,274
25	1,963	8,181	61,200	1,457	85	22,698	321,555	2,405,400	57,271
26	2,124	9,203	68,842	1,639	86	23,235	333,038	2,491,299	59,317
27	2,290	10,306	77,094	1,836	87	23,779	344,792	2,579,219	61,410
28	2,463	11,494	85,981	2,047	88	24,328	356,818	2,669,184	63,552
29	2,642	12,770	95,527	2,274	89	24,885	369,121	2,761,217	65,743
30	2,827	14,137	105,753	2,518	90	25,447	381,704	2,855,341	67,984
31	3,019	15,599	116,685	2,778	91	26,016	394,569	2,951,581	70,276
32	3,217	17,157	128,345	3,056	92	26,590	407,720	3,049,959	72,618
33	3,421	18,817	140,758	3,351	93	27,172	421,161	3,150,499	75,012
34	3,632	20,580	153,946	3,665	94	27,759	434,893	3,253,225	77,458
35	3,848	22,449	167,932	3,998	95	28,353	448,921	3,358,160	79,956
36	4,072	24,429	182,742	4,351	96	28,953	463,247	3,465,327	82,508
37	4,301	26,522	198,397	4,724	97	29,559	477,875	3,574,750	85,113
38	4,536	28,731	214,922	5,117	98	30,172	492,807	3,686,453	87,773
39	4,778	31,059	232,340	5,532	99	30,791	508,048	3,800,459	90,487
40	5,027	33,510	250,675	5,968	100	31,416	523,599	3,916,792	93,257
41	5,281	36,087	269,949	6,427	101	32,047	539,465	4,035,475	96,083
42	5,542	38,792	290,187	6,909	102	32,685	555,647	4,156,531	98,965
43	5,809	41,630	311,412	7,415	103	33,329	572,151	4,279,984	101,904
44	6,082	44,602	333,648	7,944	104	33,979	588,978	4,405,858	104,901
45	6,362	47,713	356,918	8,498	105	34,636	606,131	4,534,176	107,957
46	6,648	50,965	381,245	9,077	106	35,299	623,615	4,664,962	111,071
47	6,940	54,362	406,653	9,682	107	35,968	641,431	4,798,239	114,244
48	7,238	57,906	433,166	10,313	108	36,644	659,584	4,934,030	117,477
49	7,543	61,601	460,807	10,972	109	37,325	678,076	5,072,359	120,771
50	7,854	65,450	489,599	11,657	110	38,013	696,910	5,213,250	124,125
51	8,171	69,456	519,566	12,371	111	38,708	716,090	5,356,726	127,541
52	8,495	73,622	550,732	13,113	112	39,408	735,619	5,502,811	131,019
53	8,825	77,952	583,120	13,884	113	40,115	755,499	5,651,527	134,560
54	9,161	82,448	616,754	14,685	114	40,828	775,735	5,802,900	138,164
55	9,503	87,114	651,656	15,516	115	41,548	796,329	5,956,951	141,832
56	9,852	91,952	687,851	16,377	116	42,273	817,284	6,113,705	145,564
57	10,207	96,967	725,362	17,271	117	43,005	838,603	6,273,185	149,362
58	10,568	102,160	764,213	18,196	118	43,744	860,290	6,435,415	153,224
59	10,936	107,536	804,427	19,153	119	44,488	882,348	6,600,417	157,153
60	11,310	113,097	846,027	20,144	120	45,239	904,779	6,768,217	161,148

Note: If diameters are assumed as meters, values in columns "Surface of Sphere in Square Feet" and "Volume of Sphere—Cubic Feet" will represent Surface of Sphere in Square Meters and Volume of Sphere in Cubic Meters respectively.
Surface area of sphere = 3.141593 D^2 Square Feet.

Volume of sphere $\begin{cases} = 0.523599\ D^3 \text{ Cubic Feet.} \\ = 0.093257\ D^3 \text{ Barrels of 42 U.S. Gallons.} \end{cases}$
Number of barrels of 42 U.S. Gallons at any inch in a true sphere = $(3d-2h)\ h^2 \times .0000539681$ where d is diameter of sphere and h is depth of liquid both in inches.

Table A-4
Decimal Equivalents in Inches, Feet and Millimeters

In. Equiv. for Decimal of In.	Decimals	Millimeters Equiv. for Decimal of In.	In. Equiv. for Decimal of Ft.
1/64	.0156	0.397	3/16
1/32	.0313	0.794	3/8
3/64	.0469	1.191	9/16
1/16	.0625	1.588	3/4
5/64	.0781	1.984	15/16
3/32	.0938	2.381	1 1/8
7/64	.1094	2.778	1 5/16
1/8	.1250	3.175	1 1/2
9/64	.1406	3.572	1 11/16
5/32	.1563	3.969	1 7/8
11/64	.1719	4.366	2 1/16
3/16	.1875	4.763	2 1/4
13/64	.2031	5.159	2 7/16
7/32	.2188	5.556	2 5/8
15/64	.2344	5.953	2 13/16
1/4	.2500	6.350	3
17/64	.2656	6.747	3 3/16
9/32	.2813	7.144	3 3/8
19/64	.2969	7.541	3 9/16
5/16	.3125	7.938	3 3/4
21/64	.3281	8.334	3 15/16
11/32	.3438	8.731	4 1/8
23/64	.3594	9.128	4 5/16
3/8	.3750	9.525	4 1/2
25/64	.3906	9.922	4 11/16
13/32	.4063	10.319	4 7/8
27/64	.4219	10.716	5 1/16
7/16	.4375	11.113	5 1/4
29/64	.4531	11.509	5 7/16
15/32	.4688	11.906	5 5/8
31/64	.4844	12.303	5 13/16
1/2	.5000	12.700	6
33/64	.5156	13.097	6 3/16
17/32	.5313	13.494	6 3/8
35/64	.5469	13.891	6 9/16
9/16	.5625	14.288	6 3/4
37/64	.5781	14.684	6 15/16
19/32	.5938	15.081	7 1/8
39/64	.6094	15.478	7 5/16
5/8	.6250	15.875	7 1/2
41/64	.6406	16.272	7 11/16
21/32	.6563	16.669	7 7/8
43/64	.6719	17.066	8 1/16
11/16	.6875	17.463	8 1/4
45/64	.7031	17.859	8 7/16
23/32	.7188	18.256	8 5/8
47/64	.7344	18.653	8 13/16
3/4	.7500	19.050	9
49/64	.7656	19.447	9 3/16
25/32	.7813	19.844	9 3/8
51/64	.7969	20.241	9 9/16
13/16	.8125	20.638	9 3/4
53/64	.8281	21.034	9 15/16
27/32	.8438	21.431	10 1/8
55/64	.8594	21.828	10 5/16
7/8	.8750	22.225	10 1/2
57/64	.8906	22.622	10 11/16
29/32	.9063	23.019	10 7/8
59/64	.9219	23.416	11 1/16
15/16	.9375	23.813	11 1/4
61/64	.9531	24.209	11 7/16
31/32	.9688	24.606	11 5/8
63/64	.9844	25.003	11 13/16
1	1.0000	25.400	12

Appendix

Table A-5
Temperature Conversion Chart, Centigrade-Fahrenheit

NOTE: The numbers in boldface refer to the temperature in degrees, either Centigrade or Fahrenheit, which it is desired to convert into the other scale. If converting from Fahrenheit to Centigrade degrees, the equivalent temperature will be found in the left column; while if converting from degrees Centigrade to degrees Fahrenheit, the answer will be found in the column on the right.

Centigrade		Fahrenheit	Centigrade		Fahrenheit	Centigrade		Fahrenheit	Centigrade		Fahrenheit	Centigrade		Fahrenheit
−223.3	**−370**		−70.6	**−95**	−139	2.2	**36**	96.8	32.8	**91**	195.8	288	**550**	1022
−220.6	**−365**		−67.8	**−90**	−130	2.8	**37**	98.6	33.3	**92**	197.6	293	**560**	1040
−217.8	**−360**		−65.0	**−85**	−121	3.3	**38**	100.4	33.9	**93**	199.4	299	**570**	1058
−215.0	**−355**		−62.2	**−80**	−112.0	3.9	**39**	102.2	34.4	**94**	201.2	304	**580**	1076
−212.2	**−350**		−59.4	**−75**	−103.0	4.4	**40**	104.0	35.0	**95**	203.0	310	**590**	1094
−209.4	**−345**		−56.7	**−70**	−94.0	5.0	**41**	105.8	35.6	**96**	204.8	316	**600**	1112
−206.7	**−340**		−53.9	**−65**	−85.0	5.6	**42**	107.6	36.1	**97**	206.6	321	**610**	1130
−203.9	**−335**		−51.1	**−60**	−76.0	6.1	**43**	109.4	36.7	**98**	208.4	327	**620**	1148
−201.1	**−330**		−48.3	**−55**	−67.0	6.7	**44**	111.2	37.2	**99**	210.2	332	**630**	1166
−198.3	**−325**		−45.6	**−50**	−58.0	7.2	**45**	113.0	37.8	**100**	212.0	338	**640**	1184
−195.6	**−320**		−42.8	**−45**	−49.0	7.8	**46**	114.8	43	**110**	230	343	**650**	1202
−192.8	**−315**		−40.0	**−40**	−40.0	8.3	**47**	116.6	49	**120**	248	349	**660**	1220
−190.0	**−310**		−37.2	**−35**	−31.0	8.9	**48**	118.4	54	**130**	266	354	**670**	1238
−187.2	**−305**		−34.4	**−30**	−22.0	9.4	**49**	120.2	60	**140**	284	360	**680**	1256
−184.4	**−300**		−31.7	**−25**	−13.0	10.0	**50**	122.0	66	**150**	302	366	**690**	1274
−181.7	**−295**		−28.9	**−20**	−4.0	10.6	**51**	123.8	71	**160**	320	371	**700**	1292
−178.9	**−290**		−26.1	**−15**	5.0	11.1	**52**	125.6	77	**170**	338	377	**710**	1310
−176.1	**−285**		−23.3	**−10**	14.0	11.7	**53**	127.4	82	**180**	356	382	**720**	1328
−173.3	**−280**		−20.6	**−5**	23.0	12.2	**54**	129.2	88	**190**	374	388	**730**	1346
−170.6	**−275**		−17.8	**0**	32.0	12.8	**55**	131.0	93	**200**	392	393	**740**	1364
−167.8	**−270**	−454	−17.2	**1**	33.8	13.3	**56**	132.8	99	**210**	410	399	**750**	1382
−165.0	**−265**	−445	−16.7	**2**	35.6	13.9	**57**	134.6	100	**212**	414	404	**760**	1400
−162.2	**−260**	−436	−16.1	**3**	37.4	14.4	**58**	136.4	104	**220**	428	410	**770**	1418
−159.4	**−255**	−427	−15.6	**4**	39.2	15.0	**59**	138.2	110	**230**	446	416	**780**	1436
−156.7	**−250**	−418	−15.0	**5**	41.0	15.6	**60**	140.0	116	**240**	464	421	**790**	1454
−153.9	**−245**	−409	−14.4	**6**	42.8	16.1	**61**	141.8	121	**250**	482	427	**800**	1472
−151.1	**−240**	−400	−13.9	**7**	44.6	16.7	**62**	143.6	127	**260**	500	432	**810**	1490
−148.3	**−235**	−391	−13.3	**8**	46.4	17.2	**63**	145.4	132	**270**	518	438	**820**	1508
−145.6	**−230**	−382	−12.8	**9**	48.2	17.8	**64**	147.2	138	**280**	536	443	**830**	1526
−142.8	**−225**	−373	−12.2	**10**	50.0	18.3	**65**	149.0	143	**290**	554	449	**840**	1544
−140.0	**−220**	−364	−11.7	**11**	51.8	18.9	**66**	150.8	149	**300**	572	454	**850**	1562
−137.2	**−215**	−355	−11.1	**12**	53.6	19.4	**67**	152.6	154	**310**	590	460	**860**	1580
−134.4	**−210**	−346	−10.6	**13**	55.4	20.0	**68**	154.4	160	**320**	608	466	**870**	1598
−131.7	**−205**	−337	−10.0	**14**	57.2	20.6	**69**	156.2	166	**330**	626	471	**880**	1616
−128.9	**−200**	−328	−9.4	**15**	59.0	21.1	**70**	158.0	171	**340**	644	477	**890**	1634
−126.1	**−195**	−319	−8.9	**16**	60.8	21.7	**71**	159.8	177	**350**	662	482	**900**	1652
−123.3	**−190**	−310	−8.3	**17**	62.6	22.2	**72**	161.6	182	**360**	680	488	**910**	1670
−120.6	**−185**	−301	−7.8	**18**	64.4	22.8	**73**	163.4	188	**370**	698	493	**920**	1688
−117.8	**−180**	−292	−7.2	**19**	66.2	23.3	**74**	165.2	193	**380**	716	499	**930**	1706
−115.0	**−175**	−283	−6.7	**20**	68.0	23.9	**75**	167.0	199	**390**	734	504	**940**	1724
−112.2	**−170**	−274	−6.1	**21**	69.8	24.4	**76**	168.8	204	**400**	752	510	**950**	1742
−109.4	**−165**	−265	−5.6	**22**	71.6	25.0	**77**	170.6	210	**410**	770	516	**960**	1760
−106.7	**−160**	−256	−5.0	**23**	73.4	25.6	**78**	172.4	216	**420**	788	521	**970**	1778
−103.9	**−155**	−247	−4.4	**24**	75.2	26.1	**79**	174.2	221	**430**	806	527	**980**	1796
−101.1	**−150**	−238	−3.9	**25**	77.0	26.7	**80**	176.0	227	**440**	824	532	**990**	1814
−98.3	**−145**	−229	−3.3	**26**	78.8	27.2	**81**	177.8	232	**450**	842	538	**1000**	1832
−95.6	**−140**	−220	−2.8	**27**	80.6	27.8	**82**	179.6	238	**460**	860	566	**1050**	1922
−92.8	**−135**	−211	−2.2	**28**	82.4	28.3	**83**	181.4	243	**470**	878	593	**1100**	2012
−90.0	**−130**	−202	−1.7	**29**	84.2	28.9	**84**	183.2	249	**480**	896	621	**1150**	2102
−87.2	**−125**	−193	−1.1	**30**	86.0	29.4	**85**	185.0	254	**490**	914	649	**1200**	2192
−84.4	**−120**	−184	−0.6	**31**	87.8	30.0	**86**	186.8	260	**500**	932	677	**1250**	2282
−81.7	**−115**	−175	0.0	**32**	89.6	30.6	**87**	188.6	266	**510**	950	704	**1300**	2372
−78.9	**−110**	−166	0.6	**33**	91.4	31.1	**88**	190.4	271	**520**	968	732	**1350**	2462
−76.1	**−105**	−157	1.1	**34**	93.2	31.7	**89**	192.2	277	**530**	986	760	**1400**	2552
−73.3	**−100**	−148	1.7	**35**	95.0	32.2	**90**	194.0	282	**540**	1004	788	**1450**	2642
												816	**1500**	2732

The formulas at the right may also be used for converting Centigrade or Fahrenheit degrees into the other scales.

Degrees Cent., $C° = \frac{5}{9}(F° + 40) - 40$

Degrees Fahr., $F° = \frac{9}{5}(C° + 40) - 40$

Degrees Kelvin, $K° = C° + 273.18$

Degrees Rankine, $R° = F° + 459.72$

Source: Chicago Bridge and Iron Company.

Table A-6
Specific Gravity and Weights of Various Liquids

Liquid	At Temp. of °F	Specific Gravity	Weight in Lbs. per U.S. Gal.	Weight in Lbs. per Cu. Ft.
Acetaldehyde	64.4	0.783	6.52	49
Acetic Acid	68.0	1.049	8.74	65
Acetic Anhydride	68.0	1.083	9.02	68
Acetone	68.0	0.792	6.60	49
Aniline	68.0	1.022	8.51	64
Asphaltum	68.0	1.1–1.5	9.2–12.5	69–94
Bromine	68.0	3.119	25.98	195
Carbon Disulfide	68.0	1.263	10.52	79
Carbon Tetrachloride	68.0	1.595	13.28	100
Castor Oil	59.0	0.969	8.07	60
Caustic Soda, 66% Solution	68.0	1.70	14.16	106
Chloroform	68.0	1.489	12.40	93
Citric Acid	68.0	1.542	12.84	96
Cocoanut Oil	59.0	0.926	7.71	58
Colza Oil (Rape Seed Oil)	68.0	0.915	7.62	57
Corn Oil	59.0	0.921–0.928	7.67–7.73	57–58
Cottonseed Oil	60.8	0.926	7.71	58
Creosote	59.0	1.040–1.100	8.66–9.2	65–69
Dimethyl Aniline	68.0	0.956	7.96	60
Ether	77.0	0.708	5.90	44
Ethyl Acetate	68.0	0.901	7.50	56
Ethyl Chloride	42.8	0.917	7.64	57
Ethyl Ether	77.0	0.712–0.714	5.93–5.95	44–45
Formaldehyde	68.0	1.139	9.49	71
#1 Fuel Oil	60.0	0.80–0.85	6.7–7.1	50–53
#2 Fuel Oil	60.0	0.81–0.91	6.7–7.6	51–57
#4 Fuel Oil	60.0	0.84–1.00	7.0–8.3	52–62
#5 Fuel Oil	60.0	0.91–1.06	7.6–8.8	57–66
#6 Fuel Oil	60.0	0.92–1.08	7.7–9.0	57–67
Furfural	68.0	1.159	9.65	72
Gasoline (Motor Fuel)	60.0	0.70–0.76	5.8–6.3	44–47
Glucose	77.0	1.544	12.86	96
Glycerin	32.0	1.260	10.49	79
Hydrochloric Acid, 43.4% Sol.	60.0	1.213	10.10	76
Kerosene	68.0	0.82	6.83	51
Lactic Acid	59.0	1.249	10.40	78
Lard Oil	59.0	0.913–0.915	7.60–7.62	57
Linseed Oil—Raw	68.0	0.93	7.8	58
Linseed Oil—Boiled	59.0	0.942	7.84	59
Mercury	68.0	13.595	113.23	849
Molasses	68.0	1.47	12.2	92
Naphthalene	68.0	1.145	9.54	71
Neatsfoot Oil	59.0	0.913–0.918	7.60–7.65	57
Nitric Acid, 91% Solution	68.0	1.502	12.51	94
Olive Oil	59.0	0.915–0.920	7.62–7.66	57
Peanut Oil	59.0	0.917–0.926	7.64–7.71	57
Phenol	77.0	1.071	8.92	73
Pitch	68.0	1.07–1.15	8.91–9.58	67–72
Rosin Oil	68.0	0.98	8.61	61
Soy Bean Oil	59.0	0.924–0.927	7.70–7.72	58
Sperm Oil	59.0	0.878–0.884	7.31–7.36	55
Sulfer Dioxide	80.0	1.363	11.35	85
Sulfuric Acid, 87% Solution	64.4	1.834	15.27	114
Tar	68.0	1.2	10.0	75
Tetrachloroethane	68.0	1.596	13.29	100
Trichloroethylene	68.0	1.464	12.19	91
Tung Oil	59.0	0.939–0.949	7.82–7.90	59
Turpentine	68.0	0.87	7.25	54
Water (Sea)	59.0	1.025	8.54	64
Water (0° C)	39.2	1.00	8.34	62.4
Water (20° C)	68.0	0.998	8.32	62.3
Whale Oil	59.0	0.917–0.924	7.64–7.70	57

Source: Chicago Bridge and Iron Company.

Appendix

Table A-7
Inches of Mercury, Weight in Pounds and Inches of Water Equivalents

In. of Mercury (0°C)	Lbs. per Sq. In.	In. of Water (4°C)
1.	.49115	13.5955
1.5	.737	20.393
2.	.982	27.191
2.5	1.228	33.989
3.	1.473	40.786
3.5	1.719	47.584
4.	1.965	54.382
4.5	2.210	61.180
5.	2.456	67.977
5.5	2.701	74.775
6.	2.947	81.573
6.5	3.193	88.371
7.	3.438	95.168
7.5	3.684	101.966
8.	3.929	108.764
8.5	4.175	115.562
9.	4.420	122.359
9.5	4.666	129.157
10.	4.912	135.955
10.5	5.157	142.753
11.	5.403	149.550
11.5	5.648	156.348
12.	5.894	163.146
12.5	6.139	169.944
13.	6.385	176.741
13.5	6.631	183.539
14.	6.876	190.337
14.5	7.122	197.134
15.	7.368	203.932
15.5	7.613	210.730
16.	7.858	217.528
16.5	8.104	224.325
17.	8.350	231.123
17.5	8.595	237.921
18.	8.841	244.719
18.5	9.086	251.516
19.	9.332	258.314
19.5	9.578	265.112
20.	9.823	271.910
20.5	10.069	278.707
21.	10.314	285.505
21.5	10.560	292.303
22.	10.805	299.101
22.5	11.051	305.898
23.	11.297	312.696
23.5	11.542	319.494
24.	11.788	326.292
24.5	12.033	333.089
25.	12.279	339.887
25.5	12.524	346.685
26.	12.770	353.482
26.5	13.016	360.280
27.	13.261	367.078
27.5	13.507	373.876
28.	13.752	380.673
28.5	13.998	387.471
29.	14.243	394.269
29.5	14.490	401.067
30.	14.735	407.864

Source: Chicago Bridge and Iron Company

Table A-8
Inches to Millimeters and Feet to Meters

_____ Inches _____						Decimals[a]	Milli-meters	_____ Inches _____						Decimals[a]	Milli-meters
1/2s	1/4s	1/8s	1/16s	1/32s	1/64s			1/2s	1/4s	1/8s	1/16s	1/32s	1/64s		
					1	0.0156 25	0.397						33	0.515 625	13.097
				1	2	0.031 25	0.794					17	34	0.531 25	13.494
					3	0.046 875	1.191						35	0.546 875	13.891
			1	2	4	0.062 5	1.588				9	18	36	0.562 5	14.288
					5	0.078 125	1.984						37	0.578 125	14.684
				3	6	0.093 75	2.381					19	38	0.593 75	15.081
					7	0.109 375	2.778						39	0.609 375	15.478
		1	2	4	8	0.125 0	3.175[a]			5	10	20	40	0.625 0	15.875[a]
					9	0.140 625	3.572						41	0.640 625	16.272
				5	10	0.156 25	3.969					21	42	0.656 25	16.669
					11	0.171 875	4.366						43	0.671 875	17.066
			3	6	12	0.187 5	4.762				11	22	44	0.687 5	17.462
					13	0.203 125	5.159						45	0.703 125	17.859
				7	14	0.218 75	5.556					23	46	0.718 75	18.256
					15	0.234 375	5.953						47	0.734 375	18.653
	1	2	4	8	16	0.250 0	6.350[a]		3	6	12	24	48	0.750 0	19.050[a]
					17	0.265 625	6.747						49	0.765 625	19.447
				9	18	0.281 25	7.144					25	50	0.781 25	19.844
					19	0.296 875	7.541						51	0.796 875	20.241
			5	10	20	0.312 5	7.938				13	26	52	0.812 5	20.638
					21	0.328 125	8.334						53	0.828 125	21.034
				11	22	0.343 75	8.731					27	54	0.843 75	21.431
					23	0.359 375	9.128						55	0.859 375	21.828
		3	6	12	24	0.3750	9.525[a]			7	14	28	56	0.875 0	22.225[a]
					25	0.390 625	9.922						57	0.890 625	22.622
				13	26	0.406 25	10.319					29	58	0.906 25	23.019
					27	0.421 875	10.716						59	0.921 875	23.416
			7	14	28	0.437 5	11.112				15	30	60	0.937 5	23.812
					29	0.453 125	11.509						61	0.953 125	24.209
				15	30	0.468 75	11.906					31	62	0.968 75	24.606
					31	0.484 375	12.303						63	0.984 375	25.003
1	2	4	8	16	32	0.500 0	12.700[a]	2	4	8	16	32	64	1.000 0	25.400[a]

Table A-8 (continued on facing page)

Table A-8 (concluded)

Inches →	0	1	2	3	4	5	6	7	8	9
					Millimeters[a]					
0	—	25.4	50.8	76.2	101.6	127.0	152.4	177.8	203.2	228.6
10	254.0	279.4	304.8	330.2	355.6	381.0	406.4	431.8	457.2	482.6
20	508.0	533.4	558.8	584.2	609.6	635.0	660.4	685.8	711.2	736.6
30	762.0	787.4	812.8	838.2	863.6	889.0	914.4	939.8	965.2	990.6
40	1016.0	1041.4	1066.8	1092.2	1117.6	1143.0	1168.4	1193.8	1219.2	1244.6
50	1270.0	1295.4	1320.8	1346.2	1371.6	1397.0	1422.4	1447.8	1473.2	1498.6
60	1524.0	1549.4	1574.8	1600.2	1625.6	1651.0	1676.4	1701.8	1727.2	1752.6
70	1778.0	1803.4	1828.8	1854.2	1879.6	1905.0	1930.4	1955.8	1981.2	2006.6
80	2032.0	2057.4	2082.8	2108.2	2133.6	2159.0	2184.4	2209.8	2235.2	2260.6
90	2286.0	2311.4	2336.8	2362.2	2387.6	2413.0	2438.4	2463.8	2489.2	2514.6
100	2540.0	—	—	—	—	—	—	—	—	—

Feet →	0	1	2	3	4	5	6	7	8	9
					Meters[a]					
0	—	0.3048	0.6096	0.9144	1.2192	1.5240	1.8288	2.1336	2.4384	2.7432
10	3.0480	3.3528	3.6576	3.9624	4.2672	4.5720	4.8768	5.1816	5.4864	5.7912
20	6.0960	6.4008	6.7056	7.0104	7.3152	7.6200	7.9248	8.2296	8.5344	8.8392
30	9.1440	9.4488	9.7536	10.0584	10.3632	10.6680	10.9728	11.2776	11.5824	11.8872
40	12.1920	12.4968	12.8016	13.1064	13.4112	13.7160	14.0208	14.3256	14.6304	14.9352
50	15.2400	15.5448	15.8496	16.1544	16.4592	16.7640	17.0688	17.3736	17.6784	17.9832
60	18.2880	18.5928	18.8976	19.2024	19.5072	19.8120	20.1168	20.4216	20.7264	21.0312
70	21.3360	21.6408	21.9456	22.2504	22.5552	22.8600	23.1648	23.4696	23.7744	24.0792
80	24.3840	24.6888	24.9936	25.2984	25.6032	25.9080	26.2128	26.5176	26.8224	27.1272
90	27.4320	27.7368	28.0416	28.3464	28.6512	28.9560	29.2608	29.5656	29.8704	30.1752
100	30.4800	—	—	—	—	—	—	—	—	—

[a] Exact figure

Table A-9
Millimeters to Inches and Meters to Feet

Milli-meters	Inches Nearest 1/16″	Inches Nearest 1/64″	Decimals	Milli-meters	Inches Nearest 1/16″	Inches Nearest 1/64″	Decimals
1	1/16	3/64	0.03937	51	2	2 1/64	2.00787
2	1/16	5/64	0.07874	52	2 1/16	2 3/64	2.04724
3	1/8	1/8	0.11811	53	2 1/16	2 3/32	2.08661
4	3/16	5/32	0.15748	54	2 1/8	2 1/8	2.12598
5	3/16	13/64	0.19685	55	2 3/16	2 11/64	2.16535
6	1/4	15/64	0.23622	56	2 3/16	2 13/64	2.20472
7	1/4	9/32	0.27559	57	2 1/4	2 1/4	2.24409
8	5/16	5/16	0.31496	58	2 5/16	2 9/32	2.28346
9	3/8	23/64	0.35433	59	2 5/16	2 21/64	2.32283
10	3/8	25/64	0.39370	60	2 3/8	2 23/64	2.36220
11	7/16	7/16	0.43307	61	2 3/8	2 13/32	2.40157
12	1/2	15/32	0.47244	62	2 7/16	2 7/16	2.44094
13	1/2	33/64	0.51181	63	2 1/2	2 31/64	2.48031
14	9/16	35/64	0.55118	64	2 1/2	2 33/64	2.51969
15	9/16	19/32	0.59055	65	2 9/16	2 9/16	2.55906
16	5/8	5/8	062992	66	2 5/8	2 19/32	2.59843
17	11/16	43/64	0.66929	67	2 5/8	2 41/64	2.63780
18	11/16	45/64	0.70866	68	2 11/16	2 43/64	2.67717
19	3/4	3/4	0.74803	69	2 11/16	2 23/32	2.71654
20	13/16	25/32	0.78740	70	2 3/4	2 3/4	2.75591
21	13/16	53/64	0.82677	71	2 13/16	2 51/64	2.79528
22	7/8	55/64	0.86614	72	2 13/16	2 53/64	2.83465
23	7/8	29/32	0.90551	73	2 7/8	2 7/8	2.87402
24	15/16	15/16	0.94488	74	2 15/16	2 29/32	2.91339
25	1	63/64	0.98425	75	2 15/16	2 61/64	2.95276
26	1	1 1/32	1.02362	76	3	2 63/64	2.99213
27	1 1/16	1 1/16	1.06299	77	3 1/16	3 1/32	3.03150
28	1 1/8	1 7/64	1.10236	78	3 1/16	3 5/64	3.07087
29	1 1/8	1 9/64	1.14173	79	3 1/8	3 7/64	3.11024
30	1 3/16	1 3/16	1.18110	80	3 1/8	3 5/32	3.14961
31	1 1/4	1 7/32	1.22047	81	3 3/16	3 3/16	3.18898
32	1 1/4	1 17/64	1.25984	82	3 1/4	3 15/64	3.22835
33	1 5/16	1 19/64	1.29921	83	3 1/4	3 17/64	3.26772
34	1 5/16	1 11/32	1.33858	84	3 5/16	3 5/16	3.30709
35	1 3/8	1 3/8	1.37795	85	3 3/8	3 11/32	3.34646
36	1 7/16	1 27/64	1.41732	86	3 3/8	3 25/64	3.38583
37	1 7/16	1 29/64	1.45669	87	3 7/16	3 27/64	3.42520
38	1 1/2	1 1/2	1.49606	88	3 7/16	3 15/32	3.46457
39	1 9/16	1 17/32	1.53543	89	3 1/2	3 1/2	3.50394
40	1 9/16	1 37/64	1.57480	90	3 9/16	3 35/64	3.54331
41	1 5/8	1 39/64	1.61417	91	3 9/16	3 37/64	3.58268
42	1 5/8	1 21/32	1.65354	92	3 5/8	3 5/8	3.62205
43	1 11/16	1 11/16	1.69291	93	3 11/16	3 21/32	3.66142
44	1 3/4	1 47/64	1.73228	94	3 11/16	3 45/64	3.70079
45	1 3/4	1 49/64	1.77165	95	3 3/4	3 47/64	3.74016
46	1 13/16	1 13/16	1.81102	96	3 3/4	3 25/32	3.77953
47	1 7/8	1 27/32	1.85039	97	3 13/16	3 13/16	3.81890
48	1 7/8	1 56/64	1.88976	98	3 7/8	3 55/64	3.85827
49	1 15/16	1 59/64	1.92913	99	3 7/8	3 57/64	3.89764
50	1 15/16	1 31/32	1.96850	100	3 15/16	3 15/16	3.93701

Milli-meters	Feet-Inches Nearest 1/16″	Feet-Inches Nearest 1/64″	Decimal Inches
100	0-3 15/16	0-3 15/16	3.93701
200	0-7 7/8	0-7 7/8	7.87402
300	0-11 13/16	0-11 13/16	11.81102
400	1-3 3/4	1-3 3/4	15.74803
500	1-7 11/16	1-7 11/16	19.68504
600	1-11 5/8	1-11 5/8	23.62205
700	2-3 9/16	2-3 9/16	27.55906
800	2-7 1/2	2-7 1/2	31.49606
900	2-11 7/16	2-11 7/16	35.43307
1000	3-3 3/8	3-3 3/8	39.37008

Meters	Feet-Inches Nearest 1/16″	Decimal Feet
1	3- 3 3/8	3.2808
2	6- 6 3/4	6.5617
3	9-10 1/8	9.8425
4	13- 1 1/2	13.1234
5	16- 4 7/8	16.4042
6	19- 8 1/4	19.6850
7	22-11 9/16	22.9659
8	26- 2 15/16	26.2467
9	29- 6 5/16	29.5276
10	32- 9 11/16	32.8084
11	36- 1 1/16	36.0892
12	39- 4 7/16	39.3701
13	42- 7 13/16	42.6509
14	45-11 3/16	45.9318
15	49- 2 9/16	49.2126
16	52- 5 15/16	52.4934
17	55- 9 5/16	55.7743
18	59- 0 11/16	59.0551
19	62- 4 1/16	62.3360
20	65- 7 3/8	65.6168
21	68-10 3/4	68.8976
22	72- 2 1/8	72.1785
23	75- 5 1/2	75.4593
24	78- 8 7/8	78.7402
25	82- 0 1/4	82.0210
26	85- 3 5/8	85.3018
27	88- 7	88.5827
28	91-10 3/8	91.8635
29	95- 1 3/4	95.1444
30	98- 5 1/8	98.4252
31	101- 8 1/2	101.7060
32	104-11 13/16	104.9869
33	108- 3 3/16	108.2677
34	111- 6 9/16	111.5486
35	114- 9 15/16	114.8294

Index

Accounting meter, 152
Accumulator, 119
Actuator, 164
AGA, 152
ANSI, 2
ASTM, 2
Air cooler, 33

Bell and spigot joint, 76
Bends, 93
Block plot plan, 39
Bond, 26
Bootleg, 125
Bridle, 125
By-pass, 152, 164

Cages, 135
Capillary tubing, 151
Cast iron soil pipe, 74,75
Cast iron water pipe, 75
Catalyst, 119
Catch basins, 73, 75
Cleanouts, 81
Clearances, 57
Compressors, 70
Cone roof tanks, 39
Control buildings, 68
Control valve
 actuators, 164
 angle type, 169
 cooling fins, 164
 definition, 164
 location, 164
Cooling towers, 68
Cooling water, 82

Cracking, 142
Cut-backs, 93

Davits, 125, 128, 142
Dead legs, 84, 89, 124
Desalter, 26
Dike
 construction, 50
 drainage, 54
 freeboard, 51
 sizing, 50
Double random length, 5
Downcomer, 129
Draw-off nozzle, 142
Drop area, 142
Dual instruments, 149
Duriron pipe, 75

Equilibrium liquid
 definition, 27
 piping, 28
Equipment plot plan, 57
Excavation plan, 66

Ferrous metals, 1
Fired heaters, 68
Fire plugs, 82
Fire water systems, 82
Fittings, 5, 6-11, 93
Flanges, 5
Flare systems, 88
Flash zone, 142
Floating roof tanks, 41
Floor plate, 135

Flow diagram transposition, 66
Flow instruments, 152
Fluid definition, 1
Foam system, 55
Foundation location plan, 66
Fractionation
 definition, 27
 towers, 68, 119, 129

Galvanizing, 135
Gasoline, 142
Glycol systems, 84
Grating, 135

Handjacks, 164
Head, 34
Hillside nozzle, 125
Hinge
 detail, 128
 stop, 125
Hot vapor by-pass, 33
Hydraulics, 34
Hydrocarbon, 26
Hydrotest, 35

Indicator, 149
Inspection openings, 124
Instrument
 connections, 135
 functions, 149
Instrumentation, 149
Invert elevation, 74
Isometric spool, 93

Joint efficiency, 2
Jute, 76, 79

Ladders, 135
Level controller
  installation, 154
  location, 121
  types, 156
Level gage
  installation, 160
  location, 121
  types, 160
Line loss, 121

Manhole
  davits, 125, 128
  hinges, 125, 128
  underground, 79
Mechanical joints, 76
Mercaptan, 26
Meter runs, 152
Mill, tolerance, 2
Miters, 93
Monitors, 82, 83

Nonferrous metals, 1
Nozzle orientation, 121

Oakum, 76, 79
Offset platforms, 135
Orient, 129
Orifice flanges, 152
Orthographic spool, 93
Overhead product, 119

Piping
  at tanks, 54
  classes, 1
  drawing index, 66
  lengths, 5
  materials, 1
  thickness tolerance, 2

Platforms, 120, 135
Plot plan dimensioning, 64
Polymerization, 142
Process flow diagram, 59
Process unit flow diagram, 59
Product, 142
Pump-out systems, 84
Pump piping, 119, 129

Random length, 5
Reactor, 61
Reboilers, 70
Recorder, 149
Reflux
  accumulator, 119
  liquid, 121, 134
Reforming, 142
Relief valve
  parts, 169
Retention time, 121
Ribs, 129

Saddles, 123, 129
Sample connections, 135
Separators, 119
Sewer
  flow diagrams, 77
  tail pipe, 90
  terms, 75
Shop fabrication, 92
Site data, 37
Slide plates, 124
Specific gravity, 34
Spools, 92
Straightening vanes, 153
Static head, 34
Steam
  saturated, 35
  superheated, 35
  tracing, 84
Storage tanks, 39

Tank
  dikes, 46
  materials, 51
  spacing, 43
Taps, 151, 153
Temperature gradient, 134
Thermal reliefs, 90
Thermocouple, 150
Thermowell, 123, 150
Thread engagement, 93
Thrust blocks, 76
Tracing, 84
Transite piping, 81
Transmitters, 149
Trays, 129
Triangle solving, 93
Tube pulling area, 68
Two-phase flow, 31

Underground systems, 73

Valves, 24
Venturi tube, 153
Vertical meter runs, 153
Vessel
  davits, 142
  definition, 119
  disengaging, 121
  horizontal, 119
  pipe supports, 142
  vertical, 129
Visible glass, 160
Vitrified clay, 75

Welding, 92
Winterizing, 85

Yellowback, 150

PLEASE RETURN

PROPERTY
SYNCRUDE CANADA LTD.
OPERATIONS LIBRARY